Embedded Core Design with FPGAs

About the Author

Dr. Zainalabedin Navabi is a professor of electrical and computer engineering at Northeastern University. Dr. Navabi is the author of several textbooks and computer based trainings on VHDL, Verilog and related tools and environments. Dr. Navabi's involvement with hardware description languages begins in 1976 when he started the development of a register-transfer level simulator for one of the very first Hardware Description Languages (HDLs). In 1981 he completed the development of a synthesis tool that generated MOS layout from an RTL description. Since 1981, Dr. Navabi has been involved in the design, definition and implementation of HDLs. He has written numerous papers on the application of HDLs in simulation, synthesis and test of digital systems. He started one of the first full HDL courses at Northeastern University in 1990. Since then he has conducted many short courses and tutorials on this subject in the United States and abroad. In addition to being a professor, he is also a consultant to Electronic Design Automation (EDA) companies. Dr. Navabi received his M.S. and Ph.D. from the University of Arizona in 1978 and 1981, and his B.S. from the University of Texas at Austin in 1975. He is a senior member of IEEE, and a member of IEEE Computer Society, ASEE, and ACM. Dr. Navabi is the author of six books on various aspects of digital system design automation.

Embedded Core Design with FPGAs

Zainalabedin Navabi, Ph.D.

Professor of Electrical and Computer Engineering
Northeastern University
Boston, Massachusetts

McGraw-Hill

New York Chicago San Francisco Lisbon London Madrid
Mexico City Milan New Delhi San Juan Seoul
Singapore Sydney Toronto

3 4 5 6 7 8 9 10 IBT/IBT 1 9 8 7 6 5 4 3 2 1 0

ISBN-13: Book p/n 978-0-07-148470-1 and CD p/n 978-0-07-148471-8
         parts of
         Set ISBN 978-0-07-147481-8

ISBN-10: Book p/n 0-07-148470-1 and CD p/n 0-07-148471-X
         parts of
         Set ISBN 0-07-147481-1

*The sponsoring editor for this book was Wendy Rinaldi, the editorial
supervisor was Jody McKenzie, and the production supervisor was
Jim Kussow. It was set in Century Schoolbook by Zain Navabi. The
art director for the cover was Margaret Webster-Shapiro.*

Printed and bound by IBT Global.

McGraw-Hill books are available at special quantity discounts to
use as premiums and sales promotions, or for use in corporate
training programs. For more information, please write to the Director of Special Sales, Professional Publishing, McGraw-Hill, Two
Penn Plaza, New York, NY 10121-2298. Or contact your local book-
store.

In the memory of my sister, Shahla (Fami) Navabi.

CONTENTS

APPENDIX

PREFACE

The topic of this book is embedded system design with FPGAs. Design of an embedded system involves design of functions that may be implemented in hardware or may be implemented as software running on an embedded processor. Because of this multi-disciplinary nature of embedded systems, they include the concepts of digital systems, computer architectures, software development, computer systems and microprocessor based design. In addition, implementing embedded systems with FPGAs requires the additional knowledge of programmable devices and the corresponding design tools and languages (e.g., VHDL or Verilog).

While it is clear that thoroughly covering all these topics in one book is not possible, it must be said that in-depth knowledge of each of these topics is not necessary for becoming an embedded system designer. An embedded system designer looks at the digital system design from a system point of view; a basic understanding of all the subjects mentioned is all that such a designer requires.

Embedded Core Design with FPGAs provides all the information that is needed for designing complex embedded systems and cores. The first chapter defines what embedded systems are and how the knowledge of various digital design aspects becomes useful in design of such systems. Next we cover logic design with an advanced flavor of Register Transfer Level (RTL) design. We then discuss Verilog at the RT level, and show applications of this language for RTL simulation and synthesis. After discussion of RT level hardware and the corresponding design methodologies, we move to computer hardware and software. While the RT level material is useful for the hardware side of embedded system designs, the computer system and architecture material is intended for the software side of such designs. After covering the basics, we show how these topics are put together into the design of a complete system with hardware and software cores. Throughout this presentation, tools for RT level and advanced embedded system design are introduced and utilized.

This book can be used by hardware design practitioners who are already familiar with the basics of logic design and want to move into the arena of automated system level design. For this audience, the book includes a recap of digital design topics, computer architectures, and software programming. It also contains examples showing the use of Verilog and compilers and assemblers for embedded system designs. In addition, for an industrial setting, we show how existing hardware and software design components and library cores are used in upper level designs. Using Altera's development boards, this book gives a hands-on knowledge of the topics covered.

In an educational setting, the book can be used in an upper level technical elective course for electrical and computer engineering students as well as students in other fields of engineering. Using Altera's development boards with this book helps students see their designs being implemented and tested, and thereby get a nuts-and-bolts understanding of how things work. For students in other fields of engineering, like mechanical and chemical engineering, the book is a useful tool for design and implementation of controllers and interfaces.

Zainalabedin Navabi
navabi@ece.neu.edu
Boston, Massachusetts
July, 2006

INTRODUCTION

Embedded system design, which has become the new trend in hardware design, uses embedded cores and processors as components of a digital system. An embedded system designer uses a mix of high-level software programs and RT level descriptions to describe various parts of his or her design. The abstract communication between various parts of a system has also become a design issue that a hardware designer must be aware of. Furthermore, it is important that a system designer knows about all tools that are available for design and implementation of hardware.

The early transistor level design of digital systems gave way to the gate-level design, and in the late '80s RT level design started becoming the dominant digital design methodology. We are now seeing that for today's complex designs, RT level design is too detailed and upper level abstractions are required. Electronic System Level (ESL) is this next level of abstraction. In this level, designers describe their hardware components at a very high functional level, and with the aid of design tools, they translate their descriptions into more detailed RT or gate level descriptions.

As designs become more complex and hardware descriptions become more abstract, interconnections of components become more sophisticated. While simple wires were used for interconnecting transistors, logical signals, with more functional meaning, became gate interconnections for gate-level designs. Interconnections became more complex (i.e., busses) when we moved from gate-level to the RT level. Continuing this trend, we are now seeing that system level interconnections are becoming even more complex and are themselves consisting of complex RTL components. Simple interconnecting busses of RT level have become intelligent system busses or switch fabrics that can handle block data transfers, arbitration, and various forms of master-slave communications. Understanding this interconnection methodology is crucial for today's system level designers.

Developing software programs to implement hardware functions, understanding processor architectures that the programs runs on, design of hardware at the RT level, and describing interconnection of system components are required of today's hardware designer.

Embedded Core Design with FPGAs covers RTL, system level design methodology, FPGAs, and tools and environments that are available to a system level designer. This book can be used in an academic or industrial setting by students or engineers. In either case it assumes a general knowledge of logic design. After a review of this topic, it builds upon gate level logic design techniques to cover RTL. The first five chapters cover the main concepts of digital design with field programmable devices from a practical point of view. The remaining chapters show environments for core design and implementation of system level designs using hardware and software cores.

Chapter 1 discusses the general flow of a system level design and the role of compilers and synthesis tools. The focus is to show what is needed to become an embedded system design engineer.

Chapter 2 discusses RT level logic design from a practical point of view. Mainly, topics used in an automated RTL design are discussed here.

Chapter 3 introduces Verilog. Synthesizable RT level Verilog is emphasized, but for a complete HDL based design, testbenches and language utilities for this purpose are also discussed.

Chapter 4 discusses computer systems, computer architectures, and high level C programming. This chapter shows how processor hardware and software interact.

Chapter 5 details programmable devices. The approach we take is showing how original ROMs evolved into today's complex FPGAs.

Chapter 6 discusses tools we use for design validation, synthesis, device programming and prototyping. We discuss the use of Quartus II, ModelSim HDL simulator and DE2 and UP3 development boards.

Chapter 7 shows several interface designs. This presentation demonstrates how cores are created and utilized.

Chapter 8 shows the elements of a complete embedded system design that involves hardware and software. We show an embedded implementation of an FIR filter. To demonstrate the details, we will not take advantage of design tools and environments that are available for embedded system designs.

Chapter 9 shows how a complete and complex system that has hardware and software parts is designed and implemented. For this design we show utilization of all design aids, software tools and design automation tools that are available to an embedded system designer.

ACKNOWLEDGMENTS

Several people helped me with preparation of this manuscript. My former students Ms. Shahrzad Mirkhani and Dr. Saeed Safari wrote sections of Chapter 4 on programming and hardware design. Ms. Mirkhani was very helpful in reviewing the manuscript and making useful recommendations.

My former and present students Mr. Armin Alaghi, Ms. Elnaz Ansari, and Ms. Parisa Razaghi developed the embedded designs shown in Chapters 8 and 9. Their thoroughness and emphasis on the details were useful in generating and implementing these designs.

As with all my other publishing works, Ms. Fatemeh Asgari helped me with the preparation of the manuscript. She worked with me on the initial planning of this work, distribution of tasks during the project, and final assembly of this book. Her planning and organization has always been a key to successful completion of such projects.

Instrumental in the original proposal and arrangement of this book was Mr. Mike Phipps of Altera. His guidelines in making this book useful for students and practitioners were helpful in the organization of the book. I thank him for his support and special attention to computer engineering education.

I also thank my wife, Irma Navabi, for help, encouragement, and understanding of my working habits. Such an intensive work could not be done without the support of my wife and two sons, Arash and Arvand. I thank them for this and my other scientific achievements.

1 Elements of Embedded Design

An embedded system is a digital system with at least one processor that implements a hardware function that is part or all of the digital system. The processor(s) of an embedded system is (are) called the embedded processor(s). Embedded systems facilitate design of digital systems by giving designers the opportunity to use a C or C++ program for description and design of complex hardware functions. The high level program replaces detailed design of hardware that would normally be done by writing synthesizable HDL code or by use of hardware library components.

It must be noted that an embedded system design, as defined above, is not very different from design of microcontrollers. Embedded processors are used for hardware implementation the same way microcontrollers are. The main difference is that embedded systems offer more flexibilities and design customization. Furthermore, embedded systems offer higher level design methods for integrating those parts of a system that are regarded as hardware components with those parts of the system that are implemented with an embedded processor.

To be able to take advantages of flexibilities and high level design aids offered by embedded systems, a new methodology of hardware design must be learned. This methodology includes the use of hardware and software in the same integrated design environment. This chapter highlights elements of an embedded design. We discuss the methodology, role of software, role of hardware description languages (HDLs), integration of hardware and software, and tools and environments that are available for design of embedded systems.

1.1 Abstraction Levels

Design of digital systems has evolved from transistor level, to gate level, and to Register Transfer Level (RTL). Although HDLs and configurable library components have provided for an RTL designer ways of achieving designs quickly, a higher abstraction level of design is needed for implementation of today's complex hardware systems.

1.1.1 Transistors to Programs

Digital design started with putting transistors to implement a given hardware function. Obviously this handcrafted method of design and flexibilities offered in choice of transistor size and routing of wires, achieves an optimum design for a given function.

On the other hand, as designs become more complex, this level of design had to change to allow design of large circuits. In an evolutionary process, gate level designs replaced transistor level designs. With this move to an upper abstraction level, compromise for timing, silicon utilization, and power consumptions had to be made. In addition, design tools were developed to help designers with utilization of gates verification of designs, and translation to the transistor level.

As designs became more complex, another higher abstraction level evolved that include even less detail than the gate level. The main focus of this level of abstraction is how transfer of data happens between registers, logic units, and busses; and because of this, it is referred to as register transfer level, or RTL. As in the move from transistor level to gate level, moving from gates to RT level carries with it compromises and tradeoffs. Furthermore, this higher level of abstraction requires use of tools and various software and hardware packages to aid the designer in the design process. As in the gate level, RT level tools include those for design capture, verification, and translation from RT level to the lower abstraction level, i.e., gate level synthesis.

For the same reasons that design had to go up from gate to RT level, the time of sole RT level design had to expire, and this level of abstraction had to give way to an upper level of abstraction, which for now, we refer to as electronic system level (ESL) or just system level. At the system level, a designer is only concerned with the functionality of the system being designed, and describes the algorithm that is going to be implemented. The algorithm is described using a procedural language like the C language. The description at this level does not contain clock or gate level timing.

System level tools include design, entry tools, simulators, and, of course, hardware generation programs. Hardware generation from a system level description can be done in one of two possible ways. As in other abstraction levels, one way of generating hardware is to translate a system level description to a lower level of abstraction, i.e., RTL. Alternatively, a system level procedural description can be compiled to run on a given processor. This alternative is possible at the system level because the description is procedural and a software language like C can be used for it.

The above mentioned method of hardware generation from a system level description is what has become embedded system design. The former method, i.e., translation from system to RTL, is often referred to as C synthesis, or system level synthesis. C synthesis refers to generation of hardware from a C program, or a procedural description. Figure 1.1 shows abstraction levels discussed here.

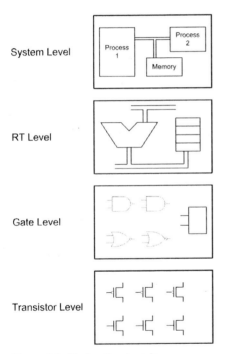

System Level

RT Level

Gate Level

Transistor Level

Figure 1.1 Abstraction Levels

1.1.2 Mixed Level Hardware

Although design at a higher level of abstraction is easier than a lower level and more tools are provided for it, designers always use a mixture of various design levels. Going from one level to another, is determined by design constraints and the effort that a designer has to put into the design.

A gate level designer goes down to transistors for especial logic functions, or when an optimized design of a cell is required. Similarly, very often in RT level, a designer finds the need for putting a few gates together for an especial RT level function, or as glue logic.

The situation with system level is no different than those at the lower abstraction levels. Often, a system level design consists of a mixture of hardware components that are described and implemented with embedded processors, as well as several RT level components. RT level components can be more handcrafted than embedded processors, and are more optimized in terms of chip area utilization, timing, and power consumption.

A system level hardware designer must be able to use a mix of RT level and system level tools and design methodologies.

1.1.3 Design Specification

The way a design is described varies based on its level of abstraction. At the transistor and gate levels, the main form of design entry is schematic entry and most of the design tools are for facilitating this form of design specification. At the RT level, hardware description languages provide an unambiguous and compact form of describing hardware. In spite of this, still block diagram schematics are used for high level interconnection of components.

At the system level, the C/C++ language is the most common format of specifying a system. At the same time, graphical tools are used for specification of interconnection of components at the top level. Components of such a block diagram may be described at the system level, RT level, or described as other block diagrams.

1.2 Embedded System Design Flow

Figure 1.2 shows the design flow for an embedded system. This flow consists of implementation of hardware functions in hardware and software, and then merging the results into one hardware realization. The subsections below describe the details of this block diagram.

1.2.1 Hardware/Software Partitioning

The first step in design of an embedded system is to decide what parts are to be implemented using hardware packages, HDL programs, or gate structures, and which parts are to be implemented with a program running on a processor. This decision is referred to as hardware/software partitioning. This is a manual (or semi-manual) process, and is perhaps the most difficult system design phase.

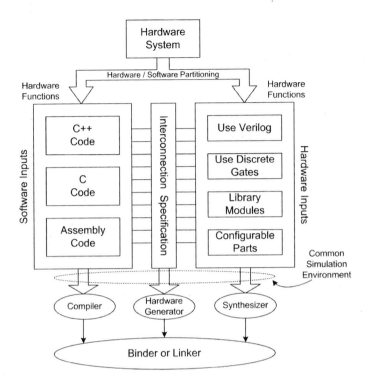

Figure 1.2 Embedded System Design Flow

The hardware part becomes a description of various hardware modules that are described in an HDL or are available as predefined hardware modules. The software part is a high level C/C++ program that after being compiled becomes the memory contents of processor that runs the program.

1.2.2 Hardware Part

The hardware part (right flow in Figure 1.2) of a complete hardware/software system may be composed of components that are described in Verilog, available in a library, or gate level parts. Using tools and design environments, a hardware designer chooses to code parts of his or her design in Verilog, or use parts from a library of predefined modules.

Often a design environment provides intellectual property (IP) cores that designers can use and integrate in their designs. Hardware design environments also include configurable parts for commonly used components such as arithmetic functions, register banks, and counters.

1.2.3 Software Part

The left flow in Figure 1.2 shows the implementation of the software part of a system. The part of a design that is to be implemented in software must become a machine language program in a given processor. The designer may choose to code this part in a high level language and compile it, or directly code it in assembly or machine language.

All the necessary software tools and compilers are available to a designer who uses a supported processor core. In this case, use of C/C++ for describing the software part of a system is the most logical choice. This way, compilation and debugging tools are provided for the designer. On the other hand, if a designer uses his or her own processor or a processor core that does not have a strong support, the designer is responsible for generating the machine language of the program he or she is implementing.

Regardless of how the programming task is done, after the completion of the design of the software part, this part looks like any hardware block with inputs and outputs. The inputs and outputs are either external to the system being designed, or they are to interconnect the hardware and software parts.

1.2.4 Interconnection Specification

The middle of diagram of Figure 1.2 shows a block that specifies interconnection of software and hardware parts. This block may be a simple shared bus, interconnection wires, or a complex switch structure. Usually, embedded system design environments have their own bus structures. Handshaking, timers, block transfer hardware, and other high level transactions take place in this bus. Such issues and detailed design of this bus are transparent to the high-level system designer.

1.2.5 Common Hardware/Software Simulation

A design that is part hardware, part software, and part switch transactions and data communication, must be simulated in a common environment for design and timing verifications.

Before a design is turned into hardware gates and netlist, the hardware part is simulated at the RT level, and the software part at the instruction level. HDL simulators provide RT level simulation, and ISS (Instruction Set Simulators) provide instruction level simulation. Usually an embedded design environment provides a co-simulation link for verifying all parts of a complete hardware / software system.

1.2.6 Hardware Synthesis

The part of a system that is described using hardware description methods (right side in Figure 1.2) is synthesized to produce a netlist of gates and primitive logical blocks. As shown in Figure 1.3, in addition to the hardware description, a synthesis program requires the synthesis target specification. Target specification tells the synthesis tool what resources, in terms of gates and logical blocks, are available for implementing our hardware specification. The resources depend on whether we are using an FPGA, an ASIC, or a custom IC for implementing our hardware.

The synthesis tool generates a netlist of components of the target library. This is usually given in an internal netlist language, VHDL, Verilog, or other netlist formats. This netlist together with hardware details of the target library components form the complete post-synthesis hardware description of the hardware part of our design. This description can be simulated for post-synthesis description, or it can be used for FPGA device programming or ASIC layout generation.

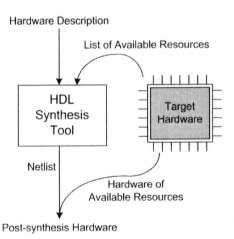

Figure 1.3 Hardware Synthesis

1.2.7 Software Compilation

For the synthesis of the hardware part of a system we must define a specific synthesis target. Similarly, compilation of the software part of a system must be done for a specific processor. Compiling our software part for the embedded processor that we are using produces a machine language program for that chosen processor. This machine language program, together with the hardware of the processor it runs on, produces the complete implementation of the software part of our system design. Figure 1.4 shows the compilation process.

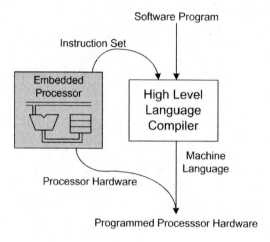

Figure 1.4 Software Compilation

1.2.8 Interconnection Hardware Generation

Between the software and hardware parts of a system in Figure 1.2 is the interconnection block. As discussed, this block is put in there by the embedded system design environment, and its hardware is generated automatically. The situation may be different if a designer does not take advantage of the facilities of the embedded design environment, and uses his or her own embedded processor and/or bussing logic. In this case, the bus hardware must be designed like any other hardware function, and synthesized to generate appropriate target device netlist.

1.2.9 Design Integrator

The last phase of an embedded system design is the integration of various hardware formats that belong to the hardware part, software part, and the bus structure. The integrator box shown in Figure 1.2 generates a complete netlist of cells of the target device. The netlist output of this phase along with the memory contents of the embedded processor are mapped into the target device for realizing our embedded system on an FPGA or ASIC.

1.3 Design Tools

The flow described in the preceding section requires a complete and integrated set of design tools and utilities. This section outlines some of the tools that are essential for any embedded system design environment.

1.3.1 Block Diagram Description

At the top level, a system is described as an interconnection of components that may be implemented in hardware or software. The block diagram tool allows hierarchical specification of system components. A component may be as large as a processor/memory system or as small as a single logic gate.

1.3.2 HDL and Other Hardware Simulators

For the verification of the hardware part of a system, HDL simulators and simulation programs for simulating library components are needed. Often, library components are available in some HDL language. Furthermore configurable library parts get translated to a standard HDL, i.e., VHDL or Verilog. Therefore, usually, an HDL simulation is all that is needed for simulation of the hardware part of a system.

1.3.3 Programming Language Compilers

Programming language compilers provide high level execution and functional verification of the software program that runs on the embedded processor. A program that describes the software part of an embedded system must be compiled and run to check and verify the correct operation of this part.

1.3.4 Netlist Simulator

A netlist simulator provides detailed simulation of post-synthesis description of the hardware part of an embedded system. This simulation uses timing files (e.g., SDF files) that are generated by the synthesis process, and generates accurate, close to actual, timing results. Because most synthesis tools generate Verilog or VHDL netlists, an HDL simulator can serve as a post-synthesis netlist simulator.

1.3.5 Instruction Set Simulator

An instruction set simulator (ISS) is a program that is aware of the instruction set of the embedded processor, and simulates the compiler output accordingly. The relation between C simulation and ISS simulation of the software part of an embedded system is like that of pre-synthesis simulation and post-synthesis simulation of the hardware part of the embedded system.

1.3.6 Hardware Synthesis Tool

The hardware synthesis tool uses the hardware description of the hardware part of an embedded system as input, and generates a netlist of the target device. As discussed in Section 1.2, from the view of the hardware part, synthesis of an HDL code with a synthesis tool is analogous to compiling the C code of the software part with a C language compiler.

1.3.7 Compiler for Machine Language Generation

As discussed in Section 1.3.3, software compilers are used for verification of the software part of a system. In addition, compilers have the important role of machine language generation just like synthesis tools generate netlists.

1.3.8 Software Builder and Debugger

Embedded design environments include a software development and debugger for aiding the designer in design of the software part of an embedded system. Editors, code development aids, and debuggers are usually included in this tool set.

1.3.9 Embedded System Integrator

A complete embedded system that includes hardware and software parts is put together using an integrator tool. Different design environments refer to such a tool differently. But, all embedded system design environments have some type of an integrator as part of their tools.

1.4 New Hardware Design Trends

The advent of embedded systems has brought new topics and subjects that a hardware designer must be aware of. This section familiarizes readers with some of the terminologies that are used in relation with an embedded system design.

1.4.1 Configurable Processors

Instead of having a fixed processor, embedded systems use configurable processors so that they can be adjusted for best performance for the applications they are running. Embedded design tools provide facilities for configuring their supported processors.

1.4.2 Standard Bus Structure

The interaction of hardware and software parts of an embedded system requires well defined protocols and their corresponding hardware implementations. Embedded systems provide their own bus structures and they automatically generate their corresponding hardware. This way, designers can focus on functionality of their designs, and not the communication mechanisms.

1.4.3 Software Programming

An embedded system designer is continuously involved in deciding what functionalities go in hardware and what goes into software. For such a designer, knowledge of software and programming languages is essential.

1.4.4 Software Utilities

Compilers, assemblers, and instruction set simulators have become essential tools that a hardware designer must be aware of. Configurable compilers that can be modified to compile programs for a variety of processors can be very useful in design of embedded systems.

1.5 Summary

This chapter introduced the concept of embedded systems. We first discussed digital design abstraction levels and how this is affected by embedded systems. We then discussed a hardware/software design flow and its implementation as an embedded system. This discussion enables us to address tools that are needed for designing embedded systems. The last section in this chapter discussed new trends and highlighted some of the topics that new hardware designers must be aware of. The rest of this book focuses on methods and tools for hardware and software parts of embedded systems.

2 Logic Design Concepts

A thorough understanding of basic logic design concepts is essential for proper use of hardware design languages and proper utilization of related tools. Although some of the mathematical concepts of logic design and elaborate minimization and simplification methods are not as crucial as they used to be a few years ago, it is still important that a hardware designer has a good understanding of logic design concepts. This includes those hardware designers designing at the gate level using schematic capture tools, or those who design hardware at the system level using hardware-software tools.

This chapter gives a review of logic design concepts. The purpose is to highlight only those topics that are essential for HDL-based design. We discuss topics with enough depth for a designer wanting to use hardware design tools targeting programmable devices as the target technology. The orientation of the material is geared towards using FPGA. For this purpose, we will not cover details that are covered in a basic logic design course, and knowledge of the theoretical concepts, and much of the background concepts are assumed here.

The chapter begins with a review of number systems, and basic logic gates. FPGA specific primitives are discussed here. Basics of combinational circuits such as Boolean algebra and Karnaugh maps and design of these combinational circuits are discussed next. We show several RT level combinational components and their designs. We will then focus on memory elements and sequential circuit design. Because of importance of state machines in RT level designs, special attention is given to these circuits in this chapter. The last section of this chapter puts all topics discussed in its preceding sections into a complete design example.

2.1 Number Systems

The transistor is the basic element of all digital electronic circuits. A transistor in a digital circuit behaves as an on-off switch. Because all elements are based on this on-off switch, they only take two distinct values. These values can be (ON, OFF), (TRUE, FALSE), (3V, 0V), or (**1, 0**).

Because of this two-value system, all numbers in a computer are in base-2 or binary system. On the other hand, we use the decimal system in our every day life. To be able to understand what happens inside a digital system, we have to be able to convert between base-10 (Decimal) and base-2 (Binary) systems.

2.1.1 Binary Numbers

A decimal number has n digits and the weight of each digit is 10^i, where i is the position of digits counting from the right hand side and starting with 0. For example, 3256 is evaluated as:

$$(3256)_D = 6*10^0 +5*10^1 +2*10^2 +3*10^3 = 3256$$

A number in base-2 is evaluated similarly, except that the weights in decimal are 2^i instead of 10^i. For example 10110 is evaluated as:

$$(10110)_B = 0*2^0 +1*2^1 +1*2^2 +0*2^3 +1*2^4 = 22$$

By considering the weights in decimal and multiplying **bi**nary digits (bit) by their weights a binary number is converted to its equivalent decimal.

For conversion from decimal to binary, a decimal number is broken into necessary 2^i parts. Corresponding to the i values for which the decimal number has a 2^i part, there is a **1** in the equivalent binary number. For example $(325)_D$ can be broken into:

256 that is 2^8,
64 that is 2^6,
4   that is 2^2, and
1   that is 2^0.

Therefore, the equivalent binary number has **1**s in positions 0, 2, 6, and 8, which makes the binary equivalent of $(325)_D$ to become $(101000101)_B$.

Methods described above for decimal to binary and binary to decimal also apply to fractional numbers. In this case the weight of digits on the right hand side of the decimal point are 10^{-1}, 10^{-2}, 10^{-3}, ... Similarly, the weights of binary digits on the right hand side of the binary point of a fractional binary number are 2^{-1}, 2^{-2}, 2^{-3}, ...

For example, $(1101.011)_B$ in binary is $(13.375)_D$ in decimal, and decimal $(19.7)_D$ translates to $(10011.101)_B$. When converting from decimal to binary, for keeping the same precision as in the decimal number, a fractional decimal digit translates to 3 fractional bits.

2.1.2 Hexadecimal Numbers

A number in binary requires many bits for its representation. This makes, writing, documenting, or entering them into a computer very error-prone. A more compact way of representing numbers, while keeping a close correspondence with binary numbers, is Hexadecimal representation.

Table 2.1 Hexadecimal Digits

Hex	Decimal	Binary
0	0	0000
1	1	0001
2	2	0010
3	3	0011
4	4	0100
5	5	0101
6	6	0110
7	7	0111
8	8	1000
9	9	1001
A	10	1010
B	11	1011
C	12	1100
D	13	1101
E	14	1110
F	15	1111

Table 2.1 shows Hexadecimal digits and their equivalent Decimal and Binary representations. As shown, a base-16 digit translates to exactly 4 bits. Because of this, conversion from (to) a binary number to its (from) hex (hexadecimal) equivalent are straightforward processes. Therefore, we can use Hex numbers as a compact way of writing binary numbers. Several examples are shown below:

$(10011.101)_B = (13.A)_H$
$(11101100)_B = (EC)_H$

2.2 Binary Arithmetic

In general, binary arithmetic is done much the same way as it is in the decimal system. In straight arithmetic, binary arithmetic is even simpler than decimal because it only involves 1s and 0s.

2.2.1 Signed Numbers

As we discussed earlier, everything inside a digital system is represented by 1s and 0s. This means that we have no way of representing plus (+) or minus (-) signs for signed numbers other than using 1s and 0s. Furthermore, unlike writing on paper and being able to use as many digits as we like, representing numbers inside a digital system is limited by the width of busses, storage units, and lines. Because of these, a binary number in a digital system uses a fixed width, and the left most bit of the number is reserved for its sign.

A simple signed number system is sign and magnitude (S&M) in which a **0** in the left-most position of the number represents a positive and a **1** represents a negative number. For example +25 in 8-bit S&M system is **00011001** and -25 is **10011001**. Note here that enough **0**s are put between the sign-bit and the magnitude of the number to complete 8 bits.

2.2.2 Binary Addition

As mentioned before, binary addition is very similar to decimal addition, and even easier. Adding two numbers starts from the right-hand side and with addition of every two bits a carry is generated. The carry is added to the addition of the next higher order bits. An example binary addition is shown below.

Carry Bits:	0	0	0	1	1	1	1	
A:	0	1	0	0	1	0	1	1
B:	0	0	1	0	1	1	0	1
Result:	0	1	1	1	1	0	0	0

Addition is done in slices (bit positions) and with every add operation, there is a sum and a carry. The sum bit is the add result of the slice being added, and the carry is carried over to the next higher slice. The right-most bit result is the least-significant bit and is calculated first, and the sign-bit is calculated last.

2.2.3 Binary Subtraction

We can perform subtraction in binary using borrows from higher bits. This is similar to the way subtraction is done in decimal. However, this requires a different process from binary addition, which means that a different hardware is needed for its implementation.

2.2.4 Two's Complement System

As an alternative procedure for adding and subtracting, we can write numbers in the 2's complement number system and perform subtractions the same way we add. This signed number representation system is used to simplify signed number arithmetic.

Unlike the S&M system, in the 2's complement system just changing the sign-bit is not enough to change a positive number to a negative number or vice-versa. In this system, to change a positive (negative) number to a negative (positive) number, all bits must be complemented and a **1** must be added to it. For example -25 is calculated as shown below:

```
00011001    (=25)
11100110    (complementing all bits)
00000001    (adding a 1)
11100111    (-25)
```

When subtracting, instead of performing *A-B*, subtraction is done by *A+(-B)*, in which *(-B)* is the two's complement of *B*. As an example consider subtraction of 25 from 93. First, 25 is turned into its two's complement negative representation that is **11100111** (as shown above). Then +93 that is **01011101** and -25 are added together as shown below:

```
  1  1  1  1  1  1  1
  0  1  0  1  1  1  0  1   (+93)
  1  1  1  0  0  1  1  1   (-25)
──────────────────────────
✗ 0  1  0  0  0  0  1  0  0
```

When adding a positive and negative number that results in a positive number, or adding two negative numbers that results in a negative number, a last carry (as in the above example) is generated that is ignored.

2.2.5 Overflow

In the 2's complement arithmetic if adding two positive (negative) numbers, (i.e., numbers whose sign bits are **0** (**1**)), results in a number that has a **1** (**0**) in its sign-bit position, an overflow has occurred. This means that the result cannot fit in the same word length as the operands. The following addition is an overflow and the result is invalid. As shown, the last bit beyond the sign-bit is dropped, as is done in 2's complement arithmetic. The final result of adding two negative numbers is a positive number that cannot be correct.

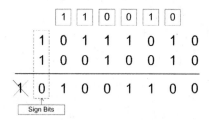

The case of overflow can be corrected by allocating more bits to the numbers involved in the two's complement arithmetic. A 2's complement number can be extended to occupy more bits by extending its sign-bit to the left. For example, 10111010 in 8-bit 2's complement system becomes 1111111110111010 in 16-bit 2's complement system. The overflow example shown above can be corrected if performed in 16-bit system as shown below:

In the above example, two negative numbers are added and a negative result is obtained, no over-flow occurs here.

2.2.6 Fixed Point Numbers

A fixed point number has integer and fractional parts. As with other binary representations, the total number of bits used for this representation is given as part of the definition of a fixed-point number. Furthermore, the position of the binary point (radix point) where the integer part ends and the fractional part begins is fixed. Shown below is a 16-bit fixed point. Radix point of this number is assumed to be to the right of its 5th bit from the right. The decimal value of this number is 971.78125.

16-bit fixed-point number with 5-bit fractional part

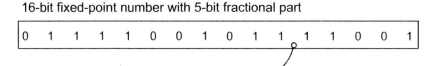

Assumed Binary Point

2.2.7 Floating Point Numbers

Fixed-point numbers are limited in the precision they represent. A more accurate form is the floating-point number representation. A floating point number in a given Radix has a Mantissa, an Exponent and a Sign. The value of the number depends on its Sign, Mantissa, Radix and Exponent, according to the following expression.

Sign Mantissa × *Radix^{Exponent}*

In the binary system *Sign* is **0** for positive numbers and **1** for the negative numbers. *Mantissa* is a normalized integer. Normalization is done such that the integer representing the Mantissa has no leading zeros resulting in more bits for representing the value of the number, and thus a better precision. Furthermore, since normalization removes all leading zeros, the left-most bit of a normalized number is always a **1**. To save space, the left-most **1** of the Mantissa is assumed and never explicitly specified. With this arrangement, a normalized 8-bit Mantissa of value 5 is represented by:

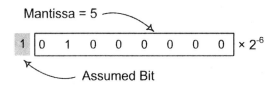

The *Radix* of a floating-point number is always 2 and not explicitly specified. An n-bit *Exponent* is a signed number that is added by $2^{n-1}-1$. The Exponent part of a floating-point number that has an 8-bit Exponent is calculated by taking the actual exponent, e.g., -6 in the above example, and adding 127 to it. Exponent value of -6 in an 8-bit exponent is represented as shown below:

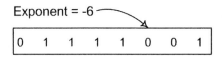

In an 8-bit exponent, all **0**'s represent exact 0 and all **1**'s represent infinity.

For uniformity between digital systems processing floating-point numbers, IEEE has established two standard formats, a single-precision 32-bit format and a double-precision 64-bit format. These formats are shown in Figure 2.1. The 32-bit format has a Sign bit, an 8-bit Exponent, and a 23-bit Mantissa. The 64-bit format has a Sign bit, an 11-bit Exponent and a 52-bit Mantissa.

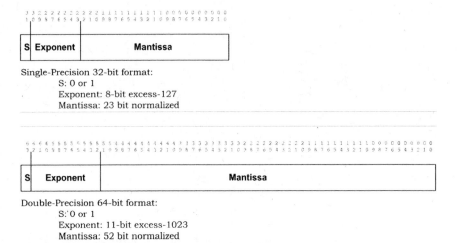

Single-Precision 32-bit format:
    S: 0 or 1
    Exponent: 8-bit excess-127
    Mantissa: 23 bit normalized

Double-Precision 64-bit format:
    S: 0 or 1
    Exponent: 11-bit excess-1023
    Mantissa: 52 bit normalized

Figure 2.1 IEEE Floating-point Formats

2.3 Basic Logic Gates and Structures

The transistor is the basic element for all digital logic components. However, for a design with several million transistors, designers cannot think at the transistor level. Therefore, transistors are put together into more abstract components, called gates, so that designers thinking at the high behavioral level can better relate to such abstract components. Later we will see that even gate structures are not abstract enough and designers need design specification at a higher level of abstraction. For this chapter, however, we concentrate on gates and gate-level designs.

2.3.1 Logic Value System

The **(0, 1)** logic value system is a simple representation for voltage levels in a digital circuit. However, this logic value system fails to represent many situations that are common in digital circuits. For example if a line is connected to neither Gnd nor Vdd, it is neither **0** nor **1**. Or a line that is both driven by logic **0** and logic **1**, is neither a **0** nor a **1**. A more complete system for representation of logic values is the four-value system, shown in Table 2.2.

Table 2.2 Four-Value Logic System

Value	Description
0	Forcing 0 or Pulled 0
1	Forcing 1 or Pulled 1
Z	Float or High Impedance
X	Uninitialized or Unknown

In logic simulations, a line that is not driven through a pull-up or a pull-down assumes **Z**. A line or a wire that is driven both by a pull-up and a pull-down structure appears as **X** in the simulation report.

2.3.2 Logic Function Representation

A logic function can be represented in a variety of ways. In many cases a name that describes the functionality of a logic function may be associated with it. The simple tabular form of a function listing input and output values is referred to as a truth table. This format is easy to read, but is limited in the size of function it can represent. An algebraic form of representation is a more compact way of representing functions and allows manipulation of a function and combining several. Often, a graphical notation may also be associated with a logical function.

Figure 2.2 shows truth table, expression, and a block diagram symbol for a *majority* function. Rows of a truth table may be numbered according to the decimal equivalent of the input combinations when the inputs are treated as a binary number with the left most input being the most significant bit of the binary number. Among various forms of the algebraic representations, a form that is referred to as the *minterm list* (for reasons that will be described later in this chapter) lists all those truth table rows for which the output is **1**. Other algebraic forms correspond to those that can be written in a computer language or those used in printed texts.

The example logic function in Figure 2.2 is a three-input one-output function. The inputs are a, b, and c, and the output is w. The output of this function is **1** when the majority of its inputs are **1**. In what follows, we will show hardware structures used for implementation of logical functions.

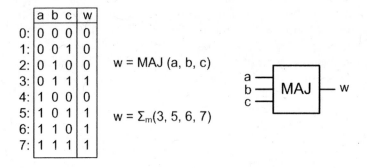

	a b c	w
0:	0 0 0	0
1:	0 0 1	0
2:	0 1 0	0
3:	0 1 1	1
4:	1 0 0	0
5:	1 0 1	1
6:	1 1 0	1
7:	1 1 1	1

$w = MAJ(a, b, c)$

$w = \Sigma_m(3, 5, 6, 7)$

Figure 2.2 Logic Function Representations

2.3.3 Transistors

The CMOS technology uses two types of transistors called NMOS and PMOS. These transistors act like on-off switches with the Gate input controlling connection (current flow) between Drain and Source terminals.

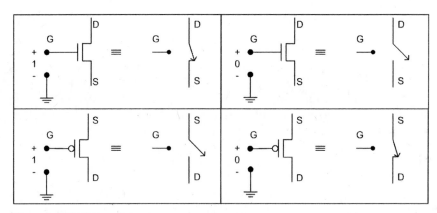

Figure 2.3 MOS Transistors

As shown in Figure 2.3, an NMOS transistor conducts when logic **1** representing a high-voltage level drives its Gate. The conduction path allows current to flow between its Source and Drain terminals. Driving the Gate of an NMOS transistor with logic **0** (low voltage value) causes an open between Source and Drain terminals, which causes no current to flow through the transistor in either way.

As shown in Figure 2.3, opposite to the way an NMOS transistor works, the PMOS transistor conducts when its gate is driven by **0**, and is open when its gate is driven by logic **1** (or high voltage value).

2.3.4 CMOS Inverter

An inverter (also referred to as NOT gate) is a logic gate with an output that is the complement of its input. Transistor level structure of this gate, its logic symbol, its algebraic notations, and its truth table are shown in Figure 2.4.

In the transistor structure shown, if a is **0**, the upper transistor conducts and w becomes **1**. If a is **1**, there will be a conduction path from w to Gnd which makes it **0**. The table shown in Figure 2.4 is called the truth table of the inverter and lists all possible input values and their corresponding outputs. The inverter symbol is a bubble that can be placed on either side of a triangle representing a buffer.

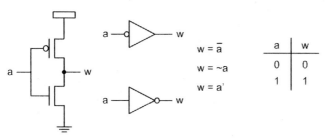

Figure 2.4 CMOS Inverter (NOT gate)

2.3.5 CMOS NAND

A CMOS NAND gate uses two series NMOS transistors for pull-down and two parallel PMOS transistors in its pull-up structure. Figure 2.5 shows structure and notations used for this gate.

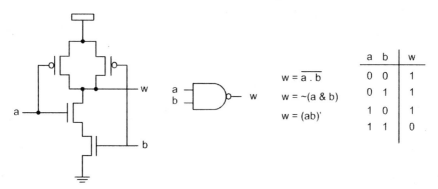

Figure 2.5 CMOS NAND

In the structure shown in Figure 2.5, if a and b are both **1**, there will be a conduction path from w to Gnd, making w **0**. Otherwise, the pull-up structure, instead of the pull-down structure, conducts that forces supply current to flow to w, making this output **1**.

2.3.6 CMOS NOR

A CMOS NOR gate uses two parallel NMOS transistors in its pull-down structure and two series PMOS transistors in its pull-up. Figure 2.6 shows structure and notations used for this gate. For the output of a NOR gate to become **1**, the pull-up structure must conduct. This means that both a and b must be **0**.

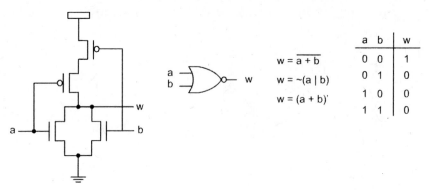

a	b	w
0	0	1
0	1	0
1	0	0
1	1	0

$w = \overline{a + b}$

$w = \sim(a \mid b)$

$w = (a + b)'$

Figure 2.6 CMOS NOR

2.3.7 AND and OR gates

Figure 2.7 shows symbolic notations, algebraic forms and truth tables for AND and OR gates. These gates are realized using inverters on the outputs of NAND and NOR gates.

$w = a \,.\, b$

$w = a \,\&\, b$

a	b	w
0	0	0
0	1	0
1	0	0
1	1	1

$w = a + b$

$w = a \mid b$

a	b	w
0	0	0
0	1	1
1	0	1
1	1	1

Figure 2.7 AND and OR gates

2.3.8 XOR gate

In addition to gates discussed above, several other logic structures become useful for realization of logic functions. One such gate is the XOR gate (Exclusive-OR) of Figure 2.8. This gate is similar to the OR gate except that its output is **1** when only one of its inputs is **1**.

A 2-input XOR gate produces a **1** output when only one of its inputs is **1**. This gate can also be considered as a 1-bit comparator that produces a 1 output when its two inputs are different. Furthermore, a 2-input XOR can be considered a controlled inverter that complements its data input (e.g., *a*) when its control input (e.g., *b*) is **1**, and keeps its unchanged when the control input is **0**. An n-input XOR gate produces a **1** output when the number of **1**'s on its inputs is odd.

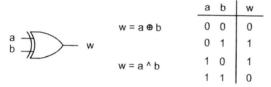

a	b	w
0	0	0
0	1	1
1	0	1
1	1	0

$$w = a \oplus b$$

$$w = a \wedge b$$

Figure 2.8 Exclusive-OR

2.3.9 MUX gate

A very useful logic structure, particularly in the FPGAs, is the multiplexer that selects one of its inputs depending on the value of its select (*s*) input. Shown in Figure 2.9, the *a* input of the MUX appears on its output when *s* is **0**. HDL expression of the MUX and its truth table are shown in Figure 2.9. The right hand side of the equation shown reads as: *if (s is 1) then (b) else (a)*. This is a convenient conditional expression that is used in the C language and Verilog.

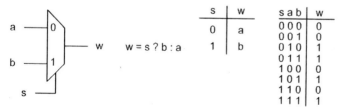

$$w = s\,?\,b:a$$

s	w
0	a
1	b

s	a	b	w
0	0	0	0
0	0	1	0
0	1	0	1
0	1	1	1
1	0	0	0
1	0	1	1
1	1	0	0
1	1	1	1

Figure 2.9 Multiplexer

A multiplexer can be used as a switch that selects one of its inputs depending on the binary value of its select input. For example, the 4-to-1 MUX of Figure 2.10 selects *a*, *b*, *c*, or *d* for when s_1s_0 is 0, 1, 2, or 3.

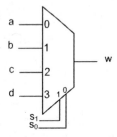

Figure 2.10 A 4-to-1 Multiplexer

A useful feature of the multiplexer is in the ability to turn it into any of the standard gates described above. Wiring s and a of a 2-to-1 multiplexer generates an AND, and wiring s and b turns a 2-to-1 multiplexer into an OR gates. Figure 2.11 shows these two configurations of the multiplexer. For the AND function, since a and s are wired together they will always be the same and, therefore, those rows of the truth table with different a and s values are crossed out. Considering only those rows that are not crosses out, the w output is the AND function of a and b. A similar analysis shows that an OR can be built by tying b and s inputs of a multiplexer.

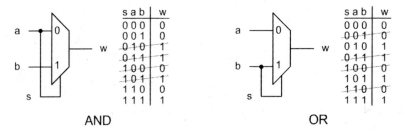

Figure 2.11 Multiplexer Configurations

Multiplexers can be cascaded to build higher order ones. Figure 2.12 shows a 4-to-1 multiplexer built out of three 2-to-1 multiplexers. In this diagram, a, b, c, or d is selected when $s_1 s_0$ is 0, 1, 2, or 3.

2.3.10 Three-State Gates

All gates discussed so far generate a **1** or a **0** on their outputs depending on the values on their inputs. A three-state (also referred to as *tri-state*) buffer (or gate) has a data input (a) and a control input (c). Depending on c, it either passes a to its output (when c is **1**) or it floats the output (when c is **0**). As previously discussed, a float wire is represented by **Z**. Figure 2.13 shows a three-state buffer with true-value output and active-high control input. A truth-table and an algebraic representation are also shown for this structure.

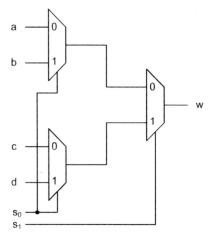

Figure 2.12 Cascading Multiplexers

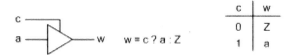

$$w = c\,?\,a\,:Z$$

c	w
0	Z
1	a

Figure 2.13 Three-State Gate

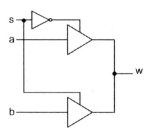

Figure 2.14 Building a Multiplexer with Tri-state Gates

Wiring two tri-state gates as shown in Figure 2.14 generates a 2-to-1 multiplexer. Similarly, a 4-to-1 multiplexer can be built by wiring four tri-state gates.

Other gate structures can be built by use of transistors arranged into complementary NMOS pull-down and PMOS pull-up structures. Furthermore, more complex functions can be built by use of gates discussed above.

2.3.11 Look-up Tables (LUT)

A Lookup table (LUT) structure is a small memory block with n inputs, one output, and 2^n entries. The memory block can be programmed to implement any function of n inputs and one output. This structure is usually implemented by a programmable fuse structure and a program memory and used in programmable devices such as FPGAs.

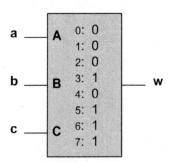

Figure 2.15 LUT Example

An example LUT that implements a majority function is shown in Figure 2.15. LUT inputs are a, b, c, and d, and its output is w. To indicate the function that the LUT is implementing, its contents are indicated by their locations similar to the way a truth-table is expressed (see Figure 2.2).

2.4 Designing Combinational Circuits

Primitive gates and structures discussed in the previous section form a set of structures with which any digital circuit can be designed. Methods of utilizing these parts for implementation of combinational functions are discussed here.

2.4.1 Boolean Algebra

When a design is being done, a designer thinks of the functionality of the design and not its gate structure. To facilitate the use of logic gates and to make a correspondence between logic gates and design functions, Boolean algebra is used.

Boolean variables used in Boolean algebra take **1** or **0** values only. This makes Boolean algebraic rules different from the algebra that is based on decimal numbers.

Boolean algebra operations are AND, OR and NOT and their algebraic notations are ., + and ¬. The AND operator between two operands can be eliminated if no ambiguities occur. An over-bar also represents the NOT operator. Basic rules used for transformation of functions into gates are discussed below. These are Boolean algebra postulates and theorems.

1. Involution Law
   - $\overline{\overline{a}} = a$

2. Identity Laws
   - $a + 0 = a$
   - $a \cdot 1 = a$

3. Null Elements
   - $a + 1 = 1$
   - $a \cdot 0 = 0$

4. Idempotent Laws
   - $a + a = a$
   - $a \cdot a = a$

5. Complementary Laws
   - $a + \overline{a} = 1$
   - $a \cdot \overline{a} = 0$

6. Commutative Laws
   - $a + b = b + a$
   - $a \cdot b = b \cdot a$

7. Associative Laws
   - $a + (b + c) = (a + b) + c$
   - $(a \cdot b) \cdot c = a \cdot (b \cdot c)$

8. Distributive Laws
   - $a + b \cdot c = (a + b) \cdot (a + c)$
   - $a \cdot (b + c) = a \cdot b + a \cdot c$

9. Absorption Laws
   - $a + a \cdot b = a$
   - $a \cdot (a + b) = a$

10. Extended Absorption Laws
    - $a + \overline{a} \cdot b = a + b$
    - $a \cdot (\overline{a} + b) = a \cdot b$

11. Duality
   - If E is true, changing AND (.) to OR (+), OR (+) to AND(.), 1 to 0, and 0 to 1 results in E_D that is also true.

12. DeMorgan's Law
   - $\overline{a \cdot b} = \overline{a} + \overline{b}$
   - $\overline{a + b} = \overline{a} \cdot \overline{b}$

Once designers obtain functionality of their designs, they translate this functionality into a set of Boolean expressions. Using the above rules, this functionality can be manipulated, minimized, and put into a form that can be realized using gates described in Section 2.3.

As an example, consider the overflow situation that may arise in two's complement addition. Consider the sign bits of the operands and the result, a7, $b7$ and $s7$. Overflow (v) occurs if $a7$ is **1**, $b7$ is **1**, $s7$ is **0** or if $a7$ is **0**, $b7$ is **0**, and $s7$ is **1**. This statement can be written as the following Boolean expression

$$v = a7 \cdot b7 \cdot \overline{s7} + \overline{a7} \cdot \overline{b7} \cdot s7$$

Applying Rule 12 (DeMorgan's Law), described above, v becomes:

$$v = \overline{\overline{a7 \cdot b7 \cdot \overline{s7}} \cdot \overline{\overline{a7} \cdot \overline{b7} \cdot s7}}$$

This expression is realized using NAND and NOT gates as shown in Figure 2.16.

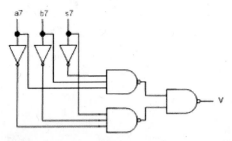

Figure 2.16 An Overflow Detector

As another example of using the above postulates and theorems for minimization of a Boolean expression consider the three-variable function $f(a, b, c)$:

$$f = a \cdot b + \overline{a} \cdot c + b \cdot c$$

According to the Complementary Laws (Rule Set 5) any expression ORed with its complement is **1**, i.e., a ORed with its complement is **1**. Also, according to the Identity Laws (Rule Set 2), any expression can be ANDed with **1** without changing its value, i.e., $b \cdot c \cdot 1$ is the same as $b \cdot c$. Applying these rules to the third product term of f, it can be rewritten as shown below:

$$f = a \cdot b + \overline{a} \cdot c + b \cdot c \cdot \left(a + \overline{a}\right)$$

In the above, using the second Distributive Law (Rule Set 8) removes the set of parenthesis and results in the product terms shown below. Note that in forming these product terms the variables are rearranged according to the second Associate Law (Rule Set 7).

$$f = a \cdot b + \overline{a} \cdot c + a \cdot b \cdot c + \overline{a} \cdot b \cdot c$$

We apply the second Distributive Law to the first and third product terms of the above expression, and the same to the second and fourth terms. After this factorization, the above expression becomes as shown below.

$$f = a \cdot b \cdot \left(1 + c\right) + \overline{a} \cdot c \cdot (1 + b)$$

According to Null Elements (Rule Set 3), in the above expression $1+c$ is 1 and $1+b$ is 1. After these replacements, and after application of the second of the Identity Laws (Rule Set 2), function f becomes as shown below.

$$f = a \cdot b + \overline{a} \cdot c$$

The expression shown above is equivalent to the function we started with, with the difference of using fewer gates for its gate level implementation, having less delays, and consuming less power. It is important to be able to reduce functions as we did above, but as in the above manipulations, it is not always obvious which rules to apply and in what order these rules have to be used. The section below describes a visual method that can help application of Boolean algebra rules.

2.4.2 Karnaugh Maps

Application of rules of Boolean algebra and expressing a hardware function with Boolean expressions is not always as easy as it is in the overflow example above. Karnaugh maps present a graphical method of representing Boolean functions. Karnaugh maps have close correspondence with tabular list of function output values, and at the same time present a visual method of applying Boolean algebra rules.

Figure 2.17 shows a 3-variable truth table and its corresponding karnaugh map (k-map). The truth table shows the listing of output values of a function in a list, and a k-map shows this information in a two-dimensional table.

A Boolean expression can be obtained for function f by reading rows of its truth table. As shown, function f is **1** for four combinations of a, b, and c. In Row #3, f is **1** if a is **0**, b is **1** and c is **1**. This means that the complement of a ANDed with b and ANDed with c make f become **1**. Therefore if $\overline{a} \cdot b \cdot c$ is **1** f becomes **1**. This term is called a product term and since it contains all variables of function f it is also called a minterm of this function. Corresponding to every row of f in which function f is **1** there is a minterm. Function f is **1** if any of its minterms are true. Therefore function f can be written by ORing its four minterms, as shown below:

$$f = \overline{a} \cdot b \cdot c + a \cdot \overline{b} \cdot c + a \cdot b \cdot c + a \cdot b \cdot \overline{c}$$

This form of representing a function is called sum of products, and since the product terms are all minterms, this representation is the Standard Sum Of Products (SSOP).

As shown in Figure 2.17, the same expression could be written by reading the Karnaugh map shown. For this, a product term corresponds to every box of the Karnaugh map that contains a **1**. However, the k-map has certain properties that we can use to come up with a more reduced form of sum of products.

Row	a	b	c	f
0	0	0	0	0
1	0	0	1	0
2	0	1	0	0
3	0	1	1	1
4	1	0	0	0
5	1	0	1	1
6	1	1	0	1
7	1	1	1	1

b c \ a	0	1
00	0	0
01	0	1
11	1	1
10	0	1

f

Figure 2.17 A 3-Variable K-map

For discussion of Karnaugh map properties, we define Boolean and physical k-map adjacency as follows:

Boolean Adjacency: Two product terms are adjacent if they consist of the same Boolean variables and only one variable appears in its true form in one and complement in another (v in one, v-bar in another).

Physical Adjacency: Two k-map boxes are adjacent if they are horizontally or vertically next to each other.

Numbering k-map rows and columns are arranged such that input combinations corresponding to adjacent boxes in the map are only different in one variable. This means that two *Physical Adjacent* boxes are also *Boolean Adjacent*. The main idea in the k-map is that two minterms that are different in only one variable can be combined to form one product term that does not include the variable that is different in the two minterms.

In the k-map of Figure 2.17, $\overline{a} \cdot b \cdot c$ and $a \cdot b \cdot c$ that are Boolean adjacent can be combined into one product term as shown below:

$$\overline{a} \cdot b \cdot c + a \cdot b \cdot c = b \cdot c$$

In the resulting product term, variable a that appears as a in one product term and a-bar in another is dropped. Because of adjacency in the k-maps, the same can be resulted without having to perform the above Boolean manipulations. Figure 2.18 shows minimization of function f using k-map grouping of terms.

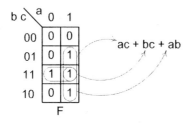

Figure 2.18 Minimizing Function *f*

This figure shows that instead of writing the ORing of minterms $\overline{a} \cdot b \cdot c$ and $a \cdot b \cdot c$ and then using Boolean algebra to reduce them to $b \cdot c$, we can directly write $b \cdot c$ by reading the k-map. Since the two 1's corresponding to these product terms are physically adjacent on the k-map, they are also Boolean adjacent. Therefore, the product term that corresponds to these two adjacent **1**'s is one that includes all the variables except the variable that appears as its true and its complement in the two adjacent k-map boxes (variable a).

The following Boolean manipulations correspond to the mappings shown in Figure 2.18:

$$f = \overline{a} \cdot b \cdot c + a \cdot \overline{b} \cdot c + a \cdot b \cdot c + a \cdot b \cdot \overline{c}$$
$$f = \overline{a} \cdot b \cdot c + a \cdot b \cdot c + a \cdot \overline{b} \cdot c + a \cdot b \cdot c + a \cdot b \cdot \overline{c} + a \cdot b \cdot c$$
$$f = (b \cdot c) \cdot (\overline{a} + a) + (a \cdot c) \cdot (b + \overline{b}) + (a \cdot b) \cdot (c + \overline{c})$$
$$f = (b \cdot c) \cdot (1) + (a \cdot c) \cdot (1) + (a \cdot b) \cdot (1)$$
$$f = b \cdot c + a \cdot c + a \cdot b$$

As shown above, the term $a \cdot b \cdot c$ is repeated 3 times. This is according to Boolean algebra Rule 4 of Section 2.4.1 that states ORing an expression with itself is the same as the original expression. In the k-map, application of this rule is implied by using the k-map box with a **1** that corresponds to *abc=111* in as many mappings as we need (here in 3 mappings). For another example, we use a 4-variable map.

a	b	c	d	w
0	0	0	0	1
0	0	0	1	0
0	0	1	0	1
0	0	1	1	0
0	1	0	0	0
0	1	0	1	0
0	1	1	0	0
0	1	1	1	0
1	0	0	0	1
1	0	0	1	0
1	0	1	0	1
1	0	1	1	1
1	1	0	0	0
1	1	0	1	0
1	1	1	0	0
1	1	1	1	1

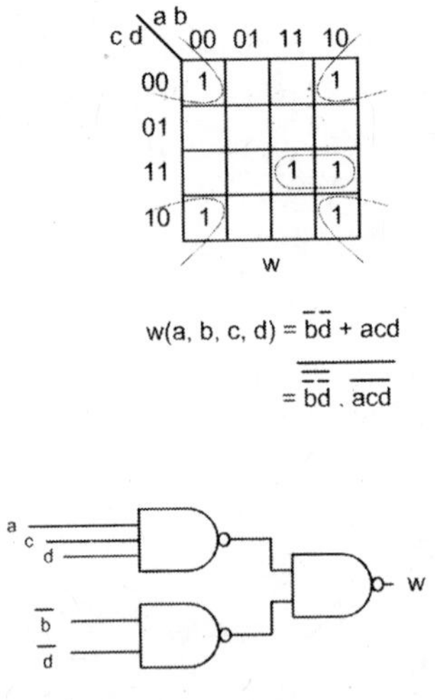
ab
cd
00 01 11 10
00
01
11
10
1
1
1
1
1
1
w
w(a, b, c, d) = $\overline{b}\,\overline{d}$ + acd
= $\overline{\overline{\overline{b}\,\overline{d}} \cdot \overline{acd}}$
a
c
d
$\overline{b}$
$\overline{d}$
w

Figure 2.19 Minimizing a 4-variable Function

A four-variable function, its k-map, its minimal Boolean realization, and its gate level implementation are shown in Figure 2.19. To make a correspondence between Boolean adjacency and k-map physical adjacency, we visualize a k-map as a spherical map in which, in the back of the sphere, the sides of the map and its four corners are adjacent.

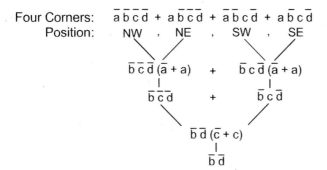

Figure 2.20 Combining Four Corners of a 4-variable Map

With this interpretation, the four corners of the k-map of Figure 2.19 form two product terms that are themselves adjacent. The complete mapping of the four corners of the map results in only one product term. By use of Boolean algebra rules, Figure 2.20 shows justification for combining the four corners of the k-map into one product term. In this diagram *Position* indicates North West, North East, South West and South East of the map.

For implementation of this function another product term is needed to cover minterms $a \cdot b \cdot c \cdot d$ and $a \cdot \overline{b} \cdot c \cdot d$. Because of adjacency of these two terms, variable b drops in the resulting product term. Figure 2.19 shows the minimal realization of function w. After a minimal SOP is obtained, it is converted to an expression using NAND operations by application of DeMorgan's theorem.

2.4.3 Don't Care Values

In some designs, certain input values never occur or if they do occur, their outputs are not important. For example, consider a design that takes a one-digit BCD number (Binary Coded Decimal) as input and generates a 1 output when the input is divisible by 3. The 4-bit input includes combinations ranging from 0 to 15 binary. However, **1010** through **1111** combinations are not valid BCD numbers and are not expected to appear on the circuit inputs.

a	b	c	d	w
0	0	0	0	1
0	0	0	1	0
0	0	1	0	0
0	0	1	1	1
0	1	0	0	0
0	1	0	1	0
0	1	1	0	1
0	1	1	1	0
1	0	0	0	0
1	0	0	1	1
1	0	1	0	-
1	0	1	1	-
1	1	0	0	-
1	1	0	1	-
1	1	1	0	-
1	1	1	1	-

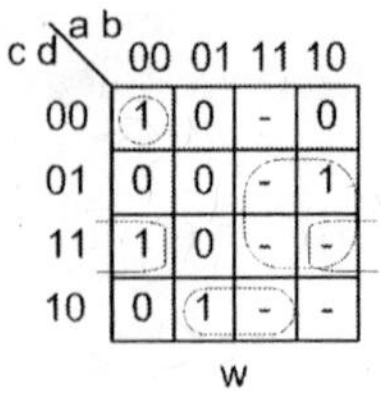

c d \ a b	00	01	11	10
00	1	0	-	0
01	0	0	-	1
11	1	0	-	-
10	0	1	-	-

w

$w(a, b, c, d) = \bar{a}.\bar{b}.\bar{c}.\bar{d} + \bar{b}.c.d + b.c.\bar{d} + a.d$

Figure 2.21 Using Don't Care Values

When we are designing this circuit with a k-map, we have to decide what to do with the k-map boxes that correspond to the invalid BCD numbers. If we fill them with all 1s, we will end up mapping unnecessary 1s. However, if we fill them with all 0s, mapping function minterms may become too limited. The alternative is to leave them as undecided or (Don't Care) values and let the mapping decide what values these invalid cases take.

We use a dash (-), or d or X for showing a Don't Care k-map value. When mapping for a minimal realization, we only use the Don't Care values if they help us form larger maps. This way, those mapped Don't Care values are used as 1s and the rest are 0s.

The solution to the problem stated above is shown in Figure 2.21. Note here that of the 6 Don't Care values 4 are used for forming larger maps and 2 are not mapped.

2.4.4 Minimal Coverage

In the above examples, we showed how minters formed product terms and how product terms were used to form a minimum implementation of a logic function. In these examples we only had one way of selecting best maps, and no decisions as to what maps to select needed to be made. An example Karnaugh map in which the choice of best product terms to choose is not as clear is shown in Figure 2.22.

We will show that in the function shown in this figure several choices exist and we need a procedural method for finding a minimal coverage of the function. For this purpose the following definitions are needed.

2.4.4.1 Implicant. An Implicant of a function is any product term that implies the function. In other words, if product term p becoming 1 causes function f to become 1, then p is an Implicant of f. In Figure 2.22, any product term consisting of any number of minterms is considered to be an Implicant of w.

2.4.4.2 Prime Implicant. A Prime Implicant (PI) of a function is an Implicant that is not completely covered by any other Implicant but itself. In Figure 2.22, all product terms circled in the second Karnaugh map are Prime Implicants of w.

2.4.4.3 Essential Prime Implicant. An Essential Prime Implicant (EPI) of a function is a PI that has at least one minterm that is not covered by any other PI. In Figure 2.22, product terms circled in bold are Essential Prime Implicants of w.

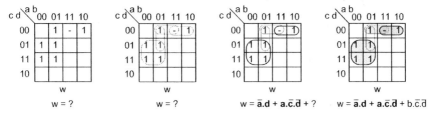

Figure 2.22 Finding Minimal Coverage

Using the above definitions, the below procedure finds the minimal coverage of a function.

2.4.4.4 Finding Minimal Coverage. Using the above definitions, steps shown below find the minimal coverage of a function, f.

1. Of all Implicants of f, list all PIs of f. This is shown in the second map of Figure 2.22.
2. Of all PIs of f, identify EPIs by looking for those PIs that have at least a 1 in the Karnaugh map that is not covered by any other PI. This step is shown in the third map of Figure 2.22.
3. If after the above step all minterms of function f are covered, then the minimal coverage is found. Otherwise choose from non-essential PIs to complete the coverage of f.

4. For the above step, choose those PIs that cover the most number of uncovered PIs. In Figure 2.22, after choosing EPIs of w (third map), minterm number 4 (abcd=0100) remains uncovered. We have two choices for covering this minterm, of which we have selected the PI the shown last in the expression of w in the fourth map.

Generally, Karnaugh maps are used for minimization of basic and simple functions. Other complementary methods are discussed in the sections that follow.

2.4.5 Iterative Hardware

Boolean minimization of functions by use of Boolean rules or, indirectly, by use of k-maps is only practical for small functions. Partitioning based on regularity of a structure, or based on independent functionalities, help in breaking a circuit into smaller manageable circuits.

For example, consider a 4-bit comparator that generates a **1** when its 4-bit A input is greater than its 4-bit B input (Figure 2.23). One way of designing this circuit is to come up with its minimal realization by doing an 8-variable k-map. Obviously, this is not practical.

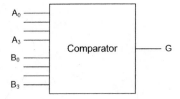

Figure 2.23 A 4-bit Comparator

Alternatively, we can design the comparator by first comparing the most significant bits of its two inputs and working our way into the least significant bit.

The G output of Figure 2.23 becomes **1** if A_3 is greater than B_3. Logically, this means that the $A_3 \cdot \overline{B_3}$ product term forms an AND gate that is an input for an OR gate that generates G. Next, we compare A_2 and B_2 only if the decision for putting a **1** on G cannot be made by A_3 and B_3. This means that the decision based on A_2 and B_2 can only be made if A_3 and B_3 are equal. Therefore the $A_2 \cdot \overline{B_2}$ product term can only cause the G output to become **1** when A_3 and B_3 are equal ($\overline{A_3 \oplus B_3} = 1$). Repeating this logic for all bits of the two inputs from bits 3 down to 0, we will cover all logics that cause G to become

1 when we reach A_0 and B_0. Figure 2.24 shows the resulting hardware for our 4-bit comparator. This hardware has a repeating part, and can easily be extended for larger magnitude comparators.

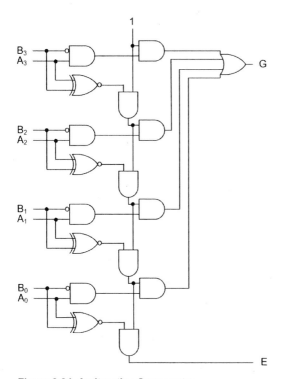

Figure 2.24 An Iterative Comparator

The OR gate shown in this figure can be broken into four 2-input OR gates so that each stage of the circuit uses its own gate, and all stages become exact instances of the same structure. Figure 2.25 shows a comparator bit that is structured such that it can be repeated any number of times for forming a comparator of any size.

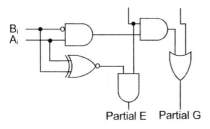

Figure 2.25 A Comparator Bit for a Regular Structure

As another example of iterative hardware, consider the design of an 8-bit adder. Adding two 8-bit numbers is shown in Figure 2.26. As shown in this figure, adding is done bit-by-bit starting from the right-hand side. The process of adding repeats for every bit position. This process uses a carry-in from its previous position ($i-1$), adds it to A_i and B_i, and generates S_i as well as a carry-out for the next position. Therefore, hardware for the 8-bit adder uses eight repetitions of a one-bit adder that is called a Full-Adder (FA).

	0	1	0	1	1	1	1	

A:	0	0	1	0	1	0	1	1
B:	0	0	1	0	1	1	0	1

	0	1	0	1	1	0	0	0

Figure 2.26 Adding Two 8-bit Numbers

The FA hardware has 3 inputs (Carry-in (c_i), bit i of A (a), and bit i of B (b)) and two outputs (Carry-out (c_o) and bit i of sum (s)). Figure 2.27 shows the design of an FA using k-maps. Also shown in this figure is an 8-bit adder using eight full-adders. This adder design is referred to as "ripple-carry" since the carry ripples from one FA to another.

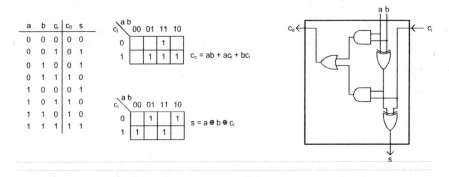

a	b	c_i	c_o	s
0	0	0	0	0
0	0	1	0	1
0	1	0	0	1
0	1	1	1	0
1	0	0	0	1
1	0	1	1	0
1	1	0	1	0
1	1	1	1	1

$c_o = ab + ac_i + bc_i$

$s = a \oplus b \oplus c_i$

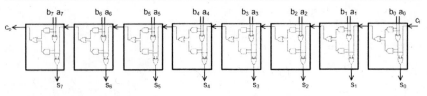

Figure 2.27 An 8-bit Ripple Carry Adder

Hardware components like the comparator and the adder described above are iterative, cascadable, extendable, and in many cases configurable. In designing digital systems it is important to have a library of such packages available. Instead of designing from scratch, a digital designer uses these packages and configures them to meet his or her design requirements.

Discrete logic gates used to match inputs and outputs of various packages are referred to as "Glue Logic".

2.4.6 Multiplexers and Decoders

Other packages that become useful in many high level designs include multiplexers and decoders. Use of multiplexers as primitive structures was discussed in Section 2.3.9; this section discusses multi-bit multiplexers that are often used for high-level RTL design.

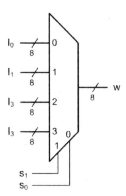

Figure 2.28 An 8-bit 4-to-1 Mux

A multiplexer is like an n-position switch that selects one of its n inputs to appear on the output. A multiplexer with n inputs is called an *n-to-1 Mux*. The number of bits of the inputs (b) determines the size of the multiplexer. A multiplexer with n data inputs requires $s=log_2(n)$ number of select lines to select one of the n inputs; i.e., $2^s=n$.

For example, a multiplexer that selects one of its four $(n=4)$ 8-bit $(b=8)$ inputs is called an 8-bit 4-to-1 Mux. This multiplexer needs 2 select lines $(s=2)$. Schematic diagram of this multiplexer is shown in Figure 2.28. This circuit can be built using an array of AND-OR gates or three-state gates wired to implement a wired-OR logic. Figure 2.29 shows the gate level design of a 1-bit 4-to-1 Mux.

Multiplexers are used for data selection, bussing, parallel-to-serial conversion, and for implementation of arbitrary logical functions. A 1-bit 2-to-1 Mux can be wired to implement NOT, AND, and OR gates. Together with a NOT, a 2-to-1 Mux can be used for implementation of most primitive gates. Because of this property, many FPGA cells contain multiplexers for implementing logic functions.

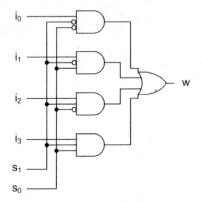

Figure 2.29 A 4-to-1 Mux

Another part that is often used in high level designs is a decoder. Generally, a combinational circuit that takes a certain code as input and generates a different code is referred to as a decoder. For example, a circuit that takes as input a 4-bit BCD (Binary Coded Decimal) and generates outputs for display on a Seven Segment Display (SSD) is called a BCD to SSD decoder.

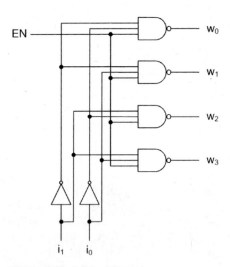

Figure 2.30 A 2-to-4 Decoder

A more accurate definition is that a decoder has as many outputs as it has combinations of inputs. For every combination of values on its inputs a certain output of the decoder becomes active. For example a 2-to-4 binary decoder has 2 inputs forming four combinations. Its four outputs become active for input combinations, **00**, **01**, **10**, and **11**. The gate level design of this decoder is shown in Figure 2.30.

The selected output in Figure 2.30 becomes **0** and all others are **1**. The circuit also has an enable input, *EN*. For the decoder to become operational, this input must be **1**. The enable input is useful for cascading decoders.

2.4.7 Activity Levels

Activity levels for input and output ports of digital circuits refer to the way that these ports function. For example an active-low output (like the decoder described above) is **1** when not active and it becomes **0** when active. An active-low enable input of a circuit makes the circuit operational when it is **0**. When such an input is **1**, circuit outputs become inactive. The *EN* input of the decoder described above is an active-high enable input.

A NAND gate can be looked at as an AND gate with an active-low output and active-high inputs. A NAND gate can also be looked at as an OR gate with active-low inputs and active-high output (see Figure 2.31). The following Boolean expressions justify these views of a NAND gate:

$$\overline{a \cdot b} = \neg(a \cdot b)$$
$$\overline{a \cdot b} = (\overline{a} + \overline{b})$$

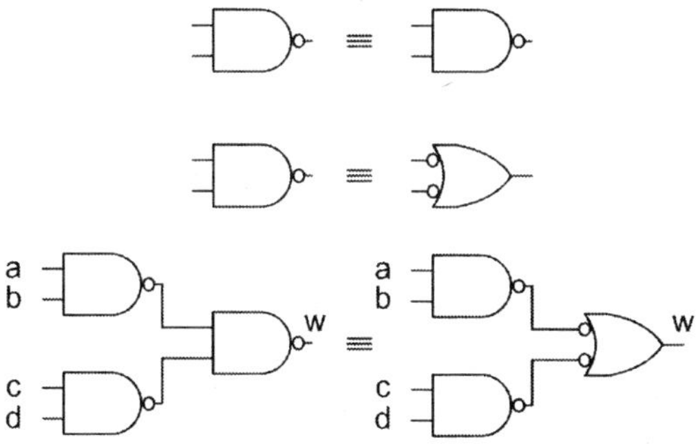

Figure 2.31 NAND Gate Activity Levels

Using correct polarities and notations with correct activity-level markings, make circuit diagrams more readable. For example in

Figure 2.31 the two circuits with w output are equivalent. The one on the left requires writing Boolean expressions to understand its functionality, but the function of the one on the right can easily be understood by inspection.

2.4.8 Enable / Disable Inputs

Many digital logic packages, like multiplexers and decoders come with enable (EN) and/or output-enable (OE) inputs. When an input is referred to as EN, it means that if this input is not active, all circuit outputs are inactive. On the other hand, an OE input is for three-state control of the output. In a circuit with an OE input, if OE is active, the outputs of the circuit are as defined by the function of the circuit. However when OE is inactive, all circuit outputs become high-impedance or float (Z value).

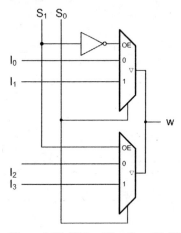

Figure 2.32 Wiring Circuits with OE Control Inputs

Circuits with three-state outputs require an OE input. Outputs of such circuits can be wired to form wired-OR logic. Figure 2.32 shows two 2-to-1 multiplexers with three-state outputs that are wired to form a 4-to-1 multiplexer. If the multiplexers of Figure 2.32 had EN inputs instead of OE inputs, forming the final output of the circuit, w, would require an OR gate.

2.4.9 A High-Level Design

In the first part of this chapter we showed that instead of using transistors in a design, we wire them to form upper-level structures (primitive gates) with easier functionalities that digital designers can relate to. In the second part, we discussed the use of gates in still higher level structures such as adders, comparators, decoders and multiplexers. With these higher-level structures, designers will be able to think at a more functional level and not have to get involved in putting thousands of gates together for a simple design.

This level of design is called RT (Register Transfer) level. In today's designs, designers think at this level and most design tools work at this level. Most design libraries include configurable RTL components for designers to use.

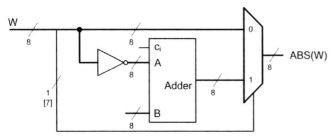

Figure 2.33 An Absolute Value Circuit

As a simple RT level design, consider an 8-bit Absolute-Value calculator. The circuit takes a positive or negative 2's complement input and generates the absolute value of its input on its 8-bit output. The circuit diagram using RT level packages is shown in Figure 2.33.

The circuit uses an array of eight NOT gates to form the complement of the input. Using the adder shown a **1** is added to this complement to generate the two's complement of the input. The multiplexer on the output uses the sign-bit of the input to select the input if it is positive or to select the 2's complement of the input if the input is a negative number.

2.5 Storage Elements

Circuits discussed so far in this chapter were combinational circuits that did not retain a history of events on their inputs. To be able to design circuits that can make decisions based on past events, we need to have circuits with memory that can remember some of what has happened on their inputs. This section discusses the use of memory elements that help us achieve this.

The past history of a memory circuit participates in determination of its present output values. Therefore, outputs of these circuits are not only a function of their inputs, but also a function of their past history. This history enters the logic structure of a memory circuit by

way of feedbacks from its outputs back to its inputs. The more lines
that are fed back means that the circuit remembers more of its past.

2.5.1 The Basic Latch

The circuit shown in Figure 2.34 is the basic latch. We will show that
this circuit has some memory. The circuit has one feedback line from
its y output to its input. One feedback line can take **0** or **1**, which
means that the circuit remembers only two things from its past.

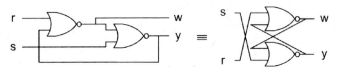

Figure 2.34 The Basic latch (Two Equivalent Circuits)

Applying the waveform shown in Figure 2.35 to the inputs of the
latch of Figure 2.34 shows that a pulse on s sets the w output to **1** and
a pulse on r sets it to **0**.

Note from of the timing diagram of Figure 2.35 that at time a
when s and r are both **0**, w is **0**, and at time b when the same exact
values appear on the circuit inputs the output is **1**. This reveals that
the output depends not only on the present input, and that the circuit
is remembering something from its past history.

An interpretation of the behavior of this circuit is that a complete positive pulse on s causes w to **s**et and a complete pulse on r
causes it to **r**eset. Because of this behavior, the circuit of Figure 2.34
is called an SR-Latch. This structure is the basic element for most
static memory structures. Alternative structures that implement this
memory behavior use NAND gates or inverters and pass-transistors.

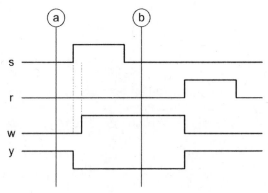

Figure 2.35 Setting and Resetting the Basic Latch

2.5.2 Clocked D Latch

The memory behavior of the SR-latch does not have a close correspondence to the way we think about storing data or saving information. The structure shown in Figure 2.36 improves this behavior. In this structure, when *clock* is **1** a **1** on D causes s to become **1** which causes Q to set to **1**, and a **0** on D causes r to become **1** to reset Q.

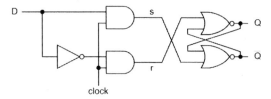

Figure 2.36 A Clocked D-Latch

This structure behaves as follows: when *clock* becomes **1**, the value of D will be stored until the next time that *clock* becomes **1**. At all times this value appears on Q and its complement on $\overline{Q}$.

This behavior that at a given time, determined by the *clock*, a value is stored until the next time we decide to store a new value, corresponds more to the way we think about memories. The circuit of Figure 2.36 is called a clocked D-latch and is used in applications for storing data, buffering data, and temporary storage of data. For storing multiple bits of data, multiple latches with a common clock can be used. Figure 2.37 shows a quad latch using a symbolic representation of a latch.

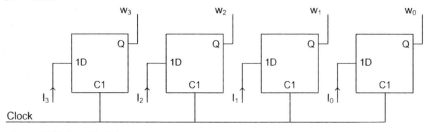

Figure 2.37 Quad Latch

In Figure 2.36 when *clock* is 1, data on D pass through the latch and reach Q and changes on D directly affect Q. Because of this, this structure is called a transparent latch. The symbolic notation of latch shown in Figure 2.37 indicates dependence of D on *clock*. This shows that control signal *1* that is the *clock* signal controls the D input.

2.5.3 Flip-Flops

The latch as discussed above is a good storage element, but because of
its transparency, it cannot be used in feedback circuits. Take for example a situation that the output Q of the latch goes through a combinational circuit and feeds back to its own inputs (see Figure 2.38).
Because latches are transparent, the feedback path stays open while
the clock signal is active. This will result in an unpredictable latch
output due to the uncontrolled number of times that data feeds back
through the logic block. In some cases the output oscillates while the
clock is active.

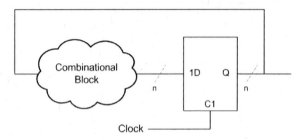

Figure 2.38 Latch Feedback Causes Unpredictable Results

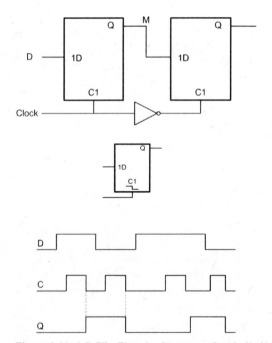

Figure 2.39 A D-Flip-Flop, Its Structure, Symbolic Notation and Waveform

To overcome the above mentioned problem, a structure without transparency must be used. Conceptually this is like the use of double doors for building entrances. At any one time only one door is open to keep the air-conditioned air inside the building and eliminate flow of air into the building.

For our case, we use two latches with inverting clocks as shown in Figure 2.39. When *Clock* is **0**, the first latch stops data on *D* from propagating to the output. When *clock* becomes **1**, data is allowed to propagate only as far as the output of the first latch (*M*). While this is happening the second latch stops data on *M* from propagating any further. As soon as *Clock* becomes **0**, *D* and *M* are disconnected and data stored in *M* propagates to *Q*. The latch on the left is called master, and the one on the right is the slave. This structure is called a master-slave D-flip-flop. At all times, input and output of this structure are isolated.

Other forms of flip-flops that have this isolation feature use a single edge of the clock to accept the input data and affect the output. Such structures are called edge-trigger flip-flops. Figure 2.40 shows a rising- and a falling-edge D-flip-flop. The triangle indicates edge triggering and the bubble on the clock input of the circuit on the right indicates negative (falling) edge triggering.

Figure 2.40 Edge Trigger Flip-flops

2.5.4 Flip-Flop Control

The initial value of a flip-flop output depends on its internal gate delays, and in most cases is unpredictable. To force an initial state into a flip-flop, set and reset control inputs should be used. Other control inputs for flip-flops are clock-enabling and three-state output control signals.

A *Set* or *Preset* control input forces a flip-flop into its **1** state, and a *Reset* or *Clear* input forces it to **0**. We refer to these signals as flip-flop initialization inputs. Such control inputs can act independent of the clock, or act like the D-input with the specified edge of the clock. In the former case, the initialization inputs must be put into the internal logic of a flip-flop and are called asynchronous control inputs. In the case that a control signal only affects the flip-flop when the flip-flop is clocked, it is called a synchronous control input. Synchronous control inputs can be added to a flip-flop by adding external logic. Figure 2.41 shows four flip-flops with various forms of synchronous and asynchronous controls. To indicate clock dependency in a flip-flop with a synchronous control, the clock identifier (number *1* on the right hand side of letter *C*) is used on the left hand side of the control signal name.

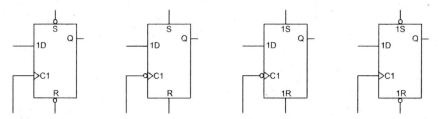

Figure 2.41 Flip-flops with Synchronous and Asynchronous Control

Another control input for flip-flops is a clock enabling input. When enabled, the flip-flop accepts its input when a clock pulse arrives, and when disabled, clocking the flip-flop does not change its state.

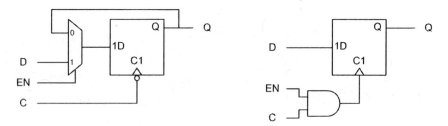

Figure 2.42 Clock Enabling

Two implementations for clock enabling are shown in Figure 2.42. The one on the left, circulates data back into the flip-flop when it is disabled (*EN* = 0). When enabled, the external data on the *D* input is clocked into the flip-flop. The structure shown on the right, uses an AND gate to actually gate the clock and stop the flip-flop from accepting data on its *D* input. This is called clock gating and because of its critical timing issues, care must be taken when using this implementation.

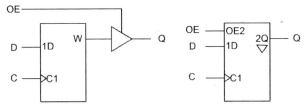

Figure 2.43 Three-State Control

Some flip-flops come with three-state outputs. In this case, a three-state buffer on the output is controlled by an *OE* (Output Enable) control input. Hardware implementation of this feature and its symbolic notation are shown in Figure 2.43. The use of a triangle on the output side of the symbolic notation is useful, but is not always used.

2.5.5 Registers

The structure formed by a group of flip-flops with a common clock signal and common control signals is called a register. As with flip-flops, registers come in different configurations in terms of their enabling, initialization and output control signals. Figure 2.44 shows an 8-bit register with an active-low three-state output control and a synchronous active low reset. A register is also said to a group of latches.

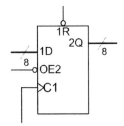

Figure 2.44 An 8-bit Register

2.6 Sequential Circuit Design

This section discusses design of circuits that have memory; such circuits are also called sequential circuits. We will first discuss the design of sequential circuits using discrete parts (gates and flip-flops) and then focus our attention on sequential packages. This approach is similar to what was done in Section 2.4 for combinational circuits.

2.6.1 Finite State Machines

A sequential circuit is a digital system that has memory and decisions it makes for a given input depend on what it has memorized. These circuits have local (inside flip-flops) or global feedbacks and the number of feedbacks determine how much of its past history it remembers.

The number of states of a sequential circuit is determined by its memory. A circuit with n memory bits has 2^n possible states. Signals or variables representing these states (n of them) are called state variables. Because sequential circuits have a finite number of states, they are also called finite-state machines, or FSM.

All sequential circuits - from a single latch to a network of high performance computers - can be regarded as an FSM. Such a machine can be modeled as a combinational circuit with feedback. If the feedback path includes an array of flip-flops with a clock for controlling the timing of data feeding back, the circuit becomes a synchronous sequential circuit. Figure 2.45 shows the Huffman model of synchronous sequential circuits. This model divides such a circuit into a combinational part and a register part.

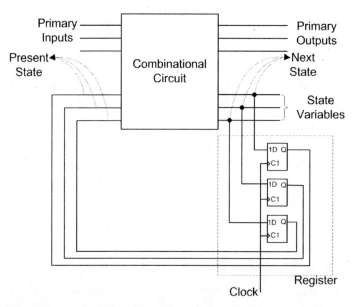

Figure 2.45 Huffman Model of a Sequential Circuit

The clock shown is the synchronization signal. Outputs that are fed back to the inputs are state variables. The inputs of the flip-flops become the present state of the machine after the circuit clock ticks.

The circuit decides on its outputs and its next state based on its inputs and its present state.

2.6.2 Designing State Machines

To show the design process for FSMs, we use a simple design with one input and one output. The circuit searches on its input for a sequence of 1s and 0s. This circuit is called a sequence detector, and the procedure used in its design applies to the design of very large FSMs.

2.6.2.1 Problem Description. A sequence detector with one input, x and one output, w, is to be designed. The circuit searches on its x input for a sequence of **1011**. If in four consecutive clocks the sequence is detected, then its output becomes **1** for exactly one clock period. The circuit continuously performs this search and it allows overlapping sequences. For example, a sequence of **1011011** causes two positive pulses on the output. Figure 2.46 shows a timing diagram example of this search.

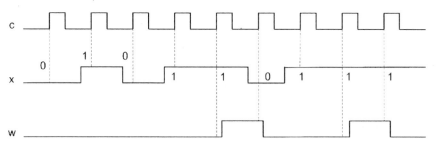

Figure 2.46 Searching for 1011

2.6.2.2 State Diagram. The above problem description is complete, but does not formally describe the machine. To design this sequence detector, a state diagram which has representations for all states of the machine must be used. A state diagram is like a flowchart and it completely describes our state machine for values that occur on its input. Input events are only considered if they are synchronized with the clock. Figure 2.47 shows the state diagram of our **1011** detector.

As shown in this state diagram, each state has a name (A through E) and a corresponding output value (w is **1** in E and **0** in the other states). There are edges out of each state for all possible values of circuit inputs.

Since we only have one input, two edges, one for x=**0** and one for x=**1** are shown for each state. Since the machine is to detect **1011**, this sequence always ends in state E, no matter what state we start from. In each state, if the input value that takes the machine one state closer to the output is not received (e.g., receiving a **0** in state *D*), the machine goes to the state that saves the most number of bits of the correct sequence. For example a **0** in state *D* takes the machine to state *C* that has a **0** output, since state *D* is the state that **101** has been detected and a **0** on *x* makes the remembered received bits **1010**. Of these remembered bits only the last **10** can be used towards a correct sequence, and therefore the machine goes to state *C* that remembers this sequence.

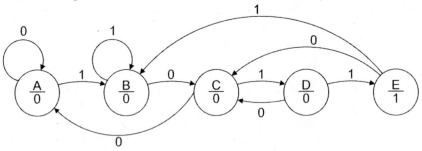

Figure 2.47   State Diagram for the 1011 Detector

2.6.2.3   State Table.   Design of digital circuits requires data and behavior of the circuit that is being designed to be represented in a tabular form. This enables us to form truth tables and/or k-maps from this behavioral description. Therefore, the next step in design of our sequence detector is to form a table from the state diagram of Figure 2.47.

State	0	1	
A	A	B	0
B	C	B	0
C	A	D	0
D	C	E	0
E	C	B	1
	State$^+$		Z

Figure 2.48   State Table of the 1011 Detector

Figure 2.48 shows the state table that corresponds to this state diagram. The first column shows the present states of the machine, *State*. Table entries are the next states of the machine for *x* values **0** and **1**. The table also shows the output of the circuit for various states of the machine. State *E* goes to state *C* for *x* of **0** and to state *B* if *x* is **1**. The value of the *w* output in this state is **1**.

2.6.2.4 State Assignment. The state table of Figure 2.48 takes us one step closer to the hardware implementation of our sequence detector, because the information is represented in a tabular form instead of the graphical form of Figure 2.47. However, hardware implementation requires all variables in a circuit description to be in binary. Obviously, in our state table, state names are not in binary.

For this binary representation, we assign a unique binary pattern (binary number) to each of the states of our state table. This step of the work is called "state assignment". Because we have five states, we need five unique binary numbers, which means that we need three bits for giving our states unique bit patterns. Figure 2.49 shows the state assignment that we have decided to use for this design.

Specific bit patterns given to the states of a state machine are not important. Binary values assigned to each state become values for y_2, y_1, and y_0. These variables are state variables of our machine.

State	y_2	y_1	y_0
A	0	0	0
B	0	0	1
C	0	1	0
D	0	1	1
E	1	0	0

Figure 2.49 State Assignment

2.6.2.5 Transition Table. Now that we have binary values for the states of our state machine, state names in the state table of Figure 2.48 must be replaced with their corresponding binary values. This will result in a tabular representation of our circuit in which all values are binary. This table is called a transition table and is shown in Figure 2.50.

			x		
y_2	y_1	y_0	0	1	
A: 0	0	0	0 0 0	0 0 1	0
B: 0	0	1	0 1 0	0 0 1	0
C: 0	1	0	0 0 0	0 1 1	0
D: 0	1	1	0 1 0	1 0 0	0
E: 1	0	0	0 1 0	0 0 1	1
			y_2^+ y_1^+ y_0^+		w

Figure 2.50 Transition Table for the 1011 Detector

A transition table shows the present values of state variables (y_2, y_1 and y_0) and their next values (y_2^+, y_1^+, y_0^+). Next state values are those that are assigned to the state variables after the circuit clock ticks once. Since only five of eight possible states are used, three combinations of state-variable value are unused. Therefore, next state and output values for these table entries are don't care values.

2.6.2.6 Excitation Tables.
So far in the design of the **1011** detector, we have concentrated on the design of the complete circuit including its combinational and register parts, as defined by the Huffman model of Figure 2.45. We have been discussing present and next state values, which obviously imply a sequential circuit.

In the next step of the design, we separate the combinational and the register parts of the design. The register part is simply an array of flip-flops with a common clock signal. The combinational part is where present values of flip-flops (their outputs) are used to generate flip-flop input values that will become their next state values.

Because a D-type flip-flop takes the value on its D input and transfers it into its output after the edge of clock, what we want to become its next state is the same as what we put on its D input. This means that the required D input values generated by the combinational part of a sequential circuit are no different than their next state values ($Q^+ = D$). Therefore, tables for values of D_2, D_1 and D_0 in our **1011** sequence detector are the same as those for y_2^+, y_1^+, and y_0^+. Flip-flop input tables are called excitation tables that are shown in Figure 2.51 for our design.

			x		
y_2	y_1	y_0	0	1	
0	0	0	0 0 0	0 0 1	0
0	0	1	0 1 0	0 0 1	0
0	1	0	0 0 0	0 1 1	0
0	1	1	0 1 0	1 0 0	0
1	0	0	0 1 0	0 0 1	1
1	0	1	- - -	- - -	-
1	1	0	- - -	- - -	-
1	1	1	- - -	- - -	-
			D_2 D_1 D_0		w

Figure 2.51 Flip-flop Excitation Tables

2.6.2.7 Implementing the Combinational Part.
Now that we have separated the combinational and register parts of our design, the next step is to complete the design of the combinational part. This part is

completely described by the table of Figure 2.51. This table includes values for $D2$, $D1$ and $D0$ in terms of x, $y2$, $y1$, and $y0$, as well as values for w in terms of $y2$, $y1$ and $y0$. Karnaugh maps shown in Figure 2.52 are extracted from the table of Figure 2.51.

   Figure 2.52 also shows Boolean expressions for the D-inputs of $y2$, $y1$ and $y0$ flip-flops. The four-input (x, $y2$, $y1$, and $y0$), four-output (w, $D2$, $D1$, and $D0$) combinational circuit is fully defined by Boolean expressions of Figure 2.52.

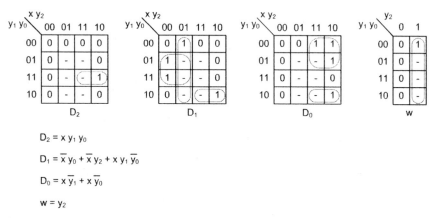

$$D_2 = x\, y_1\, y_0$$

$$D_1 = \overline{x}\, y_0 + \overline{x}\, y_2 + x\, y_1\, \overline{y_0}$$

$$D_0 = x\, \overline{y_1} + x\, \overline{y_0}$$

$$w = y_2$$

Figure 2.52 Implementing Combinational Part

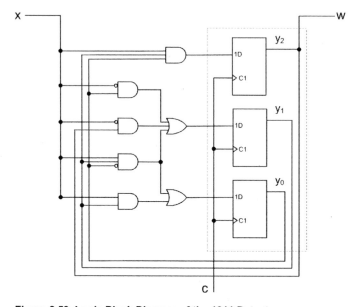

Figure 2.53 Logic Block Diagram of the 1011 Detector

2.6.2.8 Complete Implementation. The design of the **1011** sequence detector will be completed by wiring the gate-level realization of the combinational part with the flip-flops of the register part. This realization is shown in Figure 2.53.

Implementation of the **1011** detector is according to the Huffman model of Figure 2.45. The box on the left is the combinational part, and the one on the right is the register part. State variables of this circuit are y_2, y_1 and y_0 that are fed back from the outputs of the combinational part back into its inputs through the register part. The clocking mechanism and initialization of the circuit only affect the register part. For asynchronous initialization of the circuit, flip-flops with asynchronous set and/or reset inputs should be used. For synchronous initialization, AND gates on the D inputs should be used for resetting and OR gates for setting the flip-flops.

2.6.3 Mealy and Moore Machines

The design presented in the previous section produces an output that is fully synchronous with the circuit clock. In its state diagram, since the output is specified in the states of the machine, while in a given state, the output is fixed. This can also be seen in the circuit block diagram of Figure 2.53 in which the logic of the w output only uses the state variables, and does not involve x. This state machine is called a Moore machine. A more relaxed timing can be realized by use of a different machine that is referred to as a Mealy machine.

Figure 2.54 shows the Mealy state diagram of the **1011** detector. As shown, the output values in each state are specified on the edges out of the states, along with input values. This means that while in a given state, the value on x decides the value of the output. For example, in state D, if x is **0**, w is **0** and if x is **1**, w is **1**.

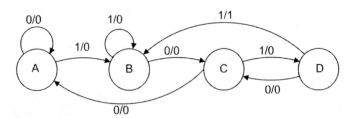

Figure 2.54 Mealy State Diagram

With this dependency, changes on x propagate to the output even if they are not accompanied by the clock. The implementation of a Mealy machine is similar to that of a Moore machine, except that the output k-map involves the inputs as well as the state variables. A sequence detector that is implemented with a Mealy machine usually requires one state less than the Moore machine that detects the same sequence. If implemented as a Mealy machine, our detector requires four states, two state variables, and three 3-variable Karnaugh maps for the two state variables and the output.

2.6.4 One-Hot Realization

Instead of going through steps discussed in Section 2.6.2 for gate-level implementation of a state machine, a more direct realization can be obtained by using one flip-flop per state of the machine. Since in a state diagram only one state is active at any one time, only one of the corresponding flip-flops becomes active. This method of state assignment is called *one-hot* assignment. This implementation uses more flip-flops than the binary state assignment discussed in Section 2.6.2, but uses fewer logic gates for activation of the flip-flops.

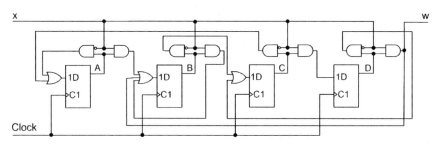

Figure 2.55 One-hot Implementation

One-hot implementation of the Mealy machine of Figure 2.54 is shown in Figure 2.55. Output of the AND gates on the outputs of the flip-flops correspond to the edges that come out of the states of the state diagram. These AND gates are conditioned by $x=0$ or $x=1$. The four flip-flops used yield 2^4 possible states. Of these 16 states only four are used (**1000**, **0100**, **0010** and **0001**). Initialization of a one-hot machine should be done such that it is put into one of its valid states. Starting the machine in **0000** is wrong because it will never get out of this state.

Some of the advantages of one-hot machines are their ease of design, regularity of their structure, and testability.

2.6.5 Sequential Packages

As there are commonly used combinational packages, like adders, decoders and multiplexers, there are commonly used sequential packages like registers, counters and shifters. An RT level designer first partitions his or her design into such predefined components, and will only resort to designing with discrete components when there are no packages that meet the design requirements.

2.6.5.1 Counters. Counters are used in many RT level designs. A counter is a sequential circuit that counts a certain sequence in ascending or descending order. An n-bit binary up-counter counts n-bit numbers in the ascending order.

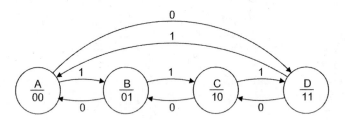

Figure 2.56 State Diagram of a 2-bit Counter

As an example we will show the design of a 2-bit up-down counter. With each clock pulse, when UD is **1** it counts up and when UD is **0** it counts down.
In the count-up mode the next count after **11** is **00**, and in the count-down mode the next count after **00** is **11**.

The state diagram for this counter is shown in Figure 2.56. Counter count outputs are shown in each state. This is a Moore state machine and the procedure discussed earlier in this chapter can be used for its design. However, because of the simple sequencing of counter circuits, many of the steps discussed in Section 2.6.2 can be skipped and we can go directly from the description of the counter to its transition tables. Furthermore, if we decide to use D-type flip-flops for our counter, excitation tables, or even D-input k-maps, can be written based on the count sequence. Figure 2.57 shows k-maps generated directly from the up and down sequences of the counter of Figure 2.56.

In the right columns of the k-maps when $UD=1$, values for $D1$ and $D0$ are set to take C_1 and C_0 through the **00, 01, 10, 11, ...** sequence. In the left columns of the k-maps, D_1 and D_0 values make the counter count the **11, 10, 01, 00, ...** sequence. Circuit shown in Figure 2.58 performs the basic up- and down- counting for our 2-bit counter.

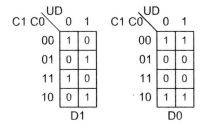

Figure 2.57 Excitation K-maps for a 2-Bit Up-Down Counter

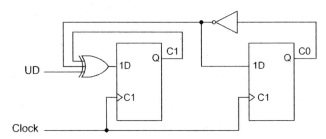

Figure 2.58 A Two-Bit Up-Down Counter

In addition to the basic counting implemented by the circuit of Figure 2.58, other features in a counter include, resetting, parallel loading, enabling, carry-in and carry-out. Resetting a counter is like resetting registers. Asynchronous resetting forces the counter into its initial state and acts independent of the clock. Synchronous resetting loads the initial state of the counter through the D-inputs of counter flip-flops, which obviously requires the proper clocking of the register.

To start counting from a given state, the counter is put into parallel-load mode and the designated start state is loaded into the flip-flops of the counter.

In this mode the counter acts just like a register. Inputs of flip-flops of a counter with parallel load feature must be available outside of the counter package.

An enable input for a counter makes it count only when this input is active. This signal controls clocking of data into the individual flip-flops of the counter.

Some counters have carry-in and carry-out input and output signals that are used for cascading several of them. Carry-out output of a modulo-n counter becomes 1 when the counter reaches its maximum count. The carry-in input of a counter (if it exists) acts just like an enable input except that it also enables the carry-out of the counter. Figure 2.59 shows a two-bit up-counter with added features of synchronous reset, enable, parallel load, carry-in and carry-out.

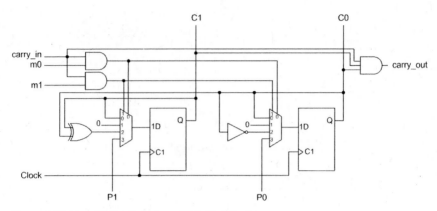

Figure 2.59 Two-Bit Up-Counter with Added Features

The m_1 and m_0 inputs of the counter shown in Figure 2.59 are its mode inputs. These inputs control data that are clocked into the flip-flops. If mode is 0 ($m_1, m_0 = 0, 0$), the counter is disabled. In mode 1 the counter resets to 0, in mode 2 the counter counts up. Mode 3 is for parallel load; in this mode P_1 and P_0 are loaded into the counter. The counter only counts if *carry_in* is **1**, otherwise it is disabled. When *carry_in* is **1** and counter reaches **11**, the *carry_out* becomes **1**. Cascading counters can be done by connecting *carry_out* of one to the *carry_in* of another.

2.6.5.2 Shifters. Shift registers are registers with the property that data shifts right or left with the edge of the clock. Shift registers are used for serial data collection, serial to parallel, and parallel to serial converters.

Figure 2.60 shows a 4-bit right shifter. With every edge of the clock data in the register moves one place to the right. Data on S_i (serial-in) starts moving into the register and data in the register moves out bit-by-bit from S_o (serial-out).

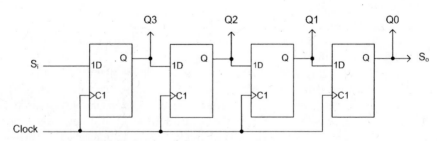

Figure 2.60 A 4-bit Shift Register

Shift-registers can be easily cascaded by connecting S_o of one to the S_i of another. Other functionalities included in these packages are left-shift, parallel load, enable, and reset. These features can be included in much the same way as in counters (Figure 2.59). Shift-registers with three-state output control use three-state gates on their output, like what is done in registers (Figures 2.43 2.44).

2.7 Memories

In their simplest form, memories are two-dimensional arrays of flip-flops, or one-dimensional arrays of registers. Flip-flops in a row of memory share read and write controls, and memory rows share input and output lines.

The number of flip-flops in a row of memory is its word-length, m. Memory words are arranged so that each word can individually be read or written into. Memory access is limited to its words. A memory of 2^n m-bit words has n address lines for addressing and enabling read and write operations into its words. The address space of such a memory is 2^n words. Input and output busses of such a memory have m bits. The block diagram of a clocked memory with a r/w (read/write) control signal is shown in Figure 2.61. The CE input shown is the Chip Enable input, which must be active for the memory to be read or written into.

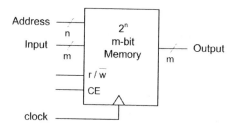

Figure 2.61 A 2^n m-bit Memory

Because accessing words in a memory can be done independent of their location in the memory array and by simply addressing them, memories are also called RAM or Random Access Memory. RAM structures come in various forms, SRAM (Static RAM), DRAM (Dynamic RAM), Pseudo-Static RAM, and many other forms that depend on their technology as well as their hardware structures.

2.7.1 Static RAM Structure

Figure 2.62 shows an SRAM that has an address space of 4, and word length of 3. The address bus for this structure is a 2-bit bus ($2^2 = 4$), and its input and output are 3-bit busses. A 2-to-4 decoder is used for decoding the address lines and giving access to the words of the memory. An external Chip-Enable disables all read and write operations when it is **0**.

The logic of the decoder shown in Figure 2.62 may be distributed inside the memory array. Other blocks in the memory shown are a read-write logic block and an IO block.

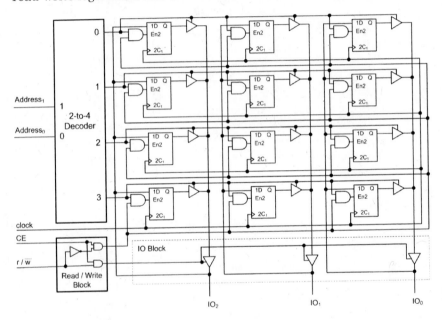

Figure 2.62 SRAM Structure

2.8 Bidirectional IO

The memory shown in Figure 2.62 has bidirectional *inout* lines used both as input and output. In the input mode, *IO* lines feed D-flip-flop inputs. In the output mode, three-state gates in the IO buffer block take the output of the addressed memory word and put it on the *IO* of the memory.

Bidirectional *inout* lines are useful for cascading memory chips and for reducing pin count of memory packages.

2.9 A Comprehensive Example: Serial Adder

In the previous section basics of combinational and sequential circuits were discussed. We showed how simple combinational parts could be designed and how state machines were defined and designed using sequential and combinational concepts. Furthermore combinational and sequential packages were defined. This section puts all the concepts into one example and shows how a complete system using sequential and combinational parts is designed.

2.9.1 Problem Statement

The example we are using is an 8-bit serial adder with two serial data inputs *ain* and *bin*, and a control input *start*. As shown in Figure 2.63, the circuit has an eight bit *result* output and a *ready* signal. After a complete pulse on *start*, operand data bits start showing up on *ain* and *bin* with every clock with least significant bits coming in first. In eight clock pulses as input data come into the circuit, they are added and the result becomes ready on *result*. At this time the *ready* signal becomes 1 and it remains 1 until a 1 is detected on the *start* input. While the circuit is performing its data collection and addition, pulses on *start* are ignored.

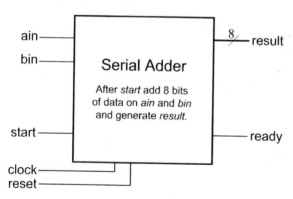

Figure 2.63 Serial Adder Block Diagram

2.9.2 Design Partitioning

The design of the serial adder has a datapath and a controller. The datapath collects data, adds them, and shifts the result into a shift-register. The controller waits for *start*, controls shifting of data into the shift-register, and issues *ready* when the addition operation is complete. Figure 2.64 shows the outline of this partitioning. In what follows, the details of the two parts of this figure will be discussed.

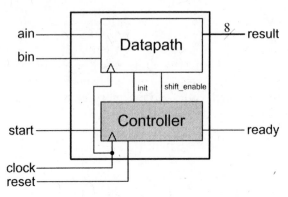

Figure 2.64 Serial Adder Data/Control Partitioning

2.9.3 Datapath Design

In the datapath of the serial adder a full-adder adds data coming in on *ain* and *bin*. With each addition, the sum is shifted into a shift-register. As data are added, the carry result of the full adder is saved in a flip-flop to be used for the addition of the next set of bits coming on *ain* and *bin*. This flip-flop must be reset for each new round of 8-bit addition.

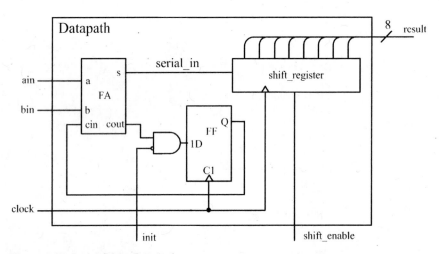

Figure 2.65 Serial Adder Datapath

Figure 2.65 shows the details of hardware of the datapath. The FA shown in this figure is the same as that shown in Figure 2.27; the FF shown is a rising-edge trigger D-type flip-flop (similar to the leftmost flip-flop of Figure 2.41, without its asynchronous control inputs). The AND gate at the input of the flip-flop provides it with a synchronous reset. This input connects to the *init* input that comes from the controller. As stated above, this flip-flop saves the carry output of a lower order bit addition for it to be used in the next upper-order bit addition.

The shift-register of the datapath is an 8-bit shift-register with a design similar to that of Figure 2.60. Our required shift-register, however, needs an enable input that does not exist in the 4-bit shift-register of Figure 2.60. This feature can easily be added by using multiplexers at the input of the flip-flops used in the shift-register. Figure 2.66 shows a shift-register bit that can be cascaded to form the required shift-register of our design. The input of the left-most input of this structure becomes the serial-input of the shift-register. The sum output from the full-adder is connected to this input. The output of the right-most bit of the shift-register is the serial output that is not needed in our design.

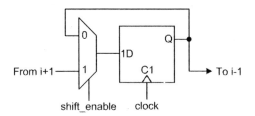

Figure 2.66 Cascadable Shift-register Bit with an Enable Input

Figure 2.67 shows the controller of our serial adder. On one side there is a state machine that waits for *start* and issues *count_enable* and *ready*. The state machine waits for the *complete* signal to be issued by the counter before it returns to its initial state that waits for another pulse on *start*. The outputs of this state machine are *ready*, *init*, *count_enable*, and *shift_enable*. The *init* and *shift_enable* outputs go out to the datapath to control initialization and shift activities.

On the other side of the controller is a counter that counts when *count_enable* is issued. Eight clock pulses after *init* resets this counter to 0, and while *count_enable* is active, the counter reaches its **111** state and issues the *complete* signal. When this signal is issued, the controller disables *count_enable*, which causes the counter to hold its last state.

Parallel with *count_enable*, the controller also issues *shift_enable* that goes out to the datapath. While this signal is active, add results from the full-adder (FA) are shifted into the datapath shift-register. Note that after eight shifts, because the shift-register is disabled, the output remains on the circuit *result* output.

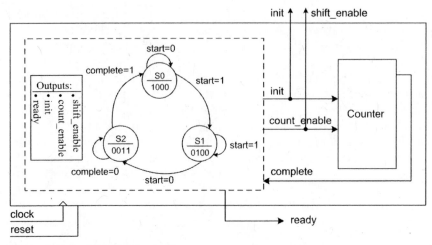

Figure 2.67 Serial Adder Controller

The state machine part of the controller can be implemented in a variety of ways. Figure 2.68 shows the one-hot implementation of this circuit. This circuit uses *start* and *complete* inputs and issues *ready*, *init*, *shift_enable* and *count_enable*. The *reset* input is the serial-adder's external reset input that sets the machine to its state *S0* by setting the first flip-flop to **1** and the rest to **0**.

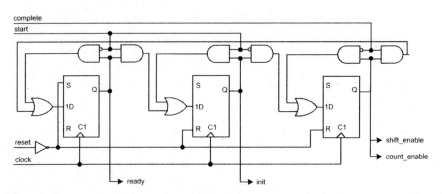

Figure 2.68 Controller State Machine Implementation

The counter part of the controller of Figure 2.67 is a simple 3-bit modulo-8 binary counter that can be implemented in a variety of ways. We have used an iterative, toggle flip-flop based structure that forms a synchronous counter. Gate level details of this structure are shown in Figure 2.69.

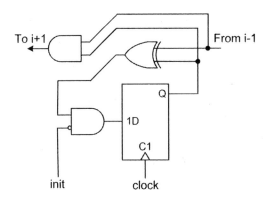

Figure 2.69 Controller Counter Implementation

The *init* input that comes from the controller state machine provides a synchronous reset for the counter of our controller. When this input is **0**, counter slices (Figure 2.69) are working in their normal count mode of operation. In each slice, the XOR feedback of the flip-flop makes it a toggle flip-flop with an enable input. When the upper input of the XOR is **0**, the output of the flip-flop circles back into its input causing its contents to remain unchanged. On the other hand, when the upper XOR input is **1** (it is enabled), it toggles with every clock pulse.

Cascading three of the structures shown in this figure generates a modulo-8 binary counter. When enabled, the right-most bit always toggles. The second bit from the right only toggles when the right-most bit is **1**. In general, cascading of the AND gates causes each bit of the counter to toggle with the clock when all bits to its right are **1**. Therefore, if, for example, the counter contents are **011**, the next clock causes the right-most **1** to toggle to **0**. Also, the second **1** toggles, because its right-most bit is **1**. And, since the two bits to the right of the **0** are **1**, this bit toggles as well. This toggling causes the next contents of the counter to become **100** which is one count above **011** in the binary system.

As stated above, the AND gate input of the right-most bit acts as the count enable input. This is because, when this bit is **0**, it propagates to all AND gates of the counter, causing all XOR inputs to be **0** which disables the toggling of the flip-flops. When the AND gates are enabled, the very last (left-most) AND gate output becomes **1** when all flip-flop outputs are **1**. This happens when the counter reaches **111** which is the final count of the counter. Therefore, as shown in Figure 2.69, the last AND output is the counter *complete* output that is used by the controller state machine.

The complete controller of the serial adder is formed by wiring the counter and the state machine together as shown in Figure 2.67.

Using the controller, the complete serial adder is formed by wiring this (Figure 2.67) and the circuit of   Figure 2.65 together to form the block diagram of Figure 2.64.

2.10 Summary

This chapter presented an overview of basic logic design. The focus was mostly on the design techniques and not on their theoretical background. We covered combinational and sequential circuits at the gate and RT levels. At the combinational gate-level, we discussed Karnaugh maps, but mainly concentrated on the use of iterative hardware and packages. In the sequential part, state machines were treated at the gate level; we also discussed sequential packages such as counters and shift-registers. The use of these packages facilitates RT level designs and use of HDLs in design. In the last section we covered a complete example illustrating a design methodology for digital system designs out of discrete gates and components. This comprehensive example showed how various design techniques could be used in a complete design.

3 RTL Design with Verilog

The level of hardware description that hardware description languages are most used for is the register transfer level (RTL). Between gate level on the low abstraction side, and system level on the high abstraction side, the RT level of abstraction is a good balance between correspondence to actual hardware and ease of description for hardware designers. At this level of abstraction, designs can be simulated with HDL simulators, they are synthesizable, and automatic generation of hardware is provided by most hardware design EDA environments.

This chapter presents Verilog at the RT level. We discuss how a design is described in Verilog for simulation and synthesis. For this purpose, only a subset of Verilog is needed and many complex language structures that are used in cell modeling and higher level non-synthesizable designs are not covered here. In order to utilize this language in a design and test environment, certain language structures that do not necessarily correspond to specific hardware structures, but are used for testing RT level designs, are also described in this chapter.

The chapter begins with a discussion of the main structures of Verilog. After this introductory presentation, we will start covering various constructs of the language using small pointed examples. The examples progressively become more complex and more constructs of the language are covered. After we present a sufficient set of constructs for design of hardware, we will turn our attention to developing testbenches for testing designs in Verilog. Several typical testbenches for the designs presented in the earlier parts of this chapter will be discussed in this part.

3.1 Basic Structures of Verilog

The basic structure of Verilog in which all hardware components and testbenches are described is called a module. Language constructs, according to Verilog syntax and semantics form the inside of a module. These constructs are designed to facilitate description of hardware components for simulation, synthesis, and specification of testbenches to specify test data and monitor circuit responses. Figure 3.1 shows a simulation model that consists of a design and its testbench in Verilog. Verilog constructs (shown by dotted lines) of the Verilog model being tested are responsible for description of its hardware, while language constructs used in a testbench are responsible for providing input data to the module being tested and analysis or display of its outputs. Simulation output is generated in form of a waveform for visual inspection or data files for machine readability.

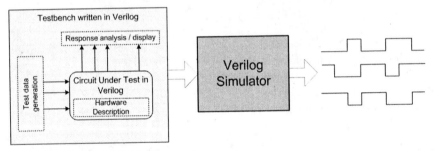

Figure 3.1 Simulation in Verilog

After a design passes basic functional validations, it must be synthesized into a netlist of components of a target library. The target library is the specification of the hardware that the design is being synthesized to. Verilog constructs used in the Verilog description of a design for its verification, or those for timing checks and timing specifications are not synthesizable. A Verilog design that is to be synthesized must use language constructs that have a clear hardware correspondence.

Figure 3.2 shows a block diagram specifying the synthesis process. Circuit being synthesized and specification of the target library are the inputs of a synthesis tool. The outputs of synthesis are a netlist of components of the target library, and timing specification and other physical details of the synthesized design.

Often synthesis tools have an option to generate this netlist in Verilog. In this case (Figure 3.3), the same testbench prepared for the pre-synthesis simulation can be used with the netlist generated by the synthesis tool. This simulation, which is often regarded as post-synthesis simulation, uses timing information generated by the synthesis tool and yields simulation results with detailed timing.

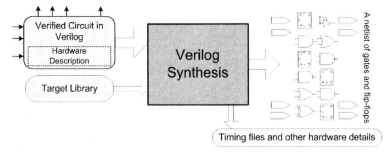

Figure 3.2 Synthesis of a Verilog Design

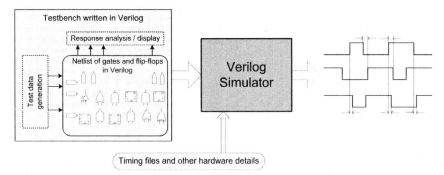

Figure 3.3 Post-synthesis Simulation in Verilog

3.1.1 Modules

The main structure used in Verilog for description of hardware components and their testbenches is a **module**. A module can describe a hardware component as simple as a transistor or a network of complex digital systems. A module that encloses a design's description can be described to test the module under design, in which case it is regarded as the testbench of the design. As shown in Figure 3.4, modules begin with the **module** keyword and end with **endmodule**. A complete design may consist of several modules. A design file describing a design takes the .v extension. For describing a system, it is usually best to include only one module in a design file.

A design may be described in a hierarchy of other modules. The top-level module is the complete design, and modules lower in the hierarchy are the design's components. Module instantiation is the construct used for bringing a lower level module into a higher level one. Figure 3.5 shows a hierarchy of several nested modules.

Design File: design1.v

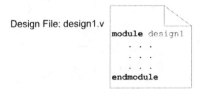

Figure 3.4 Module

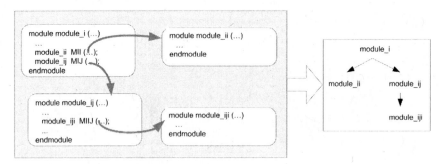

Figure 3.5 Module Hierarchy

3.1.2 Module Outline

The first part of a module description that begins with the **module** keyword and ends with a semicolon is regarded as its header. As shown in Figure 3.6, in addition to the **module** keyword, a module header includes the module name and list of its ports. Port declarations specifying the mode of a port (i.e., input, output, etc) and its length can be included in the header, or as separate declarations in the body of the module. Module declarations appear after the module header. In addition to ports not declared in the module header, this part can include declaration of signals used inside the module, or temporary variables. Specification of the operation of a module follows module declarations. In this part, various interacting statements form the description of the behavior of the module.

```
module name (ports or ports and their declarations);
    port declarations if not in the header;
    other declarations;
    . . .
    statements
    . . .
endmodule
```

Figure 3.6 Module Outline

Operation of a module can be described at the gate level, using Boolean expressions, at the behavioral level, or a mixture of various levels of abstraction. Figure 3.7 shows three ways the same operation can be described. Module *simple_1a* in uses Verilog's gate primitives, *simple_1b* uses concurrent statements, and *simple_1c* uses a procedural statement. Module *simple_1a* describes instantiation of three gate primitives of Verilog. In contracts, *simple_1b* uses Boolean expressions to describe the same functions for the outputs of the circuit. The third description, *simple_1c*, uses a conditional **if**-statement inside a procedural statement to generate proper function on one output, and uses a procedural Boolean function for forming the other circuit output.

```
module simple_1a (input i1, i2, i3, output w1, w2);
   wire c1;
   nor g1 (c1, i1, i2);
   and g2 (w1, c1, i3);
   xor g3 (w2, i1, i2, i3);
endmodule

module simple_1b (input i1, i2, i3, output w1, w2);
   assign w1 = i3 & ~(i1 | i2);
   assign w2 = i1 ^ i2 ^ i3;
endmodule

module simple_1c (input i1, i2, i3, output w1, w2);
   reg w1, w2;
   always @(i1, i2, i3) begin
      if (i1 | i2 ) w1 = 0; else w1 = i3;
      w2 = i1 ^ i2 ^ i3;
   end
endmodule
```

Figure 3.7 Module Definition Alternatives

The subsections that follow describe details of module ports and description styles. In the examples in this chapter Verilog keywords and reserved words are shown in bold. Verilog is case sensitive. It allows letters, numbers and special character "_" to be used for names. Names are used for modules, parameters, ports, variables, wires, signals, and instance of gates and modules.

3.1.3 Module Ports

In the module header, following its name is a set of parenthesis with a list of module ports. This list includes inputs, outputs and bidirectional input/output lines. Ports may be listed in any order. This ordering can only become significant when a module is instantiated, and does not affect the way its operation is described. Top-level modules used for testbenches have no ports.

Along with input and output names, in the set of parenthesis that follow the module name, sizes and types of ports may be specified. A port may be **input, output** or **inout**. The latter type is used for bi-directional input/output lines.

Size of a multi-bit port comes in a pair of numbers separated by a colon and bracketed by square brackets. The number on the left of the colon is the index of the left most bit of the vector, and that on the right is the index of the right most bit of the vector. Figure 3.8 shows an example circuit with scalar, vectored, **input, output** and **inout** ports. Ports named a, and b are one-bit inputs, and port c is a one-bit input/output. Ports av and bv are 8-bit inputs of *acircuit*. The set of square brackets that follow the **input** keyword applies to all ports that follow it. Another input/output is port cv that is an 8-bit vector. Port w of *acircuit* is declared as a one-bit output, and wv is an 8-bit output of this module.

```
module acircuit (input a, b, inout c, input [7:0] av, bv,
                 inout [7:0] cv, output w,
                 output [7:0] wv);

    .  .  .
endmodule
```

Figure 3.8 Module Notation

Alternatively, port declarations may appear as separate declaration statements in the module body, outside of its header part. Figure 3.9 shows a circuit that is identical to that of Figure 3.8, but uses the latter format.

```
module acircuit (a, b, c, av, bv, cv, w, wv);
    input a, b;
    output w;
    inout c;
    input [7:0] av, bv;
    output [7:0] wv;
    inout [7:0] cv;

    .  .  .
endmodule
```

Figure 3.9 Module Ports

3.1.4 Module Variables

In addition to port declarations, a module declarative part may also include wire and variable declarations that are to be used inside the module.

Wires (that are called **net** in Verilog) are declared by their types, **wire**, **wand** or **wor**; and variables are declared as **reg**. Wires are used for interconnections and have properties of actual signals in a hardware component. Variables are used for behavioral descriptions and are similar to variables in software languages. Figure 3.10 shows several wire and variable declarations.

```
module bcircuit (input a, b, input [7:0] av, bv,
                 output w, output [7:0] wv);
    wire d;
    wire [7:0] dv;
    reg e;
    reg [7:0] ev;
    . . .
endmodule
```

Figure 3.10 Wire and Variable Declaration

Wires represent simple interconnection wires, busses, and simple gate or complex logical expression outputs. When wires are used on the left hand sides of **assign** statements, they represent outputs of logical structures. Wires can be used in scalar or vector form. Multiple concurrent assignments to a **net** are allowed and the value that the wire receives is the resolution of all concurrent assignments to the **net**. Figure 3.11 shows several examples of wires used on the right and left hand sides of **assign** statements.

```
module vcircuit (input [7:0] av, bv, cv, output [7:0] wv);
    wire [7 :0] Iv, jv;
    assign Iv = av & cv;
    assign Iv = bv & cv;
    assign jv = av | cv;
    assign wv = iv ^ jv;
endmodule
```

Figure 3.11 Using Wires (net)

In contrast to a **net**, a **reg** variable type does not represent an actual wire and is primarily used as variables are used in a software language. In Verilog, we use a **reg** type variable for temporary variables, intermediate values, and storage of data. A **reg** type variable can only be used in a procedural body of Verilog. Multiple concurrent assignments to a **reg** should be avoided. Figure 3.12 shows several examples of **reg** type variables used in a Verilog module.

In the vector form, inputs, outputs, wires and variables may be used as a complete vector, part of a vector, or a bit of the vector. The latter two are referred to as part-select and bit-select. Examples of

part-select and bit-select on right and left hand sides of an **assign**
statement are shown in Figure 3.13.

```
module rcircuit (input sel; input [7:0] av, bv, cv,
                 output [7:0] wv, output reg [7:0] rv);
   reg [7:0] kv;
   always @(av, bv) begin
      if (sel) rv = av; else rv = bv;
       kv = av;
   end
   assign wv = kv ~^ cv;
endmodule
```

Figure 3.12 Using reg Type Variables

```
module vcircuit (input [7:0] av, bv, cv, output [7:0] wv,
                 output ys);
   assign wv [3:0] = av [7:4] & cv [7:4];
   assign ys = cv [4];
   assign wv [7] = av [0];
endmodule
```

Figure 3.13 Using Part-select and Bit-select

3.1.5 Logic Value System

Verilog uses a 4-value logic value system. Values in this system are **0**,
1, **Z**, and **X**. Value **0** is for logical **0** which in most cases represents a
path to ground (Gnd). Value **1** is logical **1** and it represents a path to
supply (Vdd). Value **Z** is for float, and **X** is used for un-initialized, un-
defined, un-driven, unknown, and value conflicts. Values **Z** and **X** are
used for wired-logic, busses, initialization values, tri-state structures,
and switch-level logic.

A gate input, or a variable or signal in an expression on the right
hand side of an assignment can take any of the four logic values.
Output of a two-valued primitive gate can only take **0**, **1** and **X**, while
output of a tri-state gate or a transistor primitive can also take a **Z**
value. A right-hand-side expression can evaluate to any of the four
logic values and can thus assign **0**, **1**, **Z**, or **X** to its left hand side **net**
or **reg**.

For more logic precision, Verilog uses strengths values as well as
logic values. Our dealing with Verilog is for design and synthesis, and
these issues will not be discussed here.

3.1.6 Wire (net) Resolutions

As discussed above, Verilog allows multiple concurrent assignments to **net** type variables. Furthermore, a **net** can be used as the output of two or more gates or components, which has the same effect as using a **net** on the left hand sides of several assignments. As an example, consider several tri-state gates with outputs connected to the same wire. In this case, we expect the resulting value to be a wiring resolution of all the driving values.

Verilog offers three types of resolutions for wired-and, wired-or, and wiring logic. For various wiring functions, we use **wire**, **wand**, **wor**, **tri**, **tri0** and **tri1 net** types. When two wires (**net**s) are connected, the resulting value depends on the two **net** values, as well as the type of the interconnecting **net**. Figure 3.14 shows **net** values for **net** types **wire**, **wand** and **wor**. The default **net** type is **wire**. The **tri net** type is the same as the **wire** type. A **net** of type **tri0** or **tri1** resolves to **0** and **1**, respectively, when driven by all **Z** values. Table shown here is referred to a **net** resolution table.

Net type	Two driving **net** values									
	0, 0	0, 1	0, Z	0, X	1, 1	1, Z	1, X	Z, Z	Z, X	X, X
wire	0	X	0	X	1	1	X	Z	X	X
wand	0	0	0	0	1	X	X	Z	X	X
Wor	0	1	X	X	1	1	1	X	X	X

Figure 3.14 "net" Type Resolutions

In the example of Figure 3.15, w and y are declared as **wand** and **wire** types. Multiple assignments to w are resolved by the "wand" labeled row of table of Figure 3.14, and multiple assignments to y are resolved by the "wire" labeled row of this table.

```
module resolving (input a, b, c, output w);

    wand w;
    wire y;

    assign w = a;
    assign w = b;
    assign y = a;
    assign y = c;

endmodule
```

Figure 3.15 Multiple Assignments Causing Resolutions

Several examples of **net** resolutions are shown in Figure 3.16. As shown here, if in the code of Figure 3.15 a is **1** and b is **Z**, since resolution of w is **wand**, the value assigned to w becomes **X**.

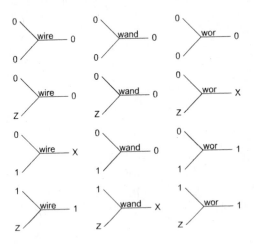

Figure 3.16 "net" Resolution Examples

3.2 Combinational Circuits

A combinational circuit can be represented by its gate level structure, its Boolean functionality, or description of its behavior. At the gate level, interconnection of its gates are shown; at the functional level, Boolean expressions representing its outputs are written; and at the behavioral level a software-like procedural description represents its functionality. This section shows these three levels of abstraction for describing combinational circuits. Examples for combining various forms of descriptions and instantiation of already described components will also be described here.

3.2.1 Gate Level Combinational Circuits

Verilog provides primitive gates and transistors. Some of the more important Verilog primitives and their logical representations are shown in Figure 3.17. In this figure w is used for gate outputs, i for inputs and c for control inputs.

Basic logic gates are **and, nand, or, nor, xor, xnor**. These gates can be used with one output and any number of inputs. The other two structures shown in the first column of this figure, are **not** and **buf**. These gates can be used with one input and any number of outputs.

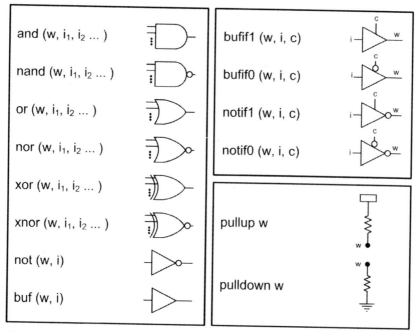

Figure 3.17 Basic Primitives

Another group of primitives shown in this figure are three-state (tri-state is also used to refer to these structures) gates. Gates shown here have w for their outputs, i for data inputs, and c for their control inputs. These primitives are **bufif1**, **notif1**, **bufif0**, and **notif0**. When control c for such gates is active (**1** for first and third, and **0** for the others), the data input, i, or its complement appears on the output of the gate. When control input of a gate is not active, its output becomes high-impedance, or **Z**.

Verilog also has primitives for unidirectional and bi-directional MOS and CMOS structures. Shown in Figure 3.18 are NMOS, PMOS and CMOS structures. Primitives in the left column are unidirectional, and those in the right column are bidirectional. These are switches that are used in switch level description of gates, complex gates, and busses. The **nmos (pmos)** primitive is a simple switch with an active high (low) control input. The **cmos** switch is usually used with two complementary control inputs. These switches behave like the tri-state gates, and they are only different from tri-state gates in their output voltage levels and drive strengths. These parameters are modeled by wire strengths and are not discussed in this book.

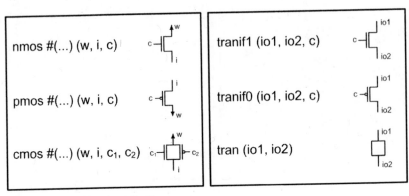

Figure 3.18 Basic MOS Primitives

3.2.1.1 Majority Example. We use the majority circuit of Figure 3.19 to illustrate how primitive gates are used in a design. The description shown in Figure 3.20 corresponds to this circuit. The module description has inputs and outputs according to the schematic of Figure 3.19.

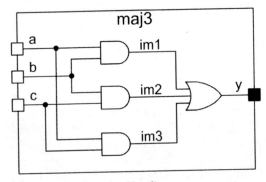

Figure 3.19 A Majority Circuit

Line 1 of the code shown is the **timescale** directive. This defines all time units in the description and their precision. For our example, *1ns/100ps* means that all numbers in the code that represent a time value are in nanoseconds and they can have up to one fractional digit (100 Ps). It is always a good practice to use this statement in every Verilog file as the first statement. This eliminates ambiguities in decisions made by Verilog simulators, and carry over of time units from one module to another. Even if a module does not use any delay parameter, if it becomes necessary to introduce a delay in the debugging process, the **timescale** directive, instead of simulator defaults, sets the time units.

The statement that begins in Line 6 and ends in Line 9 instantiates three **and** primitives. The construct that follows the primitive name specifies rise and fall delays for the instantiated primitive ($t_{plh}=2$, $t_{phl}=4$). This part is optional and if eliminated, 0 values are assumed for rise and fall delays. Line 7 shows inputs and outputs of one of the three instances of the **and** primitive. The output is *im1* and inputs are module input ports *a* and *b*. The port list on Line 7 must be followed by a comma if other instances of the same primitive are to follow, otherwise a semicolon should be used, like the end of Line 9. Line 8 and Line 9 specify input and output ports of the other two instances of the **and** primitive. Line 10 is for instantiation of the **or** primitive at the output of the majority gate. The output of this gate is *y* that comes first in the port list, and is followed by inputs of the gate. In this example, intermediate signals for interconnection of gates are *im1*, *im2*, and *im3*. Scalar interconnecting wires need not be explicitly declared in Verilog.

```
`timescale 1ns/100ps                              // Line 01

module maj3 ( a, b, c, y );
   input a, b, c;
   output y;

   and #(2,4)                                      // Line 06
      ( im1, a, b ),                               // Line 07
      ( im2, b, c ),                               // Line 08
      ( im3, c, a );                               // Line 09
   or  #(3,5) ( y, im1, im2, im3 );                // Line 10

endmodule
```

Figure 3.20 Verilog Code for the Majority Circuit

The three **and** instances could be written as three separate statements, like instantiation of the **or** primitive. If we were to specify different delay values for the three instances of the **and** primitive, we had to have three separate primitive instantiation statements.

3.2.1.2 Multiplexer Example. Figure 3.21 shows a 2-to-1 multiplexer using three-state gates. Three-state gates are instantiated in the same way as the regular logic gates. Outputs of three-state gates can be wired to form wired-and, wired-or, or wiring logic. The resolution of this **net** is determined by the type of the **net** at this node, i.e., **wire**, **wand**, **wor**, **tri**, **tri0** and **tri1 net** types.

The Verilog code of this multiplexer is shown in Figure 3.22. Lines 4 and 5 in this code instantiate two three-state gates. Their output is *y*, and since it is driven by both gates a wired-net is formed. Since *y* is not declared, its **net** type defaults to **wire**. When *s* is **1**, *bufif1* conducts and the value of *b* propagates to its output. At the same time, because *s* is **1**, *bufif0* does not conduct and its output becomes **Z**. Resolution of these values driving **net** *y* is determined by the **wire net** resolution as shown in Figure 3.14.

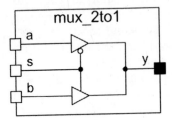

Figure 3.21 Multiplexer Using Three-state Gates

```
`timescale 1ns/100ps

module mux_2to1 ( input a, b, s, output y );
    bufif1 #(3) ( y, b, s );                    // Line 04
    bufif0 #(5) (y, a, s);                      // Line 05
endmodule
```

Figure 3.22 Multiplexer Verilog Code

3.2.1.3 CMOS NAND Example. As another example of instantiation of primitives, consider the two-input CMOS NAND gate shown in Figure 3.23.

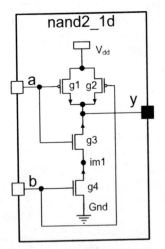

Figure 3.23 CMOS NAND Gate

The Verilog code of Figure 3.24 describes this CMOS NAND gate. Logically, NMOS transistors in a CMOS structure push **0** into the output of the gate. Therefore, in the Verilog code of the CMOS NAND, input to output direction of NMOS transistors are from *Gnd* towards *w*. Likewise, PMOS transistors push a **1** value into *w*, and therefore, their inputs are considered the *Vdd* node and their outputs are connected to the *w* node. The *im1* signal is an intermediate **net** and is explicitly declared.

```
`timescale 1ns/100ps

module cmos_nand ( input a, b, output w);

wire im1;
supply1 vdd;
supply0 gnd;
    nmos #(3, 4)
        T1 (im1, gnd, b),
        T2 (w, im1, a);

    pmos #(4, 5)
        T3 (w, vdd, a),
        T4 (w, vdd, b);
endmodule
```

Figure 3.24 CMOS NAND Verilog Description

In the Verilog code of CMOS NAND gate, primitive gate instance names are used. This naming (*T1*, *T2*, *T3*, *T4*) is optional for primitives and mandatory when modules are instantiated. Examples of module instantiations are shown in the next section.

3.2.2 Gate Level Synthesis

Gate level descriptions in Verilog are synthesizable. However, it must be noted that a designer using a gate level description cannot expect the same exact gates and interconnections to be used in the synthesized output. The hardware generated by a synthesis tool, and how gates are implemented merely depends on the target technology. Furthermore, delays used in a gate level description are always ignored by the synthesis tools.

The synthesis of *maj3* module of Figure 3.20 using Altera's Quartus II and specifying Cyclone as the target FPGA results in using a single look-up table of a logic element, as shown in Figure 3.25. As shown, the three inputs of the circuit are used in a look-up table to produce the necessary combinational output.

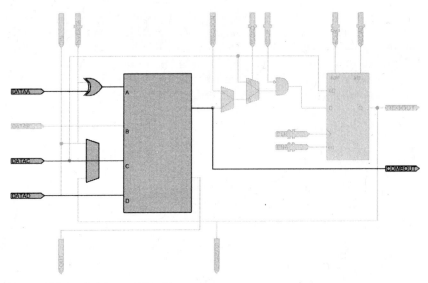

Figure 3.25 Logic Element Used for *maj3*

The RTL view of this implementation that is also produced by our synthesis tool gives a better view of the functionality of this generated hardware. Figure 3.26 shows this view of our gate level *maj3* module. As expected, the AND-OR functionality of the Majority function is obtained from the synthesis tool.

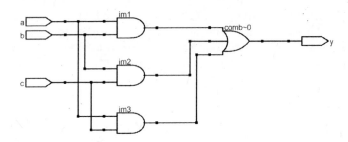

Figure 3.26 RTL (logical) View of Synthesized *maj3*

Gate level descriptions using tri-state primitive structures are also synthesizable. If the synthesis target hardware does not have tri-state structures inside the chip (such as Altera's Cyclone) regular AND-OR gates will be used for the implementation of a description that uses tri-states. For example, when synthesized, the description of the multiplexer of Figure 3.22 uses a single look-up table.

HDL synthesizers cannot synthesize MOS switch level descriptions. The code of Figure 3.24 is not accepted by synthesis tools.

3.2.3 Descriptions by Use of Equations

At a higher level than gates and transistors, a combinational circuit may be described by use of Boolean, logical, and arithmetic expressions. For this purpose the Verilog concurrent **assign** statement is used. Figure 3.27 shows Verilog operators that can be used with **assign** statements.

Bitwise Operators	&			^	~	~^	^~	
Reduction Operators	&	~&			~|	^	~^	^~
Arithmetic Operators	+	-	*	/	%		
Logical Operators	&&	||	!				
Compare Operators	<	>	<=	>=	==		
Shift Operators	>>	<<					
Concatenation Operators	{}	{n{}}					
Conditional Operator	?:						

Figure 3.27 Verilog Operators

3.2.3.1 XOR Example. As our first example for using an **assign** statement consider the description of an XOR gate as shown in Figure 3.28. The **assign** statement uses y on the left-hand-side and equates it to Exclusive-OR of a, b, and c inputs.

```
module xor3 ( input a, b, c, output y );
    assign y = a ^ b ^ c;
endmodule
```

Figure 3.28 XOR Verilog Code

Effectively, this **assign** statement is like driving y with the output of a 3-input **xor** primitive gate. The difference is that, the use of an **assign** statement gives us more flexibility and allows the use of more complex functions than what is available as primitive gates. Instead of being limited to the gates shown in Figure 3.17, we can write our own expressions using operators of Figure 3.27.

3.2.3.2 Full-Adder Example. Figure 3.29 shows another example of using **assign** statements. This code corresponds to a full-adder circuit (see Chapter 2). The s output is the XOR result of a, b and ci inputs, and the co output is an AND-OR expression involving these inputs.

A delay value of 10 ns is used for the s output and 8 ns for the co output. As with the gate outputs, rise and fall delay values can be specified for a **net** that is used on the left-hand side of an **assign** statement. This construct allows the use of two delay values. If only one value is specified, it applies to both rise and fall transitions.

```
`timescale 1ns/100ps
module add_1bit ( input a, b, ci, output s, co );
   assign #(10) s = a ^ b ^ ci;
   assign #(8) co = ( a & b ) | ( b & ci ) | ( a & ci );
endmodule
```

Figure 3.29 Full Adder Verilog Code

Another property of **assign** statements that also corresponds to gate instantiations is their concurrency. The statements in the Verilog module of Figure 3.29 are concurrent. This means that the order in which they appear in this module is not important. These statements are sensitive to events on their right hand sides. When a change of value occurs on any of the right hand side **net** or variables, the statement is evaluated and the resulting value is scheduled for the left hand side **net**.

3.2.3.3 Comparator Example. Figure 3.30 shows another example of using **assign** statements. This code describes a 4-bit comparator. The first **assign** statement uses a bitwise XOR operation on its right hand side. The result that is assigned to the im intermediate **net** is a 4-bit vector formed by XORing bits of a and b input vectors. The second **assign** statement uses the NOR reduction operator to NOR bits of im to generate the equal output for the 4-bit comparator.

The above describes the comparator using its Boolean function. However, using compare operators of Verilog, the eq output of the comparator may be written as:

```
assign eq = (a == b);
```

In this expression, $(a == b)$ results in **1** if a and b are equal, and **0** if they are not. This result is simply assigned to eq.

```
module comp_4bit (input [3:0] a, b, output eq );
   wire [3:0] im;
   assign im = a ^ b;
   assign eq = ~| im;
endmodule
```

Figure 3.30 Four-Bit Comparator

The right-hand side expression of an **assign** statement can have a conditional expression using the **?** and **:** operators. These operators are like if-then-else. In reading expressions that involve a condition operator, **?** and **:** take places of **then** and **else** respectively. The if-condition appears to the left of **?**.

3.2.3.4 Multiplexer Example.

Figure 3.31 shows a 2-to-1 multiplexer using a conditional operator. The expression shown reads as follows: **if** s is **1, then** y is *i1* **else** it becomes *i0*.

```
module mux2_1 (input [3:0] i0, i1, input s, output [3:0]y
               );
    assign y = s ? i1 : i0;
endmodule
```

Figure 3.31 A 2-to-1 Mux using Condition Operator

3.2.3.5 Decoder Example.

Figure 3.32 shows another example using the conditional operator. In this example a nesting of several ?: operations are used to describe a decoder.

```
module dcd2_4(input a, b, output d0, d1, d2, d3 );

    assign {d3, d2, d1, d0} =
                        ( {a, b} == 2'b00 ) ? 4'b0001 :
                        ( {a, b} == 2'b01 ) ? 4'b0010 :
                        ( {a, b} == 2'b10 ) ? 4'b0100 :
                        ( {a, b} == 2'b11 ) ? 4'b1000 :
                                              4'b0000 ;

endmodule
```

Figure 3.32 Decoder Using ?: and Concatenation

The decoder description also uses the concatenation operator **{ }** to form vectors from its scalar inputs and outputs. The decoder has four outputs, *d3, d2, d1* and *d0* and two inputs *a* and *b*. Input values **00, 01, 10,** and **11** produce **0001, 0010, 0100,** and **1000** outputs. In order to be able to compare *a* and *b* with their possible values, a two-bit vector is formed by concatenating *a* and *b*. The *{a, b}* vector is then compared with the four possible values it can take using a nesting of ?: operations. Similarly, in order to be able to place vector values on the outputs, the four outputs are concatenated using the **{ }** operator and used on the left-hand side of the **assign** statement shown in Figure 3.32.

This example also shows the use of sized numbers. Constants for the inputs and outputs have the general format of **n`bm**. In this format, **n** is the number of bits, **b** is the base specification and **m** is the number in base **b**. For calculation of the corresponding constant, number **m** in base **b** is translated to **n** bit binary. For example, *4`hA* becomes **1010** in binary.

3.2.3.6 Adder Example.

For another example using **assign** statements, consider an 8-bit adder circuit with a carry-in and a carry-out output. The Verilog code of this adder, shown in Figure 3.33, uses an **assign** statement to set concatenation of co on the left-hand side of s to the sum of a, b and ci. This sum results in nine bits with the leftmost bit being the resulting carry. The sum is captured in the 9-bit left-hand side of the **assign** statement in {co, s}.

```
module add_8bit ( input [7:0] a, b, input ci,
                  output [7:0] s, output co );
    assign {co, s} = a + b + ci;
endmodule
```

Figure 3.33 Adder with Carry-in and Carry-out

So far in this section we have shown the use of operators of Figure 3.27 in **assign** statements. A Verilog description may contain any number of **assign** statements and can use any mix of the operators discussed. The next example shows multiple **assign** statements.

3.2.3.7 ALU Example.

As our final example of **assign** statements, consider an ALU that performs add and subtract operations and has two flag outputs gt and $zero$. The gt output becomes **1** when input a is greater than input b, and the $zero$ output becomes **1** when the result of the operation performed by the ALU is **0**.

Figure 3.34 shows the Verilog code of this ALU. Using a conditional operation, the $addsub$ input decides whether ALU inputs should be added or subtracted. Other Verilog constructs used in this description are arithmetic, concatenation, conditional, compare and relational operations.

```
module ALU ( input [7:0] a, b, input addsub,
             output gt, zero, co, output [7:0] r );

    assign {co, s} = addsub ? (a + b) : (a - b);
    assign gt = (a>b);
    assign zero = (r == 0);
endmodule
```

Figure 3.34 ALU Verilog Code Using a Mix of Operations

3.2.4 Instantiating Other Modules

We have shown how primitive gates can be instantiated in a module and wired with other parts of the module. The same applies to instantiating a module within another. For regular structures, Verilog provides repetition constructs for instantiating multiple copies of the same module, primitive, or set of constructs. Examples in this section illustrate some of the capabilities.

3.2.4.1 ALU Example Using Adder.

The ALU of Figure 3.34 starts from scratch and implements every function it needs inside the module. If we have a situation that we need to use a specific design from a given library, or we have a function that is too complex to be repeated everywhere is it used, we can make it into a module and instantiate it when we need to use it.

Figure 3.35 shows another version of the above ALU circuit. In this new version, addition is handled by the adder circuit of Figure 3.33. On Line 5 of Figure 3.35, the *b_bbar* signal in the *ALU_Adder* module receives input *b* or its complement depending on the *addsub* input. If addition is to be done, *addsub* is **0** and *b* goes on *b_bbar*. On the other hand, if subtraction is to be done, *b_bbar* becomes the complement of *b*. This multi-bit vector is used for the input of the 8-bit adder that is instantiated on Line 4 of Figure 3.35. Since subtraction is being done in 2's complement system, *addsub* is used for the carry-in of the adder to add a **1** to the result if subtraction is to take place.

```
module ALU_Adder (input [7:0] a,b, input addsub,  // Line 01
                  output gt, zero, co, output [7:0] r );
    wire [7:0] b_bbar;
    add_8bit ADD (a, b_bbar, addsub, r, co);       // Line 04
    assign b_bbar = addsub ? ~b : b;               // Line 05
    assign gt = (a>b);
    assign zero = (r == 0);
endmodule
```

Figure 3.35 ALU Verilog Code Using Instantiating an Adder

Instantiation of a component such as *add_8bit* in the above example starts with the component name, an instance name (*ADD*) and the port connection list. The latter part decides how local variables of a module are mapped to the ports of the component being instantiated. The above example uses an ordered list, in which a local variable, e.g., *b_bbar*, takes the same position as the port of the component it is connecting to, e.g., *b*. Alternatively, a named port connection such as that shown below can be used.

```
add_8bit ADD (.a(a), .b(b_bbar), .ci(addsub),
             .s(r), .co(co));
```

Using this format allows port connections to be made in any order. Each connection begins with a dot, followed by the name of the port of the instantiated componet, e.g., *b*, and followed by a set of parenthesis enclosing the local variable that is connected to the instantiated component, e.g., *b_bbar*. This format is less error-prone than the ordered connection.

3.2.4.2 Iterative Adder Description.

Verilog uses the **generate** statement for describing regular structures that are composed of smaller subcomponents. An example is a large memory array, or a systolic array multiplier. In such cases, a cell unit of the array is described, and by use of several generate-statements it is repeated in several directions to cover the entire array of the hardware. Here we show an alternative description for our *add_8bit* module of Figure 3.33. In this new description we use eight instances of the full-adder of Figure 3.29.

The code of Figure 3.36 uses the **parameter** construct to specify a constant value for *SIZE*. In the body of this module on Line 7, a variable for generating eight instances of *add_1bit* is declared using the **genvar** declaration. The **generate** statement that begins on Line 8 loops eight times to generate a full-adder in every iteration using *a* and *b* input bits 0 to 7. Each instance of *FA* uses a carry bit from the *carry* vector, and produces its carry on the next bit position of *carry*. On lines 5 and 6 of this code, *ci* is put into *carry[0]*, and *co* is taken from *carry[8]*.

```
module add_8bit #(parameter SIZE=8)              // Line 1
                ( input [SIZE-1:0] a, b, input ci,
                  output [SIZE-1:0] s, output co );
    wire [SIZE:0] carry;
    assign carry[0] = ci;                        // Line 5
    assign co = carry[SIZE];                     // Line 6
    genvar i;                                    // Line 7
    generate for (i=0; i<SIZE; i=i+1) begin: row // Line 8
       add_1bit FA (a[i], b[i], carry[i], s[i], carry[i+1]);
    end endgenerate
endmodule
```

Figure 3.36 Adder Verilog Code Using Generate Statement

3.2.5 Synthesis of Assignment Statements

Descriptions of the previous section concentrated on using **assign** statements in description of modules. We also showed modules that instantiated other modules consisting of **assign** statements. In general, assign statements, regardless of the hierarchy they are used in and their complexity, are synthesizable.

As an example of this synthesis, consider the *ALU_Adder* of Figure 3.35. The RTL view of this circuit after being synthesized by Quartus II is shown in Figure 3.37. As shown, the generated hardware uses multiplexers for conditional complementing of the *b* input. On the front side of the circuit, the *EQUAL* and *LESS_THAN* blocks are responsible for generating *zero* and *lt* outputs. The gray box is the adder block that replaces the ADD instance in the code of Figure 3.35. The complete hardware uses 27 logic-elements of a Cyclone FPGA.

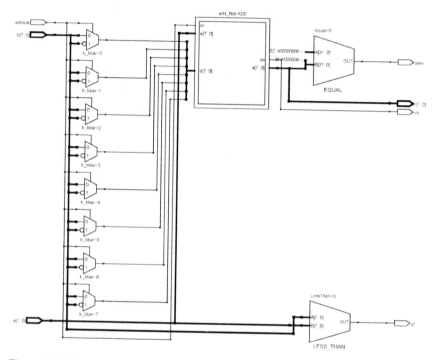

Figure 3.37 *ALU_Adder* RTL View after Synthesis

3.2.6 Descriptions with Procedural Statements

At a higher level of abstraction than describing hardware with gates and expressions, Verilog provides constructs for procedural description of hardware. Unlike gate instantiations and **assign** statements that correspond to concurrent sub-structures of a hardware component, procedural statements describe the hardware by its behavior. Also, unlike concurrent statements that appear directly in a module body, procedural statements must be enclosed in procedural blocks before they can be put inside a module.

The main procedural block in Verilog is the **always** block. This is considered a concurrent statement that runs concurrent with all other statements in a module. Within this statement, procedural statements like **if-else** and **case** statements are used and are executed sequentially. If there are more than one procedural statement inside a procedural block, they must be bracketed by **begin** and **end** keywords.

Unlike assignments in concurrent bodies that model driving logic for left hand side wires, assignments in procedural blocks are assignments of values to variables that hold their assigned values until a different value is assigned to them. A variable used on the left hand side of a procedural assignment must be declared as **reg**.

An event control statement is considered a procedural statement, and is used inside an **always** block. This statement begins with an at-sign, and in its simplest form, includes a list of variables in the set of parenthesis that follow the at-sign, e.g., *@ (v1* **or** v2 ...); .

When the flow of the program execution within an **always** block reaches an event-control statement, the execution halts (suspends) until an event occurs on one of the variables in the enclosed list of variables. If an event-control statement appears at the beginning of an **always** block, the variable list it contains is referred to as the *sensitivity list* of the **always** block. For combinational circuit modeling all variables that are read inside a procedural block must appear on its sensitivity list.

Examples that follow show various ways combinational component may be modeled by procedural blocks.

3.2.6.1 Majority Example.

Figure 3.38 shows a majority circuit described by use of an **always** block. In the declarative part of the module shown, the *y* output is declared as **reg** since this variable is to be assigned a value inside a procedural block.

```
module maj3 ( input a, b, c, output reg y );
   always @( a or b or c )
   begin
      y = (a & b) | (b &c) | (a & c);
   end
endmodule
```

Figure 3.38 Procedural Block Describing a Majority Circuit

The **always** block describing the behavior of this circuit uses an event control statement that encloses a list of variables that is considered as the sensitivity list of the **always** block. The **always** block is said to be sensitive to *a*, *b* and *c* variables. When an event occurs on any of these variables, the flow into the **always** block begins and as a result, the result of the Boolean expression shown will be assigned to variable *y*. This variable holds its value until the next time an event occurs on *a*, *b*, or *c* inputs.

In this example, since the **begin** and **end** bracketing only includes one statement, its use is not necessary. Furthermore, the syntax of Verilog allows elimination of semicolon after an event control statement. This effectively collapses the event control and the statement that follows it into one statement.

3.2.6.2 Majority Example with Delay.
The Verilog code shown in Figure 3.39 is a majority circuit with a 5ns delay. Following the **always** keyword, the statements in this procedural block are an event-control, a delay-control and a procedural assignment. The delay-control statement begins with a sharp-sign and is followed by a delay value. This statement causes the flow into this procedural block to be suspended for 5ns. This means that after an event on one of the circuit inputs, evaluation and assignment of the output value to *y* takes place after 5 nanoseconds.

Note in the description of Figure 3.39 that **begin** and **end** bracketing is not used. As with the event-control statement, a delay-control statement can collapse into its next statement by removing their separating semicolon. The event-control, delay-control and assignment to *y* become a single procedural statement in the **always** block of *maj3* code.

```
`timescale 1ns/100ps

module maj3 ( input a, b, c, output reg y );

    always @(a, b, c ) #5 y = (a & b) | (b &c) | (a & c);

endmodule
```

Figure 3.39 Majority Gate with Delay

3.2.6.3 Full-Adder Example.
Another example of using procedural assignments in a procedural block is shown in Figure 3.40. This example describes a full-adder with sum and carry-out outputs.

The **always** block shown is sensitive to *a*, *b*, and *ci* inputs. This means that when an event occurs on any of these inputs, the **always** block wakes up and executes all its statements in the order that they appear. Since assignments to *s* and *co* outputs are procedural, both these outputs are declared as **reg**.

The delay mechanism used in the full-adder of Figure 3.40 is called an intra-statement delay that is different than that of the majority circuit of Figure 3.39.

```
`timescale 1ns/100ps
module add_1bit ( input a, b, ci, output reg s, co );
    always @( a or b or ci )
    begin
        s = #5 a ^ b ^ ci;
        co = #3 (a & b) | (b &ci) | (a & ci);
    end
endmodule
```

Figure 3.40 Full-Adder Using Procedural Assignments

In the majority circuit, the delay simply delays execution of its next statement. However, the intra-statement delay of Figure 3.40 only delays the assignment of the calculated value of the right-hand side to the left-hand side variable. This means that in Figure 3.40, as soon as an event occurs on an input, the expression $a^{\wedge}b^{\wedge}c$ is evaluated. But, the assignment of the evaluated value to s and proceeding to the next statement takes 5ns.

Because assignment to co follows that to s, the timing of the former depends on that of the latter, and evaluation of the right-hand side of co begins 5ns after an input change. Therefore, co receives its value 8ns after an input change occurs. To remove this timing dependency and be able to define a statement timing independent of its previous one, a different kind of assignment must be used.

Assignments in Figure 3.40 are of the blocking type. Such statements block the flow of the program until they are completed. A different assignment is of the non-blocking type. A different version of the full-adder that uses this construct is shown in Figure 3.41. This assignment schedules its right hand side value into its left hand side to take place after the specified delay. Program flow continues into the next statement while propagation of values into the first left hand side is still going on.

In the example of Figure 3.41, evaluation of the right hand side of s is done immediately after an input changes, and the resulting value is scheduled for s after 5ns. Evaluation of the right hand side of co also occurs at the same time as that of s. The resulting value for s is scheduled for it after 8ns. The 8ns delay makes the timing of s and co of Figure 3.41 the same as those of Figure 3.40.

Since our focus is on synthesizable coding and gate delay timing issues are not of importance, we will mostly use blocking assignments in this book.

```
`timescale 1ns/100ps
module add_1bit ( input a, b, ci, output reg s, co );

    always @( a or b or ci )
    begin
        s <= #5 a ^ b ^ ci;
        co <= #8 (a & b) | (b &ci) | (a & ci);
    end
endmodule
```

Figure 3.41 Full-Adder Using Non-Blocking Assignments

3.2.6.4 Procedural Multiplexer Example.

For another example of a procedural block, consider the 2-to-1 multiplexer of Figure 3.42. This example uses an **if-else** construct to set y to $i0$ or $i1$ depending on the value of s.

As in the previous examples, all circuit variables that participate in determination of value of y appear on the sensitivity list of the **always** block. Also since y appears on the left hand side of a procedural assignment, it is declared as **reg**.

The **if-else** statement shown in Figure 3.42 has a condition part that uses an equality operator. If the condition is true (i.e., s is equal to **0**), the block of statements that follow it will be taken, otherwise the block of statements after the **else** are taken. In both cases, the block of statements must be bracketed by **begin** and **end** keywords if there is more than one statement in a block.

```
module mux2_1 (input i0, i1, output reg s, y );
    always @( i0 or i1 or s ) begin
        if ( s==1'b0 )
            y = i0;
        else
            y = i1;
    end
endmodule
```

Figure 3.42 Procedural Multiplexer

3.2.6.5 Procedural ALU Example.

The **if-else** statement, used in the previous example, is easy to use, descriptive and expandable. However, when many choices exist, a **case**-statement which is more structured may be a better choice. The ALU description of Figure 3.43 uses a **case** statement to describe an ALU with add, subtract, AND and XOR functions.

The ALU has a and b data inputs and a 2-bit f input that selects its function. The Verilog code shown in Figure 3.43 uses a, b and f on its sensitivity list. The **case**-statement shown in the **always** block uses *f* to select one of the **case** alternatives. The last alternative is the **default** alternative that is taken when *f* does not match any of the alternatives that appear before it. This is necessary to make sure that unspecified input values (here, those that contain **X** and/or **Z**) cause the assignment of the default value to the output and not leave it unspecified.

```
module alu_4bit (input [3:0] a, b, input [1:0] f, output
                 reg [3:0] y );
   always @ ( a or b or f ) begin
      case ( f )
         2'b00 : y = a + b;
         2'b01 : y = a - b;
         2'b10 : y = a & b;
         2'b11 : y = a ^ b;
         default: y = 4'b0000;
      endcase
   end
endmodule
```

Figure 3.43 Procedural ALU

3.2.7 Combinational Rules

Completion of **case** alternatives or **if-else** conditions is an important issue in combinational circuit coding. In an **always** block, if there are conditions under which the output of a combinational circuit is not assigned a value, because of the property of **reg** variables the output retains its old value. The retaining of old value infers a latch on the output. Although, in some designs this latching is intentional, obviously it is unwanted when describing combinational circuits. With this, we have set two rules for coding combinational circuits with **always** blocks.

1. List all inputs of the combinational circuit in the sensitivity list of the **always** block describing it.
2. Make sure all combinational circuit outputs receive some value regardless of how the program flows in the conditions of **if-else** and/or **case** statements. If there are too many conditions to check, set all outputs to their inactive values at the beginning of the **always** block.

3.2.8 Synthesizing Procedural Blocks

Following combinational synthesis rules discussed above, we can easily develop Verilog designs for synthesis. We use the *alu_4bit* example of Figure 3.43 for demonstrating this synthesis. Figure 3.44 shows the RTL view of the synthesis result of this circuit. As shown, the circuit has four selectors driven by f that is first decoded and then used for selecting one of the four functions of the ALU. The logic for the four functions of a and b appear to the left of the selectors.

The complete implementation of this circuit uses 17 of the 5980 logic-elements of an EP1C6Q240C8 Cyclone FPGA.

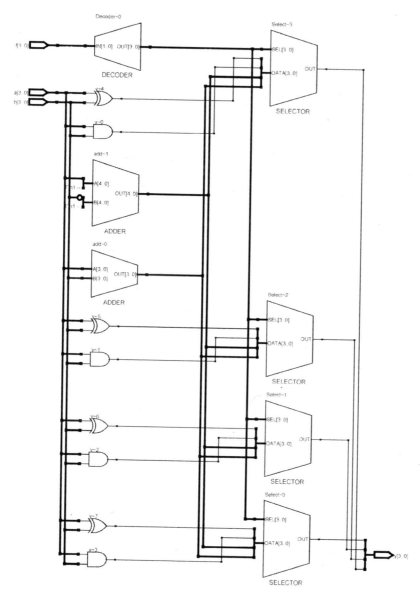

Figure 3.44 Synthesis of Procedural ALU

3.2.9 Bussing

Bus structures can be implemented by use of multiplexers or three-state logic. In Verilog, various methods of describing combinational circuits can be used for the description of a bus.

Figure 3.45 shows Verilog coding of *busout* that is a three-state bus and has three sources, *busin1*, *busin2*, and *busin3*. Sources of *busout* are put on this bus by active-high enabling control signals, *en1*, *en2* and *en3*. Using the value of an enabling signal, a condition statement either selects a bus driver or a 4-bit **Z** value to drive the *busout* output.

```
module bussing (busin1, busin2, busin3, en1, en2, en3,
                busout );
    input [3:0] busin1, busin2, busin3;
    input en1, en2, en3;
    output [3:0] busout;

    assign busout = en1 ? busin1 : 4'bzzzz;
    assign busout = en2 ? busin2 : 4'bzzzz;
    assign busout = en3 ? busin3 : 4'bzzzz;

endmodule
```

Figure 3.45 Implementing a 3-State Bus

Verilog allows multiple concurrent drivers for **net**s. However, a variable declared as a **reg** and used on a left hand side in a procedural block (**always** block), should only be driven concurrently by one source. This makes the use of **net**s more appropriate for representing busses.

3.3 Sequential Circuits

As with any digital circuit, a sequential circuit can be described in Verilog by use of gates, Boolean expressions, or behavioral constructs (e.g., the **always** statement). While gate level descriptions enable a more detailed description of timing and delays, because of complexity of clocking and register and flip-flop controls, these circuits are usually described by use of procedural **always** blocks. This section shows various ways sequential circuits are described in Verilog. The following discusses primitive structures like latch and flip-flops, and then generalizes coding styles used for representing these structures to more complex sequential circuits including counters and state machines.

3.3.1 Basic Memory Elements at the Gate Level

A clocked D-latch latches its input data during an active clock cycle. The latch structure retains the latched value until the next active clock cycle. This element is the basis of all static memory elements.

A simple implementation of the D-latch that uses cross-coupled NOR gates is shown in Figure 3.46. The Verilog code of Figure 3.47 corresponds to this D-latch circuit. This description uses primitive **and** and **nor** structures.

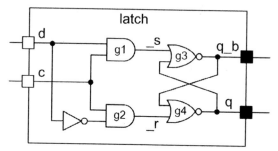

Figure 3.46 Clocked D-latch

As shown in this Verilog code, the tilde (~) operator is used to generate the complement of the d input of the latch. Using AND gates, the d input and its complement are gated to generate internal _s and _r inputs. These are inputs to the cross-coupled NOR structure that is the core of the memory in this latch.

```
`timescale 1ns/100ps
module latch ( input d, c, output q, q_b );
    wire  _s, _r;
    and   #(6) g1 ( _s, c, d ),
               g2 ( _r, c, ~d );
    nor   #(4) g3 ( q_b, _s, q ),
               g4 ( q, _r, q_b );
endmodule
```

Figure 3.47 Verilog Code for a Clocked D-latch

Alternatively, the same latch can be described with an **assign** statement as shown below.

$$\text{assign } \#(3) \; q = c \; ? \; d : q;$$

This statement simply describes what happens in a latch. The statement says that when c is **1**, the q output receives d, and when c is **0** it retains its old value. Using two such statements with complementary clock values describe a master-slave flip-flop. As shown in Figure 3.48, the *qm* **net** is the master output and *q* is the flip-flop output.

```
`timescale 1ns/100ps

module master_slave ( input d, c, output q );
    wire qm;
    assign #(3) qm = c ? d : qm;
    assign #(3) q = ~c ? qm : q;
endmodule
```

Figure 3.48 Master-Slave Flip-Flop

This code uses two concurrent **assign** statements. As discussed before, these statements model logic structures with **net** driven outputs (*qm* and q). The order in which the statements appear in the body of the *master_slave* **module** is not important.

3.3.2 Memory Elements Using Procedural Statements

Although latches and flip-flops can be described by primitive gates and **assign** statements, such descriptions are hard to generalize, and describing more complex register structures cannot be done this way. This section uses **always** statements to describe latches and flip-flops. We will show that the same coding styles used for these simple memory elements can be generalized to describe memories with complex control as well as functional register structures like counters and shift-registers.

Latches. Figure 3.49 shows a D-latch described by an **always** statement. The *q* output of the latch is declared as **reg** because it is being driven inside the **always** procedural block. Latch clock and data inputs (*c* and *d*) appear in the sensitivity list of the **always** block, making this procedural statement sensitive to *c* and *d*. This means that when an event occurs on *c* or *d*, the **always** block wakes up and it executes all its statements in the sequential order from **begin** to **end**.

```
module d_latch ( input d, c, output reg q, output q_b );
    always @ ( c or d )
        if ( c ) begin
            #4 q = d;
        end
    assign #3 q_b = ~q;
endmodule
```

Figure 3.49 Procedural Latch

The **if**-statement enclosed in the **always** block puts d into q when c is active. This means that if c is **1** and d changes, the change on d propagates to the q output. This behavior is referred to as transparency, which is how latches work. While clock is active, a latch structure is transparent, and input changes affect its output.

Any time the **always** statement wakes up, if c is **1**, it waits 4 nanoseconds and then puts d into q. When q changes and it receives its new value, its complement will become the driving value for q_b output. Note that, since q is used on the left hand side in an **always** statement it is a **reg**, but since q_b is used in an **assign** statement it is a **net**. By default, the type an output, such as q_b, is a **net** of **wire** type.

3.3.2.2 D Flip-Flop.

While a latch is transparent, a change on the D-input of a D flip-flops does not directly pass on to its output. The Verilog code of Figure 3.50 describes a positive-edge trigger D-type flip-flop.

The sensitivity list of the procedural statement shown includes **posedge** of clk. This **always** statement only wakes up when clk makes a **0** to **1** transition. When this statement does wake up, the value of d is put into q. Obviously this behavior implements a rising-edge D flip-flop.

```
`timescale 1ns/100ps
module d_ff ( input d, clk, output reg q, output q_b );
    always @( posedge clk )
    begin
        #4 q <= d;
    end
    assign #3 q_b = ~q;
endmodule
```

Figure 3.50 A Positive-Edge D Flip-Flop

Instead of **posedge**, use of **negedge** would implement a falling-edge D flip-flop. After the specified edge, the flow into the **always** block begins. In our description, this flow is halted by 4 nanoseconds by the #4 delay-control statement. After this delay, the value of d is read and put into q. Following this transaction, the flow into the **always** block goes back to its beginning waiting for another positive-edge of the clock. Although the behavior of this code would be the same if we used a blocking assignment ($q = d$) instead of the non-blocking assignment ($q <= d$), it is more customary to use the latter in describing sequential circuits in Verilog.

3.3.2.3 Synchronous Control. The coding style presented for the above simple D flip-flop is a general one and can be expanded to cover many features found in flip-flops and even memory structures. The description shown in Figure 3.51 is a D-type flip-flop with synchronous set and reset (*s* and *r*) inputs.

```
module d_ff (input d, s, r, clk, output reg q, output q_b);
    always @ ( posedge clk ) begin
        if ( s ) begin
            #4 q <= 1'b1;
        end else if ( r ) begin
            #4 q <= 1'b0;
        end else begin
            #4 q <= d;
        end
    end
    assign #3 q_b = ~q;
endmodule
```

Figure 3.51 D Flip-Flop with Synchronous Control

The description uses an **always** block that is sensitive to the positive-edge of *clk*. When *clk* makes a **0** to **1** transition, the flow into the **always** block begins. Immediately after the positive-edge, *s* is inspected and if it is active (**1**), after 4 ns *q* is set to **1**, 3 ns after which *q_b* is set to **0** (by the **assign** statement outside of the **always**). Following the positive-edge of *clk*, if *s* is not **1**, *r* is inspected and if it is active, *q* is set to **0**. If neither *s* or *r* are **1**, the flow of the program reaches the last **else** part of the **if**-statement and assigns *q* to *d*.

The behavior discussed here only looks at *s* and *r* on the positive-edge of *clk*, which corresponds to a rising-edge trigger D-type flip-flop with synchronous active high set and reset inputs. Furthermore, the set input is given a higher priority over the reset input. The flip-flop structure that corresponds to this description is shown in Figure 3.52.

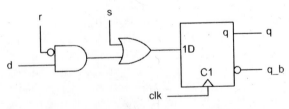

Figure 3.52 D Flip-Flop with Synchronous Control

Other synchronous control inputs can be added to this flip-flop in a similar fashion. A clock enable (*en*) input would only require inclusion of an **if**-statement in the last **else** part of the **if**-statement in the code of Figure 3.51.

3.3.2.4 Asynchronous Control. The control inputs of the flip-flop of Figure 3.51 are synchronous because the flow into the **always** statement is only allowed to start when the **posedge** of *clk* is observed. To change this to a flip-flop with asynchronous control, it is only required to include asynchronous control inputs in the sensitivity list of its procedural statement.

Figure 3.53 shows a D flip-flop with active high asynchronous set and reset control inputs. Note that the only difference between this description and the code of Figure 3.51 (synchronous control) is the inclusion of **posedge** *s* and **posedge** *r* in the sensitivity list of the **always** block. This inclusion allows the flow into the procedural block to begin when *clk* becomes **1** or *s* becomes **1** or *r* becomes **1**. The **if**-statement in this block checks for *s* and *r* being **1**, and if none are active (activity levels are high) then clocking *d* into *q* occurs.

```
module d_ff (input d, s, r, clk, output reg q, output q_b);

    always @ (posedge clk or posedge s or posedge r )
    begin
        if ( s ) begin
            #4 q <= 1'b1;
        end else if ( r ) begin
            #4 q <= 1'b0;
        end else  begin
            #4 q <= d;
        end
    end

    assign #3 q_b = ~q;

endmodule
```

Figure 3.53 D Flip-Flop with Asynchronous Control

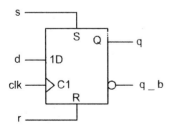

Figure 3.54 Flip-Flop with Asynchronous Control Inputs

An active high (low) asynchronous input requires inclusion of **posedge** (**negedge**) of the input in the sensitivity list, and checking its **1** (**0**) value in the **if**-statement in the **always** statement. Furthermore, clocking activity in the flip-flop (assignment of d into q) must always be the last choice in the **if**-statement of the procedural block. The graphic symbol corresponding to the flip-flop of Figure 3.53 is shown in Figure 3.54.

3.3.3 Flip-flop Synthesis

In the above discussion, flip-flop and latches described by use of procedural statements are synthesizable. The flip-flop of the logic-element used for the implementation of code of Figure 3.51 uses its lookup-table to generate a **1** or **0** for setting and resetting the flip-flop. As shown in Figure 3.55, all data (D, Set or Reset) of the flip-flop go through the D input and its asynchronous inputs are unused. Data on the D input is always controlled by the clock.

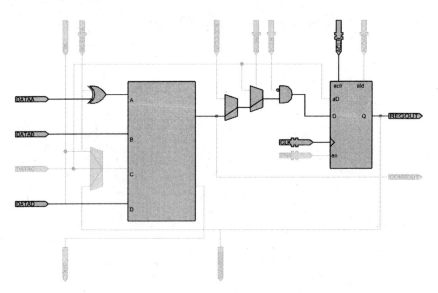

Figure 3.55 Synchronous Flip-Flop Synthesis

On the other hand, the flip-flop of the logic-element used for the implementation of code of Figure 3.53 uses its asynchronous inputs for setting and resetting. As shown in Figure 3.56 the lookup-table used for the flip-flop with asynchronous control is only used for passing the D input of the circuit to the flip-flop D input. Logic-element flip-flop inputs aD and ald are responsible for asynchronously loading a **1** into the flip-flop and the $aclr$ handles asynchronous resetting.

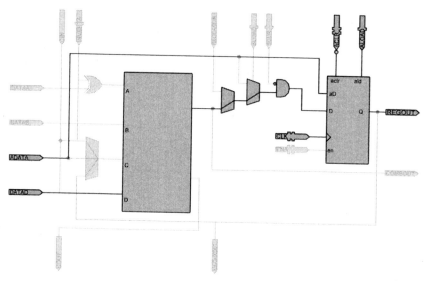

Figure 3.56 Asynchronous Flip-Flop Synthesis

RTL views of hardware generated for the two flip-flops are shown in Figure 3.57. Comparing these circuits, more clearly shows the use of the D-input of the flip-flop input for synchronous reset. The D-input of the flip-flop resulted from synthesizing Figure 3.51 has a logic block that involves d, r, and s (upper part of Figure 3.57). On the other hand, the circuit d input directly connects to the D-input of the flip-flop resulted from the synthesis of Figure 3.53 (lower part of Figure 3.57).

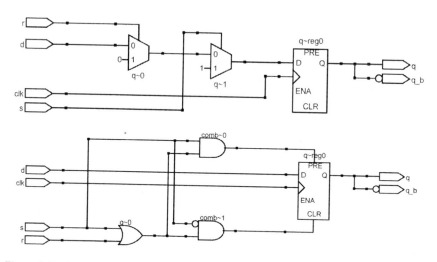

Figure 3.57 Synchronous versus Asynchronous Synthesis

3.3.4 Registers, Shifters and Counters

Registers, shifter-registers, counters and even sequential circuits with more complex functionalities can be described by simple extensions of the coding styles presented for the flip-flops. In most cases, the functionality of the circuit only affects the last **else** of the **if**-statement in the procedural statement of codes shown for the flip-flops.

3.3.4.1 Registers. Figure 3.58 shows an 8-bit register with synchronous *set* and *reset* inputs. The *set* input puts all **1**s in the register and the *reset* input resets it to all **0**s. The main difference between this and the flip-flop with synchronous control is the vector declaration of inputs and outputs.

```
module register (input [7:0] d, input clk, set, reset,
                 output reg [7:0] q);
    always @ ( posedge clk ) begin
        if ( set )
            #5 q <= 8'b1;
        else if ( reset )
            #5 q <= 8'b0;
        else
            #5 q <= d;
    end
endmodule
```

Figure 3.58 An 8-bit Register

3.3.4.2 Shift-Registers. A 4-bit shift-register with right- and left-shift capabilities, a serial-input, synchronous reset input, and parallel loading capability is shown in Figure 3.59. As shown, only the positive-edge of *clk* is included in the sensitivity list of the **always** block of this code, which makes all activities of the shift-register synchronous with the clock input. If *rst* is **1**, the register is reset, if *ld* is **1** parallel *d* inputs are loaded into the register, and if none are **1** shifting left or right takes place depending on the value of the *l_r* input (**1** for left, **0** for right). Shifting in this code is done by use of the concatenation operator { }. For left-shift, *s_in* is concatenated to the right of *q[2:0]* to form a 4-bit vector that is put into *q*. For right-shift, *s_in* is concatenated to the left of *q[3:1]* to form a 4-bit vector that is clocked into *q[3:0]*.

The style used for coding this register is the same as that used for flip-flops and registers presented earlier. In these examples, a single procedural block handles function selection (e.g., zeroing, shifting, or parallel loading) as well as clocking data *d* into the register.

```
module shift_reg (input [3:0] d, input clk, ld, rst, l_r,
                  s_in, output reg [3:0] q);
    always @ ( posedge clk ) begin
        if ( rst )
            #5 q <= 4'b0000;
        else if ( ld )
            #5 q <= d;
        else if ( l_r )
            #5 q <= {q[2:0], s_in};
        else
            #5 q <= {s_in, q[3:1]};
    end
endmodule
```

Figure 3.59 A 4-bit Shift Register

Another style of coding registers, shift-registers and counters is to use a combinational procedural block for function selection and another for clocking.

As an example, consider a shift-register that shifts *s_cnt* number of places to the right or left depending on its *sr* or *sl* control inputs (Figure 3.60). The shift-register also has an *ld* input that enables its clocked parallel loading. If no shifting is specified, i.e., *sr* and *sl* are both zero, then the shift register retains its old value.

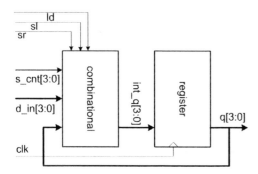

Figure 3.60 Separate Combinational/Sequential Blocks Shift Register

The Verilog code of Figure 3.61 that corresponds to this design shows two procedural blocks that are identified by *combinational* and *register*. An optional block name appears after the **begin** keyword that begins a block and is separated from this keyword by use of a colon.

The *combinational* block is sensitive to all inputs that can affect the shift register output. These include the parallel *d_in*, the *s_cnt* shift-count, *sr* and *sl* shift control inputs, and the *ld* load control input. In the body of this block an **if-else** statement decides on the value placed on the *int_q* internal variable. The value selection is based on values of *ld*, *sr*, and *sl*. If *ld* is **1**, *int_q* becomes *d_in* that is the parallel input of the shift register. If *sr* or *sl* is active, *int_q* receives the value of *q* shifted to right or left as many as *s_cnt* places. In this example, shifting is done by use of the >> and << operators. On the left, these operators take the vector to be shifted, and on the right they take the number of places to shift.

The *int_q* variable that is being assigned values in the *combinational* block is a 4-bit **reg** that connects the output of this block to the input of the *register* block.

The *register* block is a sequential block that handles clocking *int_q* into the shift register output. This block (as shown in Figure 3.61) is sensitive to the positive edge of *clk* and its body consists of a single **reg** assignment.

Note in this code that both *q* and *int_q* are declared as **reg** because they are both receiving values in procedural blocks, one in the *register* and one in the *combinational* block.

```
module shift_reg (input [3:0] d_in, input clk, input [1:0]
                  s_cnt, sr, sl, ld, output reg [3:0] q );
    reg [3:0] int_q;
    always @(d_in, s_cnt, sr, sl, ld, q) begin:combinational
        if ( ld )   int_q = d_in;
        else if ( sr )   int_q = q >> s_cnt;
        else if ( sl ) int_q = q << s_cnt;
        else int_q = q;
    end
    always @ ( posedge clk ) begin:register
        q <= int_q;
    end
endmodule
```

Figure 3.61 Shift-Register Using Two Procedural Blocks

3.3.4.3 Counters. Any of the styles described for the shift-registers in the previous discussion can be used for describing counters. A counter counts up or down, while a shift-register shifts right or left. We use arithmetic operations in counting as opposed to shift or concatenation operators in shift-registers.

Figure 3.62 shows a 4-bit up-down counter with a synchronous *rst* reset. The counter has an *ld* input for doing the parallel loading of *d_in* into the counter. The counter output is *q* and it is declared as **reg** since it is receiving values within a procedural statement.

Discussion about synchronous and asynchronous control of flip-flops and registers also apply to the counter circuits. For example, inclusion of ***posedge*** *rst* in the sensitivity list of the counter of Figure 3.62 would make its resetting asynchronous. Also, as in the other examples of clocked circuits, we are using nonblocking assignments in our clocked procedural blocks.

```verilog
module counter (input [3:0] d_in, input clk, rst, ld, u_d,
                output reg [3:0] q );
    always @ ( posedge clk ) begin
        if ( rst )
            q <= 4'b0000;
        else if ( ld )
            q <= d_in;
        else if ( u_d )
            q <= q + 1;
        else
            q <= q - 1;
    end
endmodule
```

Figure 3.62 An Up-Down Counter

3.3.5 Synthesis of Shifters and Counters

Except for the operations that are performed in the procedural blocks of the descriptions of the registers, counters, and shift registers, the above descriptions followed the same basic rules, and the same styles of coding could be used for them. The styles we presented were synthesizable, and for demonstration purposes we show the synthesis results obtained by synthesizing the shift register of Figure 3.61.

Synthesis of this shift register uses 26 logic-elements of an Altera Cyclone chip. As shown in the RTL view of Figure 3.63, the generated hardware has the same outline as the block diagram shown in Figure 3.60. The combinational logic on the left feeds the register on the right, and there is a feedback from the register to the combinational part.

3.3.6 State Machine Coding

Coding styles presented so far can be further generalized to cover finite state machines of any type. This section shows coding for Moore and Mealy state machines. The examples we will use are simple sequence detectors. These circuits represent the controller part of a digital system that has been partitioned into a data path and a controller. The coding styles used here apply to such controllers, and will be used in later chapters of this book to describe CPU and multiplier controllers.

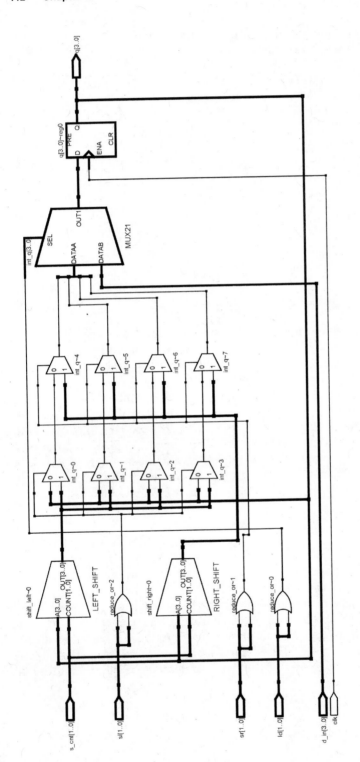

Figure 3.63 RTL View of Synthesized Shift Register

3.3.6.1 Moore Detector.

State diagram for a Moore sequence detector detecting **101** on its x input is shown in Figure 3.64. The machine has four states that are labeled, *reset, got1, got10,* and *got101*. Starting in *reset*, if the **101** sequence is detected, the machine goes into the *got101* state in which the output becomes **1**. In addition to the x input, the machine has a *rst* input that forces the machine into its *reset* state. The resetting of the machine is synchronized with the clock.

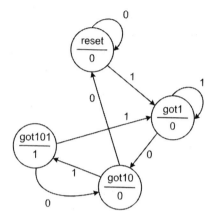

Figure 3.64 A Moore Sequence Detector

The Verilog code of the Moore machine of Figure 3.64 is shown in Figure 3.65. After the declaration of inputs and outputs of this module, **parameter** declaration declares four states of the machine as two-bit parameters. The square-brackets following the **parameter** keyword specify the size of parameters being declared. Following parameter declarations in the code of Figure 3.65, the two-bit *current* **reg** type variable is declared. This variable holds the current state of the state machine. The body of the code of this circuit has an **always** block and an **assign** statement.

The **assign** statement shown in Figure 3.65 puts a **1** on the z output when the current state of the machine, i.e., *current*, is *got101*. This statement is concurrent with the **always** block that is responsible for making the state transitions. The **always** block used in the module of Figure 3.65 describes state transitions of the state diagram of Figure 3.64. The main task of this procedural block is to inspect input conditions (values on *rst* and x) during the present state of the machine defined by *current* and set values into *current* for the next state of the machine.

```
module moore_detector (input x, rst, clk, output z);
   parameter [1:0]  reset = 0, got1 = 1,
                    got10 = 2, got101 = 3;
   reg [1:0] current;
   always @ ( posedge clk ) begin
      if ( rst ) begin
      current <= reset;
      end
      else case ( current )
         reset:  begin
            if ( x==1'b1 ) current <= got1;
            else current <= reset;
         end
         got1:  begin
            if ( x==1'b0 ) current <= got10;
            else current <= got1;
         end
         got10:  begin
            if ( x==1'b1 ) begin
               current <= got101;
            end else begin
               current <= reset;
            end
         end
         got101:  begin
            if ( x==1'b1 ) current <= got1;
            else current <= got10;
         end
         default:  current <= reset;
      endcase
   end
   assign z = (current == got101) ? 1 : 0;
endmodule
```

Figure 3.65 Moore Machine Verilog Code

The flow into the **always** block begins with the positive edge of *clk*. Since all activities in this machine are synchronized with the clock, only *clk* appears on the sensitivity list of the **always** block. Upon entry into this block, the *rst* input is checked and if it is active, *current* is set to *reset* (*reset* is a declared parameter and its value is **0**). The value put into *current* in this pass through the **always** block gets checked in the next pass with the next edge of the clock. Therefore this assignment is regarded as the next-state assignment. When this assignment is made, the **if-else** statements skip the rest of the code of the **always** block, and this **always** block will next be entered with the next positive edge of *clk*. Upon entry into the **always** block, if *rst* is not **1**, program flow reaches the **case** statement that checks the value of *current* against the four states of the machine. Figure 3.66 shows an outline of this **case**-statement.

```
case ( current )
    reset:    begin . . . end
    got1:     begin . . . end
    got10:    begin . . . end
    got101:   begin . . . end
    default:  begin . . . end
endcase
```

Figure 3.66 case-Statement Outline

The **case**-statement shown has five **case**-alternatives. A **case**-alternative is followed by a block of statements bracketed by the **begin** and **end** keywords. In each such block, actions corresponding to the active state of the machine are taken. The last **case**-alternative is the default case that is used for covering cases not covered by other alternatives. In this description, because we have four states, the use of **default** is not required for synthesis. However, for simulation purposes, in case of ambiguous values of *current* (**X** or **Z** in *current*), this alternatives sets the next state of the machine to reset. For synthesis or simulation, it is recommended to use **default** regardless of the number of states. We will elaborate on this issue in the next state machine example of this section.

Figure 3.67 shows the Verilog code of the *got10* state and its diagram from the state diagram of Figure 3.64. As shown, the **case**-alternative that corresponds to the *got10* state specifies the next values for that state. Determination of the next state is based on the value of *x*. If *x* is **1**, the next state becomes *got101*, and if *x* is **0**, the next state becomes *reset*.

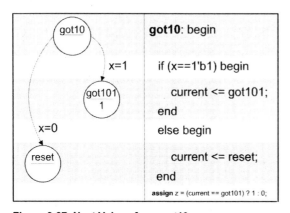

Figure 3.67 Next Values from *got10*

In this coding style, for every state of the machine there is a **case**-alternative that specifies the next state values. For larger machines, there will be more **case**-alternatives, and more conditions within an alternative. Otherwise, this style can be applied to state machines of any size and complexity.

This same machine can be described in Verilog in many other ways. We will show alternative styles of coding state machines by use of examples that follow.

3.3.6.2 A Mealy Machine Example.

Unlike a Moore machine that has outputs that are only determined by the current state of the machine, in a Mealy machine, the outputs are determined by the state the machine is in as well as the inputs of the circuit. This makes Mealy outputs not fully synchronized with the circuit clock. In the state diagram of a Mealy machine the outputs are specified along the edges that branch out of the states of the machine.

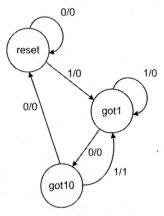

Figure 3.68 A 101 Mealy Detector

Figure 3.68 shows a **101** Mealy detector. The machine has three states, *reset*, *got1* and *got10*. While in *got10*, if the x input becomes **1** the machine prepares to go to its next state with the next clock. While waiting for the clock, its output becomes **1**. While this is happening, if the clock arrives, the machine goes out of *got10* and enters the *got1* state. This machine allows overlapping sequences. The machine has no external resetting mechanism. A sequence of two zeros on input x puts the machine into the *reset* state in a maximum of two clocks.

The Verilog code of the **101** Mealy detector is shown in Figure 3.69. After input and output declarations, a **parameter** declaration defines bit patterns (state assignments) for the states of the machine. Note here that state value 3 or 2'b11 is unused. As in the previous example, we use the *current* two-bit **reg** to hold the current state of the machine.

After the declarations, an **initial** block sets the initial state of the machine to *reset*. This procedure for initializing the machine is only good for simulation and is not synthesizable.

```verilog
module mealy_detector ( input x, clk, output z );
    parameter [1:0]
        reset   = 0,   // 0 = 0 0
        got1    = 1,   // 1 = 0 1
        got10   = 2;.  // 2 = 1 0

    reg [1:0] current;

    initial current = reset;
    always @ ( posedge clk )
    begin
        case ( current )
        reset:
            if( x==1'b1 ) current <= got1;
            else current <= reset;
        got1:
            if( x==1'b0 ) current <= got10;
            else current <= got1;
        got10:
            if( x==1'b1 ) current <= got1;
            else current <= reset;
        default: current <= reset;
        endcase
    end
    assign z= ( current==got10 && x==1'b1 ) ? 1'b1 : 1'b0;

endmodule
```

Figure 3.69 Verilog Code of 101 Mealy Detector

This example uses an **always** block for specifying state transitions and a separate **assign** statement for setting values to the z output. The **always** statement responsible for state transitions is sensitive to the circuit clock and has a **case** statement that has **case** alternatives for every state of the machine. Consider for example, the *got10* state and its corresponding Verilog code segment, as shown in Figure 3.70. The Verilog code of this state specifies its next states. Notice in this code segment that the **case** alternative shown does not have **begin** and **end** bracketing. Actually, **begin** and **end** keywords do not appear in blocks following **if** and **else** keywords either.

Verilog only requires **begin** and **end** bracketing if there is more than one statement in a block. The use of this bracketing around one statement is optional. Since the **if** part and the **else** part each only contain one statement, **begin** and **end** keywords are not used. Furthermore, since the entire **if-else** statement reduces to only one statement, the **begin** and **end** keywords for the **case**-alternative are also eliminated.

The last **case**-alternative shown in Figure 3.69 is the **default** alternative. When checking *current* against all alternatives that appear before the **default** statement fail, this alternative is taken. There are several reasons that we use this default alternative. One is that, our machine only uses three of the possible four 2-bit assignments and 2'b11 is unused. If the machine ever begins in this state, the default case makes *reset* the next state of the machine. The second reason why we use **default** is that Verilog assumes a four-value logic system that includes **Z** and **X**. If *current* ever contains a **Z** or **X**, it does not match any of the defined case alternatives, and the default case is taken. Another reason for use of **default** is that our machine does not have a hard reset and we are making provisions for it to go to the *reset* state. The last reason for **default** is that it is just a good idea to have it.

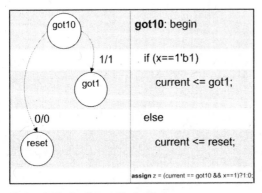

Figure 3.70 Coding a Mealy State

The last statement in the code fragment of Figure 3.70 is an **assign** statement that sets the *z* output of the circuit. This statement is a concurrent statement and is independent of the **always** statement above it. When *current* or *x* changes, the right hand side of this assignment is evaluated and a value of **0** or **1** is assigned to *z*. Conditions on the right hand side of this assignment are according to values put in *z* in the state diagram of Figure 3.68. Specifically, the output is **1** when *current* is *got10* and *x* is **1**, otherwise it is **0**. This statement implements a combinational logic structure with *current* and *x* inputs and *z* output.

3.3.6.3 Huffman Coding Style.
The Huffman model for a digital system characterizes it as a combinational block with feedbacks through

an array of registers. Verilog coding of digital systems according to the Huffman model uses an **always** statement for describing the register part and another concurrent statement for describing the combinational part.

We will describe the state machine of Figure 3.64 to illustrate this style of coding. Figure 3.71 shows the combinational and register part partitioning that we will use for describing this machine. The *combinational* block uses x and p_state as input and generates z and n_state. The *register* block clocks n_state into p_state, and reset p_state when *rst* is active.

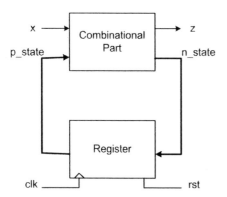

Figure 3.71 Huffman Partitioning of 101 Moore Detector

Figure 3.72 shows the Verilog code of Figure 3.64 according to the partitioning of Figure 3.71. As shown, parameter declaration declares the states of the machine. Following this declaration, n_state and p_state variables are declared as two-bit **reg**s that hold values corresponding to the states of the **101** Moore detector. The *combinational* **always** block follows this **reg** declaration. Since this is a purely combinational block, it is sensitive to all its inputs, namely x and p_state. Immediately following the block heading, n_state is set to its inactive or reset value. This is done so that this variable is always reset with the clock to make sure it does not retain its old value. As discussed before, retaining old values implies latches, which is not what we want in our combinational block.

The body of the combinational **always** block of Figure 3.72 contains a **case**-statement that uses the p_state input of the **always** block for its **case**-expression. This expression is checked against the states of the Moore machine. As in the other styles discussed before, this **case**-statement has **case**-alternatives for *reset*, *got1*, *got10*, and *got101* states.

```verilog
module moore_detector ( input x, rst, clk, output z );
   parameter [1:0] reset = 2'b00, got1 = 2'b01,
                   got10 = 2'b10, got101 = 2'b11;
   reg [1:0] p_state, n_state;

   always @ ( p_state or x ) begin : combinational
      n_state = 0;
      case ( p_state )
      reset:  begin
         if( x==1'b1 ) n_state = got1;
         else n_state = reset;
      end
      got1:  begin
         if( x==1'b0 ) n_state = got10;
         else n_state = got1;
      end
      got10:  begin
         if( x==1'b1 ) n_state = got101;
         else n_state = reset;
      end
      got101:  begin
         if( x==1'b1 ) n_state = got1;
         else n_state = got10;
      end
      default: n_state = reset;
      endcase
   end

   always @( posedge clk ) begin : register
      if( rst ) p_state = reset;
      else p_state = n_state;
   end

   assign z = (current == got101) ? 1 : 0;

endmodule
```

Figure 3.72 Verilog Huffman Coding Style

In a block corresponding to a **case**-alternative, based on input values, *n_state* is assigned values. Unlike the other styles where *current* is used both for the present and next states, here we use two different variables, *p_state* and *n_state*.

The next procedural block shown in Figure 3.72 handles the register part of the Huffman model of Figure 3.71. In this part, *n_state* is treated as the register input and *p_state* as its output. On the positive edge of the clock, *p_state* is either set to the *reset* state **(00)** or is loaded with contents of *n_state*. Together, *combinational* and *register* blocks describe our state machine in a very modular fashion.

As with other styles we presented here, a separate **assign** statement (or any other concurrent statement) is used for assignment of values to the output.

The advantage of this style of coding is in its modularity and defined tasks of each block. State transitions are handled by the *combinational* block and clocking is done by the *register* block. Changes in clocking, resetting, enabling or presetting the machine only affect the coding of the *register* block. If we were to change the synchronous resetting to asynchronous, the only change we had to make was adding *posedge* rst to the sensitivity list of the *register* block.

3.3.6.4 More Complex Outputs. Thus far in presenting state machine coding styles, we have used a simple **assign** statement for assignment of values to the output of the circuit. For a design with more input and output lines and more complex output logic, we can use an **always** block for handling assigning values to the outputs of the circuit. This block is similar to the block used for handling state transitions. For coding the output block, it is necessary to follow the rules discussed for combinational blocks in Section 3.2.7.

Figure 3.73 shows the coding of the **110-101** Mealy detector using two separate blocks for assigning values to *n_state* and the *z* output. In a situation like that of this example, in which the output logic is fairly simple, a simple **assign** statement could replace the *outputting* procedural block. In this case, *z* must be a **net** and not a **reg**.

The examples discussed above, in particular, the last two styles, show how combinational and sequential coding styles can be combined to describe very complex digital systems.

3.3.7 State Machine Synthesis

All state machine descriptions discussed above are synthesizable. For demonstration purposes we discuss synthesis results of the Mealy machine of Figure 3.69.

Except for the initial statement used in this description, everything else synthesizes. The synthesis process ignores this statement and issues a warning message. Recall that this statement was put in the code of the *mealy_detector* for initialization of the machine when being simulated.

Implementation of this state machine uses four logic-elements, three of which are registered and one only uses the lookup-table of its logic-element. Figure 3.74 shows the technology view of the synthesized hardware illustrating the use of four lookup tables. Utilization of the registered logic-elements (left-most three in Figure 3.74) is similar to the logic-element of Figure 3.55, and the right-most element is similar to the logic-element of Figure 3.25.

```verilog
module mealy_detector ( input x, en, clk, rst, output reg z
                        );
   parameter [1:0]    reset = 0,
                      got1 = 1, got10 = 2, got11 = 3;

   reg [1:0] p_state, n_state;

   always @( p_state or x ) begin : Transitions
      n_state = reset;
      case ( p_state )
      reset:
         if ( x == 1'b1 ) n_state = got1;
         else n_state = reset;
      got1:
         if ( x == 1'b0 ) n_state = got10;
         else n_state = got11;
      got10:
         if ( x == 1'b1 ) n_state = got1;
         else n_state = reset;
      got11:
         if ( x == 1'b1 ) n_state = got11;
         else n_state = got10;
      default:  n_state = reset;
      endcase
   end

   always @( p_state or x ) begin: Outputting
      z = 0;
      case ( p_state )
      reset:   z = 1'b0;
      got1:    z = 1'b0;
      got10:   if ( x == 1'b1 ) z = 1'b1;
         else z = 1'b0;
      got11:   if ( x==1'b1 ) z = 1'b0;
         else z = 1'b1;
      default:  z = 1'b0;
      endcase
   end

   always @ ( posedge clk ) begin: Registering
      if( rst ) p_state <= reset;
      else if( en ) p_state <= n_state;
   end

endmodule
```

Figure 3.73 Separate Transition and Output Blocks

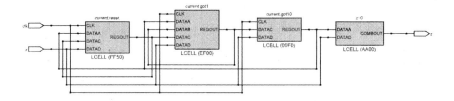

Figure 3.74 Technology View of Mealy Machine Synthesis

3.3.8 Memories

Verilog allows declaration and usage of multidimensional arrays for any of the **net** types or **reg** variables. The following declares *a_array* as a two-dimensional array of 8-bit words.

```
reg [7:0] a_array [0:1023][0:511];
```

In an array declaration, the address range (or ranges, for multi-dimensional arrays) of the elements of the array comes (or come, for multi-dimensional arrays) after the name of the array. Range specifications are enclosed in square brackets. The size and range specification of the elements of an array come after the **net** type (e.g., **wire**) or **reg** keyword. In the absence of a range specification before the name of the array, an element size of one bit is assumed. Several examples of array declarations are shown in Figure 3.75.

```
reg [7:0] Areg;            // An 8-bit vector
reg Amem [7:0];            // A memory of 8 1-bit elements
reg Dmem [7:0][0:3];       // 2-Dim mem, 1-bit elements
reg [7:0] Cmem [0:3];      // A memory of four 8-bit words
reg [2:0] Dmem [0:3][0:4]; // 2-Dim mem, 3-bit elements
reg [7:0] Emem [0:1023];   // A memory of 1024 8-bit words
```

Figure 3.75 Array Declaration Examples

3.3.8.1 Array Indexing.

Bit-select and part-select operators are used for extracting a bit or a group of bits from a declared array. Such addressing only applies to contiguous bits of an array. We use arrays declared above to demonstrate bit-select and part-select operations.

Bit-selection is done by using the addressed bit number in a set of square brackets. For example *Areg[5]* selects bit 5 of *Areg* array.

Verilog allows constant and indexed part-select. A constant part-select specifies range of bits to be selected. For example, *Areg[7:3]* selects the upper five bits of *Areg*. On the other hand an indexed part-select specifies starting index and the number of bits to be selected. For example, *Areg[3+:5]* selects the same five bits as *Areg[7:3]* does. Several examples are shown below.

```
Areg [5:3] selects bits 5, 4, and 3
Areg [5-:4] selects bits 5, 4, 3, and 2
Areg [2+:4] selects bits 5, 4, 3, and 2
```

3.3.8.2 Standard Memory. The standard format for declaring a memory in Verilog is to declare it as an array of a vector. For example *Cmem* of Figure 3.75 is a 4-word memory of 8-bit words. The address space of this memory is 4. *Emem* is a byte-oriented memory with a 10 bit address (1024 address space).

An expression can be used for addressing a memory. For example two bits of *Areg* can be used to extract an 8-bit word of *Cmem*. This is done as shown below:

```
Cmem [Areg[7:6]] // extracts Cmem word
                 // addressed by Areg[7:6]
```

A memory word can be used as an address for itself. The following example uses the 8-bit word at location 0 of *Emem* to address this memory:

```
Emem [Emem[0]] // extracts Emem word
               // addressed by Emem[0]
```

Verilog allows selection rules for accessing part of an addressed word of a memory. For this purpose a second set of square brackets to the right of those used for memory addressing are used for bit-or part-select of the accessed memory word. For example the four least significant bits of the word at location 355 of *Emem* are accessed by:

```
Emem [355][3:0] // 4 LSB of location 355
```
This operation is equivalent to:

```
Emem [355][3-:4] // 4 bits starting from 3, down
```

Specifying a range of addresses in Verilog is not allowed. For example, *Emem* locations 355 to 358 cannot be addressed as shown below.

```
Emem [355:358] // Illegal.
               // This is not a 4-word block
```

As discussed, declaring multi-dimensional memories is allowed in Verilog, e.g., *Dmem* above. For accessing such memories, simple indexings are allowed for specifying a word in the memory, and bit-select and part-select are allowed for accessing bit or bits of the addressed word. The following is an example using *Dmem*.

```
Dmem [0][2]58] // Illegal.
               // This is not a 4-word block
```

Figure 3.76 shows a memory block with separate input and output busses. Writing into the memory is clocked, while reading from it only requires *rw* to be **1**. An **assign** statement handles reading from the memory, and an **always** block performs writing into this memory.

```
module memory (input [7:0] inbus, output [7:0] outbus,
            input [9:0] addr, input clk, rw);
    reg [7:0] mem [0:1023];
    assign outbus = rw ? mem [addr] : 8'bz;
    always @ (posedge clk)
        if (rw == 0) mem [addr] = inbus;
endmodule
```

Figure 3.76 Memory Description

3.4 Writing Testbenches

Verilog coding styles discussed so far were for coding hardware structures, and in all cases synthesizability and direct correspondence to hardware were our main concerns. On the other hand, testbenches do not have to have hardware correspondence and they usually do not follow any synthesizability rules. We will see that delay specifications, and **initial** statements that do not have a one-to-one hardware correspondence are used generously in testbenches.

```
module moore_detector ( input x, rst, clk, output z );
    parameter [1:0] a=0, b=1, c=2, d=3;
    reg [1:0] current;
    always @( posedge clk )
        if ( rst )    current = a;
        else case ( current )
            a : current <= x ? b : a ;
            b : current <= x ? b : c ;
            c : current <= x ? d : a ;
            d : current <= x ? b : c ;
            default : current <= a;
        endcase
    assign z = (current==d) ? 1'b1 : 1'b0;
endmodule
```

Figure 3.77 Circuit Under Test

For demonstration of testbench coding styles, we use the Verilog code of Figure 3.77 that is a **101** Moore detector, as the circuit to be tested. This description is functionally equivalent to that of Figure 3.65. The difference is in the use of condition expressions (?:) instead of **if-else** statements. This code will be instantiated in the testbenches that follow.

3.4.1 Generating Periodic Data

Figure 3.78 shows a module that is used as a testbench that instantiates *moore_detector* and applies test data to its inputs. The first statement in this code is the **'timescale** directive that defines the time unit of this description. The testbench itself has no ports, which is typical of all testbenches. All data inputs to a circuit-under-test are locally generated in its testbench.

```
`timescale 1 ns / 100 ps
module test_moore_detector;
    reg x, reset, clock;
    wire z;

    moore_detector uut ( x, reset, clock, z );
    initial  begin
        clock=1'b0; x=1'b0; reset=1'b1;
    end
    initial #24 reset=1'b0;
    always #5 clock=~clock;
    always #7 x=~x;
endmodule
```

Figure 3.78 Generating Periodic Data

Because we are using procedural statements for assigning values to ports of the circuit-under-test, all variables mapped with the input ports of this circuit are declared as **reg**. The testbench uses two **initial** blocks and two **always** blocks. The first initial block initializes *clock*, *x*, and *reset* to **0**, **0**, and **1** respectively. The next **initial** block waits for 24 time units (ns in this code), and then sets *reset* back to **0** to allow the state machine to operate.

The **always** blocks shown produce periodic signals with different frequencies on *clock* and *x*. Each block waits for a certain amount of time and then it complements its variable. Complementing begins with the initial values of *clock* and *x* as set in the first **initial** block. We are using different periods for *clock* and *x*, so that a combination of patterns on these circuit inputs is seen. A more deterministic set of values could be set by specifying exact values at specific times.

3.4.2 Random Input Data

Instead of the periodic data on x we can use the **$random** predefined system function to generate random data for the x input. Figure 3.79 shows such a testbench.

```
`timescale 1 ns / 100 ps

module test_moore_detector;
    reg x, reset, clock;
    wire z;
    moore_detector uut( x, reset, clock, z );
    initial  begin
        clock=1'b0; x=1'b0; reset=1'b1;
        #24 reset=1'b0;
    end
    initial #165 $finish;
    always #5 clock=~clock;
    always #7 x=$random;
endmodule
```

Figure 3.79 Random Data Generation

This testbench also combines the two **initial** blocks for initially activating and deactivating *reset* into one. In addition, this testbench has an **initial** block that finishes the simulation after 165 ns.

When the flow into a procedural block reaches the **$finish** system task, the simulation terminates and exits. Another simulation control task that is often used is the **$stop** task that only stops the simulation and allows resumption of the stopped simulation run.

3.4.3 Timed Data

A very simple testbench for our sequence detector can be done by applying test data to x and timing them appropriately to generate the sequence we want, very similar to the way values were applied to *reset* in the previous examples. Figure 3.80 shows this simple testbench.

Techniques discussed in the above examples are just some of what one can do for test data generation. These techniques can be combined for more complex examples. After using Verilog for some time, users form their own test generation techniques. For small designs, simulation environments generally provide waveform editors and other tool-dependent test generation schemes. Some tools come with code fragments that can be used as templates for testbenches.

```
`timescale 1ns/100ps
module test_moore_detector;
   reg x, reset, clock;
   wire z;

   moore_detector uut( x, reset, clock, z );

   initial begin
      clock=1'b0; x=1'b0; reset=1'b1;
      #24 reset=1'b0;
   end

   always #5 clock=~clock;

   initial begin
      #7   x=1;
      #5   x=0;
      #18  x=1;
      #21  x=0;
      #11  x=1;
      #13  x=0;
      #33 $stop;
   end
endmodule
```

Figure 3.80 Timed Test Data Generation

3.5 Sequential Multiplier Specification

This section uses a comprehensive example to put all Verilog concepts and techniques presented in the preceding sections into one design. The design example is an add-and-shift sequential multiplier, with an 8-bit A and B inputs and a 16-bit result. The block diagram of this circuit is shown in Figure 3.81. This multiplier has an 8-bit bi-directional I/O for inputting its A and B operands, and outputting its 16-bit output one byte at a time.

Multiplication begins with the *start* pulse. On the clock edge that *start* is **1**, operand A is on the *databus* and in the next clock, this bus will contain operand B. The two operands appear on the bus in two consecutive clock pulses. After accepting these data inputs, the multiplier begins its multiplication process and when it is completed, it starts sending the result out on the *databus*. When the least-significant byte is placed on *databus*, the *lsb_out* output is issued, and in the next clock for the most-significant byte, *msb_out* is issued. When both bytes are outputted, *done* becomes **1**, and the multiplier is ready for another set of data. The multiplexed bi-directorial *databus* is used to reduce the total number of I/O pins of the multiplier.

3.5.1 Shift-and-Add Multiplication Process

When designing multipliers there is always a compromise to be made between how fast the multiplication process is done and how much hardware we are using for its implementation.

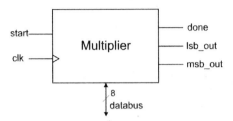

Figure 3.81 Multiplier Block Diagram

A simple multiplication method that is slow but efficient in use of hardware is the shift-and-add method. In this method, depending on bit i of operand A, either operand B is added to the collected partial result and then shifted to the right (when bit i is **1**), or (when bit i is **0**) the collected partial result is shifted one place to the right without being added to B.

This method can better be understood by considering how binary multiplication is done manually. Figure 3.82 shows manual multiplication of two 8-bit binary numbers.

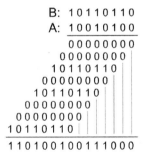

A × B: 1 1 0 1 0 0 1 0 0 1 1 1 0 0 0

Figure 3.82 Manual Binary Multiplication

We start considering bits of A from right to left. If a bit value is **0** we select **00000000** to be added with the next partial product, and if it is a **1**, the value of B is selected. This process repeats, but each time **00000000** or B is selected, it is written one place to the left with respect to the previous value. When all bits of A are considered, we add all calculated values to come up with the multiplication results.

Understanding hardware implementation of this procedure becomes easier if we make certain modifications to this procedure. First, instead of moving our observation point from one bit of *A* to another, we put *A* in a shift-register, always observe its right-most bit, and after every calculation, we move it one place to the right, making its next bit accessible. Second, for the partial products, instead of writing one and the next one to its left, we move the partial product to the right as we are writing it.

Finally, instead of calculating all partial products and adding them up at the end, we add a newly calculated partial product to the previous one and write the calculated value as the new partial result. Therefore, if the observed bit of *A* is **0**, **00000000** is to be added to the previously calculated partial result, and the new value should be shifted one place to the right. In this case, since the value being added to the partial result is **00000000**, adding is not necessary, and only shifting the partial result is sufficient. This process is called *shift*. However, if the observed bit of *A* is **1**, *B* is to be added to the previously calculated partial result, and the calculated new sum must be shifted one place to the right. This is called *add-and-shift*.

Repeating the above procedure, when all bits of *A* are shifted out, the partial result becomes the final multiplication result. We use a 4-bit example to clarify the above procedure.

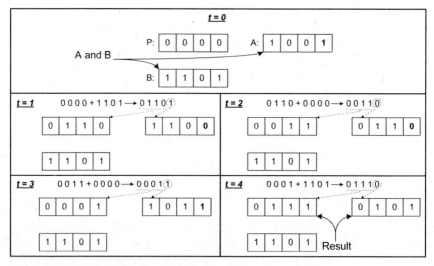

Figure 3.83 Hardware Oriented Multiplication Process

As shown in Figure 3.83, A = **1001** and B = **1101** are to be multiplied. Initially at time 0, A is in a shift-register with a register for partial results (P) on its left.

At time 0, because $A[0]$ is **1**, the partial sum of $B + P$ is calculated. This value is **01101** (shown in the upper part of time 1) and has 5 bits to consider carry. The right most bit of this partial sum is shifted into the A register, and the other bits replace the old value of P. When A is shifted, **0** moves into the $A[0]$ position. This value is observed at time 1. At this time, because $A[0]$ is **0**, **0000** + P is calculated (instead of $B + P$). This value is **00110**, the right most bit of which is shifted into A, and the rest replace P. This process repeats 4 times. At the end of the 4th cycle, the least significant 4 bits of the multiplication result become available in A and the most-significant bits in P. The example used here performed 9*13 and 117 is obtained as the result of this operation.

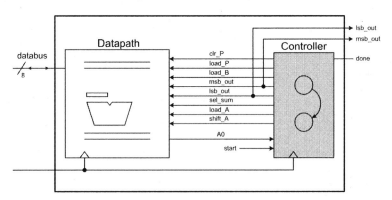

Figure 3.84 Datapath and Controller

3.5.2 Sequential Multiplier Design

The multiplication process discussed in the previous section justifies the hardware implementation that is being discussed here.

3.5.2.1 Control Data Partitioning. The multiplier has a datapath and a controller. The data part consists of registers, logic units and their interconnecting busses. The controller is a state machine that issues control signals for control of what gets clocked into the data registers.

As shown in Figure 3.84, the datapath registers and the controller are triggered with the same clock signal. On the rising edge of a clock, the controller goes into a new state. In this state, several control signals are issued, and as a result the components of the datapath start reacting to these signals. The time given for all activities of the datapath to stabilize is from one edge of the clock to another. Values that are propagated to the inputs of the datapath registers are clocked into these registers with every positive edge of the clock.

3.5.2.2 Multiplier Datapath.

Figure 3.85 shows the datapath of the sequential multiplier. As shown, P and B are outputs of 8-bit registers and A is the output of an 8-bit shift-register. These components are implemented with **always** statements in the Verilog code of the multiplier. An adder, a multiplexer and two tri-state buffers constitute the other components of this datapath. These components are implemented with **assign** statements.

Control signals that are outputs of the controller and inputs of the datapath (Figure 3.84), are named according to their functionalities like loading registers, shifting, etc. These signals are shown in the corresponding blocks of Figure 3.85 next to the data component that they control.

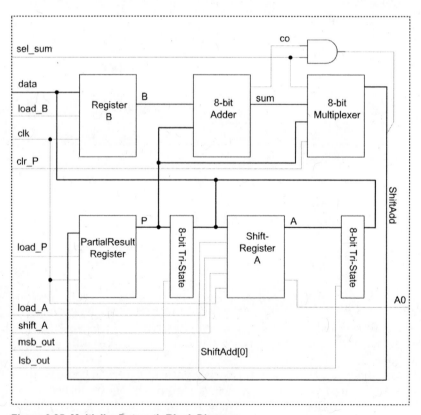

Figure 3.85 Multiplier Datapath Block Diagram

The input data bus (signal named *data* in this figure) connects to the inputs of *A* and *B* to load multiplier and multiplicand into these registers. This bi-directional bus is driven by the outputs of *P* and *A* through tri-state buffers. These tri-states become active when multiplication result is ready.

The output from *B* and *P* are added to form *co* and *sum* to be put in *P* if adding is to take place. Otherwise, *P* is put on *ShiftAdd* to be shifted, while being put back into *P*. *ShiftAdd* is the multiplexer output that selects *sum* or *P*. The *sel_sum* control input determines if *sum* or *P* is to go on the multiplexer output.

The AND function shown in Figure 3.85 selects carry-out from the adder or **0** depending on the value of *sel_sum* control input. This value is concatenated to the left of the multiplexer output to form a 9-bit vector. This vector has *P+B* or *P* with a carry to its left. The rightmost bit of this 9-bit vector is split and goes into the serial input of the shift-register that contains *A*, and the other eight bits go into register *P*. Note that concatenation of the AND output to the left of the multiplexer output and splitting the right bit from this 9-bit vector, effectively produces a shifted result that is clocked into *P*.

3.5.2.3 Datapath Description. The complete datapath Verilog description of the multiplier is shown in Figure 3.86. Verilog **assign** and **always** statements are used to describe components of the datapath. As shown here, the first two **always** statements represent registers *B* and *P* for operand *B* and the partial result, *P*. The **assign** statement that comes next in this figure represents the 8-bit adder. This adder adds *P* and *B*.

Another component of our multiplier datapath is an 8-bit shift-register for operand *A* of the multiplier. This shift-register either loads *A* with *data* (controlled by *load_A*) or shifts its contents (controlled by *shift_A*). An **always** statement that implements this shift-register is shown in Figure 3.86. Following this statement, an **assign** statement representing the multiplexer for selection of *sum* or *P* is shown in the Verilog code of the datapath. This statement puts **8'h0** on *ShiftAdd* if *clr_P* is active. We will use this enabling feature of the multiplexer for producing an eight-bit zero at its output to load into *P* for resetting the *P* partial result register at the start of the multiplication process.

The last two **assign** statements of Figure 3.86 represent two sets of tri-state buffers driving the bidirectional *data* bus of the datapath. As shown, if *lsb_out* is **1**, *A* (the least-significant byte of result) drives *data* and if *msb_out* is **1**, *P* (The most-significant byte) drives *data*.

```
module datapath ( input clk, clr_P, load_P, load_B,
                  msb_out, lsb_out, sel_sum,
                  load_A, shift_A,
                  inout [7:0] data, output A0 );
   wire [7:0] sum, ShiftAdd;
   reg [7:0] A, B, P;
   wire co;

   always @( posedge clk ) if (load_B) B <= data;

   always @( posedge clk )
      if (load_P) P <= {co&sel_sum, ShiftAdd[7:1]};

   assign { co, sum } = P + B;

   always @( posedge clk )
      case ( { load_A, shift_A } )
         2'b01 : A <= { ShiftAdd[0], A[7:1] };
         2'b10 : A <= data;
         default : A <= A;
      endcase
   assign A0 = A[0];

   assign ShiftAdd = clr_P ? 8'h0 : ( ~sel_sum ? P : sum );

   assign data = lsb_out ? A : 8'hzz;
   assign data = msb_out ? P : 8'hzz;

endmodule
```

Figure 3.86 Datapath Verilog Code

3.5.2.4 The Multiplier Controller.
The multiplier controller is a finite state machine that has two starting states, eight multiplication states, and two ending states. States and their binary assignments are shown in Figure 3.87.

```
`define   idle    4'b0000
`define   init    4'b0001
`define   m1      4'b0010
`define   m2      4'b0011
`define   m3      4'b0100
`define   m4      4'b0101
`define   m5      4'b0110
`define   m6      4'b0111
`define   m7      4'b1000
`define   m8      4'b1001
`define   rslt1   4'b1010
`define   rslt2   4'b1011
```

Figure 3.87 Multiplier Control States

```
module controller ( input clk, start, A0,
                    output reg clr_P, load_P, load_B,
                               msb_out, lsb_out, sel_sum,
                    output reg load_A, Shift_A, done );
   reg [3:0] current;

   always @ ( current, start, A0 ) begin : cmb
      clr_P = 0; load_P = 0; load_B = 0;
      msb_out = 0; lsb_out = 0;
      sel_sum = 0; load_A = 0; Shift_A = 0; done = 0;
      case ( current )
         `idle :
            if (~start) done = 1;
            else begin
               load_A = 1; clr_P = 1; load_P = 1;
            end
         `init :
            load_B = 1;
         `m1,`m2,`m3,`m4,`m5,`m6,`m6,`m7,`m8 : begin
            Shift_A = 1; load_P = 1;
            if (A0) sel_sum = 1;
         end
         `rslt1 :
            lsb_out = 1;
         `rslt2 :
            msb_out = 1;
         default: begin
            clr_P = 0; load_P = 0; load_B = 0;
            msb_out = 0; lsb_out = 0; sel_sum = 0;
            load_A = 0; Shift_A = 0; done = 0;
         end
      endcase
   end
   always @ ( posedge clk ) begin : seq
      case ( current )
         `idle : current <= ~start ? `idle : `init;
         `init : current <= `m1;
         `m1,`m2,`m3,`m4,`m5,`m6,`m6,`m7,`m8 :
            current <= current+1;
         `rslt1 : current <= `rslt2;
         `rslt2 : current <= `idle;
         default : current <= `idle;
      endcase
   end

endmodule
```

Figure 3.88 Verilog Code of Controller

In the `idle state, the multiplier waits for *start* while loading *A*. In `init, it loads the second operand *B*. In `m1 to `m8, the multiplier performs add-and-shift of *P+B*, or *P+0*, depending on *A0*. In the last two states (`rslt1 and `rslt2), the two halves of the result are put on *data* that becomes the *databus* port of the multiplier.

The Verilog code of controller is shown in Figure 3.88. This Code declares *datapath* ports, and uses two **always** blocks (*cmb* and *seq*) to issue control signals and make state transitions. At the beginning of the *cmb* **always** block all control signal outputs are set to their inactive values. This eliminates unwanted latches that may be generated by a synthesis tool for these outputs. The second **always** block, *seq*, handles state transitions.

The 4-bit *current* variable represents the currently active state of the machine. As shown in *cmb*, when *current* is `idle and *start* is **0**, the *done* output remains high. In this state if *start* becomes **1**, control signals *load_A*, *clr_P* and *load_P* become active to load *A* with *databus* and clear the *P* register. Clearing *P* requires *clr_P* to put **0**'s on the *ShiftAdd* of the datapath and loading the **0**'s into *P* by asserting *load_P*.

In `m1 to `m8 states, *A* is shifted, *P* is loaded, and if *A0* is **1**, *sel_sum* is asserted. As discussed in relation to *datapath*, *sel_sum* controls shifted *P+B* (or shifted *P+0*) to go into *P*. In the result states, *lsb_out* and *msb_out* are asserted in two consecutive clocks in order to put *A* and *P* on the *data* bus respectively.

In the *seq* block, a **case** statement, similar to and parallel to that of the *cmb* block, handles state transitions by assigning values to *current*.

```
module Multiplier ( input clk, start,
                    inout [7:0] databus,
                    output lsb_out, msb_out, done );
   wire clr_P, load_P, load_B, msb_out,
        lsb_out, sel_sum, load_A, Shift_A;

   datapath dpu( clk, clr_P, load_P, load_B,
                 msb_out, lsb_out, sel_sum,
                 load_A, Shift_A,
                 databus, A0 );

   controller cu( clk, start, A0, clr_P, load_P, load_B,
                  msb_out, lsb_out, sel_sum,
                  load_A, Shift_A, done );
endmodule
```

Figure 3.89 Top-Level Multiplier Code

3.5.2.5 Top-Level Code of the Multiplier. Figure 3.89 shows the top-level *Multiplier* **module**. The *datapath* and *controller* modules are instantiated here. The input and output ports of this unit are according to the block diagram of Figure 3.81. This description is synthesizable, and can be used in any FPGA device programming environment for synthesis and device programming.

3.5.3 Multiplier Testing

This section shows an auto-check adaptive testbench for our sequential multiplier. Several forms of data applications and result monitoring are demonstrated by this example. The outline of the *test_multiplier* **module** is shown in Figure 3.90.

```
`timescale 1ns/100ps

module test_multiplier;
    reg clk, start, error;
    wire [7:0] databus;
    wire lsb_out, msb_out, done;
    reg [7:0] mem1[0:2], mem2[0:2];
    reg [7:0] im_data, opnd1, opnd2;
    reg [15:0] expected_result, multiplier_result;
    integer indx;

    Multiplier uut ( clk, start, databus,
                     lsb_out, msb_out, done );

    initial begin: Apply_data   ... end     // Figure 3.91
    initial begin: Apply_Start  ... end     // Figure 3.92
    initial begin: Expected_Result  ... end // Figure 3.93
    always @(posedge clk)
            begin: Actual_Result  ... end   // Figure 3.94
    always @(posedge clk)
            begin: Compare_Results ... end // Figure 3.95
    always #50 clk = ~clk;
    assign databus=im_data;
 endmodule
```

Figure 3.90 Multiplier Testbench Outline

In the declarative part of this testbench inputs and outputs of the multiplier are declared as **reg** and **wire**, respectively. Since *databus* of the multiplier is a bidirectional bus, it is declared as **wire** for reading it, and a corresponding *im_data* **reg** is declared for writing into it. An **assign** statement drives *databus* with *im_data*. When writing into this bus from the testbench, the writing must be done into *im_data*, and after the completion of writing the bus must be released by writing 8'hZZ into it.

Other variables declared in the testbench of Figure 3.90 are *expected_result* and *multiplier_result*. The latter is for the result read from the multiplier, and the former is what is calculated in the test-bench. It is expected that these values are the same.

The testbench shown in Figure 3.90 applies three rounds of test to the *Multiplier* **module**. In each round, data is applied to the module under test and results are read and compared with the expected results. These are the tasks performed by this testbench:

- Read data files *data1.dat* and *data2.dat* and apply data to *databus*

- Apply *start* to start multiplication

- Calculate the expected result

- Wait for multiplication to complete, and collect the calculated result

- Compare expected and calculated results and issue error if they do not match

These tasks are timed independently. An **always** block generates a periodic signal on the *clk* input. This clock is used for clocking the multiplier as well as for synchronizing **always** statements for collection of actual results and comparing them with the expected ones.

Reading Data Files. Figure 3.91 shows the *Apply_data* **initial** block that is responsible for reading data and applying them to *im_data*, which in turn goes on *databus*.

```
initial begin: Apply_data
   indx=0;
   $readmemh ( "data1.dat", mem1 );
   $readmemh ( "data2.dat", mem2 );
   repeat(3) begin
      #300 im_data = mem1 [indx];
      #100 im_data = mem2 [indx];
      #100 im_data = 8'hzz;
      indx = indx+1;
      #1000;
   end
   #200 $stop;
end
```

Figure 3.91 Reading Data Files

Hexadecimal data from *data1.dat* and *data2.dat* external files are read into *mem1* and *mem2*. In each round of test, data from *mem1* and *mem2* are put on *im_data*. Data from *mem2* is distanced from that of *mem1* by 100 ns. This way, the latter is interpreted as data for the *A* operand and the former for the *B* multiplication operand. After placing this data, 8'hzz is put on *im_data*. This releases the *databus* so that it can be driven by the multiplier when its result is ready.

3.5.3.2 Applying Start. Figure 3.92 shows an **initial** block in which variable initializations take place, and *start* signal is issued. Using a **repeat** statement, three 100 ns pulses distanced by 1400 ns are placed on *start*.

```
initial begin: Apply_Start
    clk=1'b0; start=1'b0; im_data=8'hzz;
    #200 ;

    repeat(3) begin
        #50    start = 1'b1;
        #100   start = 1'b0;
        #1350;
    end

end
```

Figure 3.92 Initializations and Start

3.5.3.3 Calculating Expected Result. Figure 3.93 shows an **initial** block that reads data that is placed on *databus* by the *Apply_data* block (Figure 3.91), and calculates the expected multiplication result. After *start*, when *databus* is updated, the first operand is read into *opnd1*. The next time *databus* changes, *opnd2* is read. The expected result is calculated using these operands.

```
initial begin: Expected_Result
    error=1'b0;
    repeat(3) begin
        wait ( start==1'b1 );
        @( databus );
        opnd1=databus;
        @( databus );
        opnd2=databus;
        expected_result = opnd1 * opnd2;
    end
end
```

Figure 3.93 Calculating Expected Result

3.5.3.4 Reading Multiplier Output. When the multiplier completes its task, it issues *msb_out* and *lsb_out* to signal that it has readied the two bytes of the result. The **always** block of Figure 3.94 is triggered by the rising edge of the circuit clock. After a clock edge, if *msb*_out or *lsb*_out is **1**, it reads the *databus* and puts in its corresponding position in *multiplier_result*.

```
always @(posedge clk) begin: Actual_Result
   if (msb_out) multiplier_result[15:8] = databus;
   if (lsb_out) multiplier_result[7:0] = databus;
end
```

Figure 3.94 Reading Multiplier Results

3.5.3.5 Comparing Result. Figure 3.95 shows the **always** block that is responsible for comparing actual and expected multiplication results. After the active edge of the clock, if *done* is **1**, then comparing *multiplier_result* and *expected_result* takes place. If values of these variables do not match *error* is issued.

```
always @(posedge clk) begin: Compare_Results
   if (done)
      if (multiplier_result != expected_result) error = 1;
      else error = 0;
end
```

Figure 3.95 Comparing Results

The self-running testbench presented here verifies RT-level operation of our multiplier. This design is synthesizable and because of the timing used in our testbench, it can also be used for the post-synthesis description of our multiplier.

This section showed a complete design of a system with a well-defined datapath and a controller. The design demonstrates top-down design, data/control partitioning, and a complete flow from design to test of a system. This flow will be used in the sections that follow to illustrate how Verilog can be used for design of systems that access memory for instructions and data.

3.6 Synthesis Issues

Verilog constructs described in this chapter included those for cell modeling as well as those for designs to be synthesized. In describing an existing cell, timing issues are important and must be included in the Verilog code of the cell. At the same time, description of an existing cell may require parts of this cell to be described by interconnection of gates and transistors. On the other hand, a design to be synthesized does not include any timing information because this information is not available until the design is synthesized, and designers usually do not use gates and transistors for high level descriptions for synthesis.

Considering the above, knowing that the timings are ignored by synthesis tools, and only using gates when we really have to, the codes presented in this chapter all have one-to-one hardware correspondence and are synthesizable. For synthesis, a designer must consider his or her target library to see what and how certain parts can be synthesized. For example, most FPGAs do not have internal three-state structures and three-state bussings are converted to AND-OR busses.

3.7 Summary

The focus of this chapter was on RT level description in the Verilog HDL language. The chapter used complete design examples at various levels of abstraction for showing ways in which Verilog could be used in a design. We showed how timing details could be incorporated in cell descriptions. The examples that were presented had one-to-one hardware correspondence and were synthesizable. We have shown how combinational and sequential components can be described for synthesis and how a complete system can be put together using combinational and sequential blocks for it to be tested and synthesized. The Verilog HDL is far more extensive than the coverage given to it in this chapter. However, this chapter is sufficient for doing basic designs for synthesis.

4 Computer Hardware and Software

Processors play a major role in the design of embedded systems. An embedded processor may be used as the central processing unit of an embedded system, or it may just be used as a convenient and fast way of implementing a hardware function. With embedded systems, understanding how a processor works, its software, and software utilities, such as compilers and assemblers, are key topics that a hardware designer should be familiar with.

This chapter provides the basic concepts and techniques that are necessary for design and utilization of an embedded processor. The chapter begins with an introduction to computer systems, describing the role of software and hardware in a computer system. In this part, instructions, programs, instruction execution, and processing hardware will be described. After this introduction, we will describe a computer from its software point of view. This provides the necessary background for understanding the hardware of a processor. In the description of processor hardware we begin with a simple processor example that has the basic properties found in most processing units and then continue with a more realistic processor. The hardware and software of this processor, which we refer to as SAYEH, becomes the main focus of this chapter. In a top-down fashion, we will show control-data partitioning of our example processors and design and implementation of the individual parts of these machines.

4.1 Computer System

It is important to understand what it is that we refer to as a computer. This section gives this overall view. A computer is an electronic machine which performs some computations. To have this machine perform a task, the task must be broken into small instructions, and the computer will be able to perform the complete task by executing each of its comprising instructions. In a way, a computer is like any of us trying to evaluate something based on a given algorithm.

To perform a task, we come up with an algorithm for it. Then we break down the algorithm into a set of small instructions, called a program, and using these step-by-step instructions we achieve the given task. A computer does exactly the same thing except that it cannot decide on the algorithm for performing a task, and it cannot break down a task into small instructions either. To use a computer, we come up with a set of instructions for it to do, and it will be able to do these instructions much faster than we could.

Putting ourselves in place of a computer, if we were given a set of instructions (a program) to perform, we would need an instruction sheet and a data sheet (or scratch paper). The instruction sheet would list all instructions to perform. The data sheet, on the other hand, would initially contain the initial data used by the program, and it could also be used for us to write our intermediate and perhaps the final results of the program we were performing.

In this scenario, we read an instruction from the instruction sheet, read its corresponding data from the data sheet, use our brain to perform the instruction, and write the result in the data sheet. Once an instruction is complete, we go on to the next instruction and perform that. In some cases, based on the results obtained, we might skip a few instructions and jump to the beginning of a new set of instructions. We continue execution of the given set of instructions until we reach the end of the program.

For example consider an algorithm that is used to add two 3-digit decimal numbers. Figure 4.1 shows the addition algorithm and two decimal numbers that are added by this algorithm. The algorithm starts with reading the first two digits from the paper, and continues with adding them in the brain and writing the sum and output carry on the paper in their specified positions. So the paper is used to store both the result (i.e., sum), and the temporary results (i.e., carry). The algorithm continues until it reaches Step 7 of Figure 4.1.

There are similar components in a computing machine (computer). A computing system has a memory unit. The part of the memory that is used to store instructions corresponds to the instruction sheet and the part that stores temporary and final results corresponds to the scratch paper or the data sheet. The *Central Processing Unit* (*CPU*), which corresponds to the brain, sequences and executes the instructions.

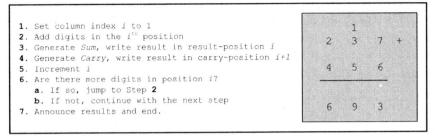

```
1. Set column index i to 1
2. Add digits in the i^{th} position
3. Generate Sum, write result in result-position i
4. Generate Carry, write result in carry-position i+1
5. Increment i
6. Are there more digits in position i?
   a. If so, jump to Step 2
   b. If not, continue with the next step
7. Announce results and end.
```

Figure 4.1 Decimal Addition

There is an important difference between storing information in these two methods. In the manual computation, the instructions are represented using natural language or some human readable guidelines and data is usually presented in decimal forms. On the other hand, in the computer, information (both instructions and data) are stored and processed in the binary form. To provide communication between the user and the computer, an *input-output (IO) device* is needed to convert information from a human language to the machine language (**0** and **1**) and vice versa. So each computer should have a CPU to execute instructions, a memory to store instructions and data, and an IO device to transfer information between the computer and the outside world.

There are several ways to interconnect these three components (Memory, CPU and IO) in a computer system. A computer with an interconnection shown in Figure 4.2 is called a *von-Neumann* computer. The CPU communicates with the IO device(s) for receiving input data and displaying results. It communicates with the memory for reading instructions and data as well as writing data.

As shown in this figure, the CPU is divided into *datapath* and *controller* parts. The datapath has storage elements (registers) to store intermediate data, handles transfer of data between its storage components, and performs arithmetic or logical operations on data that it stores. The datapath also has communication lines for transfer of data; these lines are referred to as *busses*. Activities in the datapath include reading from and writing into data registers, bus communications, and distributing control signals generated by the controller to the individual data components.

The controller commands the datapath to perform proper operation(s) according to the instruction it is executing. Control signals carry these commands from the controller to the datapath. Control signals are generated by controller state machine that, at all times, knows the status of the task that is being executed and the sort of the information that is stored in datapath registers. Controller is the thinking part of a CPU.

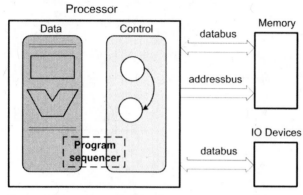

Figure 4.2 Von-Neumann Machine

4.2 Computer Software

The part of a computer system that contains instructions for the machine to perform is called its *software*. For making software of a computer available for its hardware to execute, it is put in the memory of the computer system. As shown in Figure 4.2 the memory of a system is directly accessible by its hardware. There are several ways computer software can be described.

4.2.1 Machine Language

Computers are designed to perform our commands. To command a computer, you should know the computer alphabets. As mentioned in Chapter 2, the computer alphabet is just two letters **0** and **1**. An individual command, which is presented using these two letters, is called an *instruction*. Instructions are binary numbers that are meaningful for a computer. For example, the binary number *10110010* may be a command to tell a computer to add two numbers. This number has three fields, the first field (*1011*) signifies the *add* operation, and the other two (*00* and *10*) are references to the numbers that are to be added. This binary notation is referred to as the *machine language*. This language is hardware dependent, and it is different from one machine to another. The SAYEH processor that we use in the examples of this book has its own machine language.

4.2.2 Assembly Language

The earliest programmers wrote their programs in machine language. Machine language programs are tedious and error prone to write, and difficult to understand. So the programmers used a symbolic notation closer to the human language. This symbolic language is called *assembly language* that is easier to use than the machine language. For example, the above instruction might be written as *add R0, R2*.

We need a special program, called *assembler*, to convert a program from assembly language to machine language. Because the assembly language is a symbolic representation of the machine language, each computer (e.g., SAYEH) has its own assembler.

4.2.3 High-Level Language

The development of the programming language continued and resulted in a *high-level programming language* which is closer to the programmer's problem specification. For example a user can write the *add R0, R2 instruction* mentioned above, as *R0 = R0 + R2*, which is more readable. Similarly we need a program, called *compiler*, to translate these high-level programs into the assembly language of a specific computer. Consequently the high-level languages can be used on different computers.

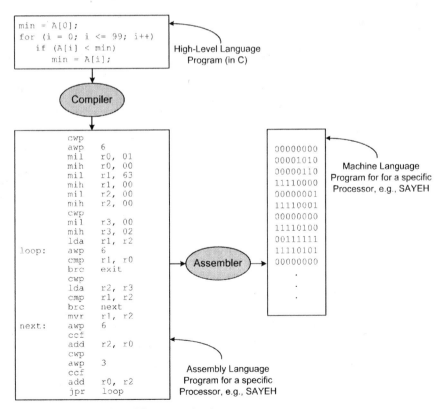

```
min = A[0];
for (i = 0; i <= 99; i++)
    if (A[i] < min)
        min = A[i];
```

High-Level Language Program (in C)

Compiler

```
        cwp
        awp    6
        mil    r0, 01
        mih    r0, 00
        mil    r1, 63
        mih    r1, 00
        mil    r2, 00
        mih    r2, 00
        cwp
        mil    r3, 00
        mih    r3, 02
        lda    r1, r2
loop:   awp    6
        cmp    r1, r0
        brc    exit
        cwp
        lda    r2, r3
        cmp    r1, r2
        brc    next
        mvr    r1, r2
next:   awp    6
        ccf
        add    r2, r0
        cwp
        awp    3
        ccf
        add    r0, r2
        jpr    loop
```

```
00000000
00001010
00000110
11110000
00000001
11110001
00000000
11110100
00111111
11110101
00000000
  .
  .
  .
```

Machine Language Program for for a specific Processor, e.g., SAYEH

Assembler

Assembly Language Program for a specific Processor, e.g., SAYEH

Figure 4.3 Translation of Programming Languages

In its first pass, a compiler reads the input program and checks the syntax of the language. If there is no syntax error in the input program, the compiler builds an internal data structure to hold the information extracted from the input program, otherwise it reports the error. In the next pass, based on the internal data structure, the target code is generated. Figure 4.3 shows a program (in the C-language) that finds the minimum value in an array. The array has 100 elements and starts from memory location 512. The compiler takes the C program as its input and generates the assembly program for the corresponding machine. The assembly program is then fed to the assembler to generate machine code for the machine that is to run this program.

4.2.4 C Programming Language

As mentioned above, there are various languages in different levels that a designer can develop a software program with. Since machine and assembly languages are processor-dependent and they are not easy to learn and use, programmers prefer to develop their software programs in higher level languages (e.g., C/C++, Pascal, and Java). A program so developed, is translated to a specific machine language by high-level language compilers. This section gives an overview of the C language that is enough for the material of the chapters that follow.

A simple C program, which is converted to SAYEH assembly language at the end of this chapter, is shown in Figure 4.4. This code reads a set of integers from a file and sorts it in the descending order. The first number in the input file determines the number of integers in the set. The name of the input file is passed to the program through the first argument of the program. We use this program to introduce language constructs of the C language.

The program of Figure 4.4 uses *preprocessor directives* on lines 01 to 05. It utilizes *functions* for a better readability (lines 08, 17, and 24). The *sort* function uses a nested *for-loop* on lines 11 and 12 and a *conditional statement* on line 13. The *get_array_size* and *get_data* functions call *file IO functions* on lines 19 and 28. Lines 37, 43, and 50 in the *main* function use *IO functions* and lines 47 and 58 utilize *memory allocation* functions for a *dynamic array*.

We will describe preprocessors (lines 01 to 05) in Section 4.2.4.1, C data types and variables (e.g., line 07) in Sections 4.2.4.2 to 4.2.4.4, conditional statements (e.g., line 13) and loops (e.g., line 12) in Sections 4.2.4.5 and 4.2.4.6. We will continue with discussing subprograms (e.g., line 08) in Section 4.2.4.7. Section 4.2.4.8 describes important points about arrays and pointers in C (e.g., line 46). This section ends with describing a few input/output functions defined in C (e.g., line 37).

```
01: #include <stdio.h>
02: #include <stdlib.h>
03: #define SWAP(s1,s2)   {g_swp = s1; s1 = s2; s2 = g_swp;}
04: #define ERR -1
05: #define OK 0
06: //global variable
07: int g_swp;
08: void sort(int *a) {//function definition
09:     int i, j, num; //local variables
10:     num = a[0]; //accessing an array
11:     for (i=1; i<num; i++)
12:         for (j=i+1; j<=num; j++)
13:             if (a[i]<a[j]) SWAP(a[i], a[j])
14:     //end of the nested loop
15:     return;
16: }
17: int get_array_size(FILE *fp) {
18:     int size;
19:     if (fread(&size, 1, sizeof(int), fp))
20:         return size;
21:     else
22:         return ERR;
23: }
24: int get_data(FILE *fp, int *a) {
25:     int entry, i;
26:     int size = a[0];
27:     for (i=1; i<=size; i++)
28:         if (fread(&entry, 1, sizeof(int), fp))
29:             a[i] = entry;
30:         else
31:             return ERR;
32:     //end of loop
33:     return OK;
34: }
35: int main(int argc, char *argv[]) {
36:     if (argc != 2)  {
37:         printf("Error in arguments.\n");
38:         return ERR;
39:     }
40:     FILE *fptr = fopen(argv[1], "rb");
41:     int sz = get_array_size(fptr);
42:     if (sz == ERR)  {
43:         printf("Error in getting the array size.\n");
44:         return ERR;
45:     }
46:     int *s_arr;
47:     s_arr = (int *)malloc((sz+1)*sizeof(int));
48:     s_arr[0] = sz;
49:     if (get_data(fptr, s_arr) == ERR)   {
50:         printf("Error in array entries.\n");
51:         return ERR;
52:     }
53:     fclose(fptr);
54:     //it is the time to sort this array
```

```
55:      sort(s_arr);
56:      //we can use the sorted array or save it to a file
57:      ...
58:      free(s_arr);
59:      exit(OK);
60: }
```

Figure 4.4 Sort Program in C Language

4.2.4.1 Preprocessors. In addition to the body of a C program, there are parts in a C code that are processed by the compiler before the syntax checking process. These parts are called preprocessors. These preprocessors begin with the '#' symbol. The main preprocessors are discussed below.

#include *filename*. This preprocessor inserts the contents of *filename* in the current position of the file that the include statement is written (lines 01 and 02 in Figure 4.4).

#define *arg1 arg2*. This preprocessor replaces *arg1* with *arg2* everywhere after the definition. This replacement can be a simple name replacement or it can be a complex macro definition. For example the following statement replaces the word *ERR* with -1 (also see lines 04 and 05 in Figure 4.4).

```
#define ERR -1
```

Alternatively, the **#define** preprocessor can be used in macro definitions. For example, in the following code the compiler searches for *SWAP* word with two arguments and replaces it with three assignments written in front of *SWAP(i, j)* in its definition.

```
#define SWAP(i, j)  {g_swp=i; i=j; j=g_swp;}
```

For example, if somewhere in code there is *SWAP(a, b)*, it will be replaced with these statements: *{g_swp = a; a = b; b = g_swp;}*.

#undef *already_defined*. This preprocessor clears the definition of an already defined expression. This means that the *already_defined* expression will be replaced only between its **#define** statement and its **#undef** statement.

#if *conditional_exp* ... #endif. The codes written between **#if** and **#endif** expressions are included in the program only if the *condi-*

tional_exp is true. Other useful conditional preprocessors are **#else**, **#ifdef**, and **#ifndef**.

4.2.4.2 Data Types in C. Like other high-level programming languages, there are various data types in C. There are simple data types as well as complex data types. Simple data types include types like **int** used for integer variables, **float** for floating point variables, and **char** for character-type (single byte) variables. A boolean data type in standard C can be modeled by an **int**, **float**, or **char** data type. In other words, **0** is interpreted as *false*, while any other value is interpreted as *true*.

In addition to the above single data types, there are complex data types in C. These complex data types include the following types.

Enumeration types. Enumeration types are a subset of integers. A simple example of an enumeration type is:

```
enum week_day {Sat, Sun, Mon, Tue, Wed, Thu, Fri};
```

The above expression simply assigns value **0** to *Sat*, **1** to *Sun*, etc. If we define variable *wd* of type *week_day*, we can assign any values between *Sat* and *Fri* to it.

Structure types. A structure, defined by the **struct** keyword, is a record with various fields. As an example, the following data type represents an instruction format.

```
struct instruction {
    char opcode;
    int operand1;
    int operand2; };
```

This type has three fields. The first field, named *opcode*, is of type **char**, while the other two fields are of type **int**. Field types used in the structures can be complex types themselves.

Union types. Union types are similar to structures in their syntax. But a C compiler will allocate memory only for the largest field of a union. For example, in the following union, if we define variable *u_var* of type *u*, the memory allocated for *u_var* would be of size **float**. Therefore, only one field has a valid value at the same time. If *u* was defined as a **struct**, there would be separate memory spaces for each of its fields (the total memory would be the size of **int** + size of **char** + size of **float**).

```
union u {
    int i_field;
    char c_field;
    float f_field; };
```

Note that the fields of a **union** or **struct** are accessible using the '.' operation. For example, to access *c_field* of *u_var*, we will write *u_var.c_field* in our program body.

4.2.4.3 Variable Definition in C. We define a variable *v_var* of a specific type *t_type* using the following syntax:

```
t_type v_var = initial_value;
```

Assigning an initial value to *v_var* is optional. For example, variable *ch* of type **char** is defined as (like lines 07, 09, and 18 in Figure 4.4):

```
char ch;
```

4.2.4.4 Variable Assignments. In C language, we can assign a value to a variable using '=' symbol (lines 10 and 41 in Figure 4.4). In the following example, the content of *ch* variable becomes 'p' and the content of *i1* becomes equal to the contents of *i2* after *i1 = i2;* statement.

```
char ch;
int i1, i2;
...
ch = 'p';
i1 = i2;
```

Unlike many high-level languages, in the C language variables of different types can be assigned to each other with the programmer's own risk. For example, you can assign the value of an **int** type variable to a **char** type variable. In this case, only the lower byte of that **int** type variable will be assigned to the **char** type variable. This is because **char** type occupies only one byte of memory.

For these incompatible-type variable assignments, programmers usually use *type-casting* (line 47 in Figure 4.4). As an example, if *i_v* is a variable of type **int** and *c_v* is a variable of type **char**, and if we want to assign *c_v* to *i_v*, we can use the following type casting:

```
i_v = (int) c_v;
```

Using type casting in the above statement makes c_v to be interpreted as an **int** type variable.

4.2.4.5 Conditional Statements. In the flow of a program, decisions must be made based on which certain tasks are to be done (lines 13, 19, 28, etc. in Figure 4.4). In C, these decisions can be made using conditional statements. These statements are discussed below.

if ... else if ... else. The format of the **if-else** conditional statement is as follows:

```
if (cond_state1) {
  ... //code block 1
} else if (cond_state2) {
  ... //code block 2
} else if {

  ...
} else {
  ... //code block n
}
```

In this format the number of **else if** sections can be zero or more. The **else** section is optional. The conditional statement, written after the **if** keyword, uses logical operations shown in Table 4.1. For example, consider the following expression:

```
((a > b) && (c <= d)) || (a == d)
```

This expression is true when a is greater than b and also c is less than or equal to d. If this condition cannot be true, the above statement can still be true if a is equal to d.

Table 4.1 Logical Operations in C Language

Logical Operation	Symbol
Logical AND	&&
Logical OR	||
Logical NOT	!
Is Equal	==
Not Equal	!=
Greater than	>
Less that	<
Greater than or Equal	>=
Less than or Equal	<=

switch ... case. If there are many branches in an **if-else** conditional statement, for readability purposes, it is easier to use **switch** statements. The format of this conditional statement is discussed here.

```
switch (a_variable) {
    case (val_1): ...//code block 1
              break;
    case (val_2): ...//code block 2
              break;
    ...
    default: ...//code block n
}
```

The above format is similar to an **if-else** statement in that each **if-else** condition is equivalent to *a_variable*==*val_i* (e.g., *a_variable*==*val_2*). The **default** section is identical to the last **else** section in an **if-else** conditional statement.

If we omit the **break** statements in the **case** sections, the statements of the next **case** sections will also be executed until they reach a **break** statement. For example, if *a_variable* is equal to *val_1* and there is no **break** statement in the first **case** section, the statements of the first two **case** sections will be executed.

4.2.4.6 Loop Statements. There are several ways to write a loop in the C programming language.

for loop. The format of the **for** loop is shown below.

```
for (statement set 1;statement set 2;statement set 3)
{
    ... //loop body
}
```

In this loop statement there can be three sets of statements between the parentheses after the **for** keyword. These sets of statements may include any number of statements of any type. However, usually *statement set 1* is used for initializing loop variables, *statement set 2* is used for the exit condition of the loop, and *statement set 3* is used for incrementing/decrementing loop variables. A general form of a **for** loop used in most C programs is shown below (see lines 11, 12, and 27 in Figure 4.4):

```
for (int i=1; i<100; i++) {...}
```

The above **for** loop begins with i equal to 1 and exits when i is equal to 100. In each iteration of the loop, i is incremented by 1 (**++** is an operation defined in C and it increments its operand by 1).

while loop. Another way of writing loops is the **while** loop that is shown below:

```
while (! exit_condition) {

    ... //loop body
}
```

The **while** loop shown above exits when the *exit_condition* becomes true. As an example, we can write our **for** loop example, shown earlier, with a **while** loop:

```
int i = 1;
while (i <100) {
    i++;
    ...
}
```

do...while loop. The **do-while** loop is similar to **while** loop, but the first iteration of the loop will be certainly executed. Our previous example for a **for** and **while** loop is shown below in **do...while** format.

```
int i = 1;
do {
    i++;
    ...
} while (i <100);
```

Note that in the loop bodies, the loop can be terminated using **break** statement. Similar to other programming languages, we can construct nested loops by inserting a loop statement inside the loop body of another loop statement (line 12 in Figure 4.4).

4.2.4.7 Subprograms. Readability and modularity of a C program can be achieved by implementing each specific task of the program in a separate subprogram. A subprogram can call other subprograms. Subprograms in C are called *functions*. Unlike languages such as Pascal, the C language does not have *procedures*. The main difference between procedures and functions is that procedures do not have a return value, while functions do. The format of a function is shown below (see also lines 08, 17, 24, and 35 in Figure 4.4):

```
ret_type func_name (type1 arg1, type2 arg2, …) {
    …//function body
}
```

To implement a procedure with functions in C, there is a special data type in this language named **void**. **void** means *nothing* in the C language. If the *ret_type* (return data type) of a function is **void**, this function acts like a procedure (line 08 in Figure 4.4). As an example, *min_max* function is shown below:

```
void min_max (int a, b) {
    if (a > b)
            printf ("a is maximum");
    else if (a < b)
            printf("a is minimum");
    else
            printf("a is equal to b");
    return;
}
```

An important point about functions in the C language is *function prototyping*. As shown below, if function *f1* is defined before function *f2* and if *f1* uses *f2*, the compiler will generate an error. This is because the compiler has not seen the definition of *f2* when it is referenced in *f1*.

```
int f1(int a1, a2) {
    …
    f2(a1, 5);
    …
}
…
void f2(int c, char d) {…}
```

A solution to this problem is that we define the prototype of function *f2* before defining *f1*, as shown below:

```
void f2(int c, char d); //f2 prototype
…
int f1(int a1, a2) {…} //f1 function
…
void f2(int c, char d) {…} //f2 function
```

Every C program has a main function from which the program execution begins. This function is called **main()** (line 35 in Figure 4.4). Since **main()** is the starting point of a program, using this function is mandatory in all C programs. This function has two arguments called *argc* and *argv* that are used for passing the program's parameters from outside to the program data structures. The former shows the number of program parameters (the count includes the program name itself), while the latter contains the name of the program arguments (including the program name). In our sort program example, shown in Figure 4.4, our program has two arguments (here *argc* must be equal to 2). The first argument contains the name of the sort program (e.g., *sort.exe*) and the second argument contains the input file name (e.g., *c:/input.srt*). Therefore, *argv[0]* contains *"sort.exe"* string and *argv[1]* contains the *"c:/input.srt"* string.

4.2.4.8 Arrays and Pointers. Like other high-level languages, you can define arrays in C. A static array is an array of a defined type with a fixed size. A precise amount of memory will be allocated when these variables are defined. For example, an array of 100 integers is defined by:

```
int int_array [100];
```

If an integer uses two bytes of memory, when the compiler reaches the above definition statement, it will allocate 200 bytes of memory for the *int_array* variable. The ith member of *int_array* can be accessed by *int_array[i-1]*, because the first index of an array in the C language is 0.

Another powerful, but potentially dangerous, capability of C language is the ability to define a *pointer* type (lines 40 and 46 in Figure 4.4). A pointer is an address that corresponds to a variable's address. Pointers are represented by a * preceding variable names. For example in the following pointer definition, *int_ptr* is a variable that points to an integer.

```
int *int_ptr;
```

In the above example *int_ptr* is the address of an integer and *int_ptr* shows the contents of this address.

Another operator that is related to pointers is the **&** (address-of) operator that is used before a variable (for example **&***int_var*) to show the address of that variable. The following code shows an example:

```
int int_var;
int_var = 1000;
int *int_ptr = &int_var;
```

In the above example **&***int_var* shows the location of *int_var* in memory. After assigning **&***int_var* to *int_ptr*, this integer pointer also points to the address of *int_var*. Therefore, **int_ptr* is equal to 1000, because it shows the contents of memory of *int_var* variable.

The C language allows definition of two (or more) dimensional pointers. For example, a pointer to another pointer is defined by ******. Pointers are complicated in C and care must be taken in using them. It is strongly recommended to study pointers in more detail before starting to use them.

Dynamic arrays are pointers, to which memory can be assigned at run-time (like *s_arr* in Figure 4.4). Allocating memory can be done by several memory allocation functions (such as *malloc* used in line 47 in Figure 4.4). It is necessary to free an allocated memory of a variable after using it (line 58 in Figure 4.4), because, unlike static arrays, it will not be de-allocated automatically. Memory leaks are resulted from improper de-allocation. A memory de-allocation function is *free* function.

Another interesting point about pointers and dynamic arrays is **void *** or a pointer to the **void** type. As stated above, **void** in C means *nothing*, but when it is used as a pointer, it means *anything*. If a variable of type **void *** is used as a function parameter, any type of pointer can be passed to that function for that parameter (for example **char ***, **int ***, etc.). In addition, if a **void *** variable is defined inside a function, memory can be allocated with any type to that variable by type casting. For example, if *v_ptr* is a **void ***, what is shown below generates an array of 100 characters:

```
v_ptr = (char *) malloc (100);
```

And an array of 50 integers is generated by the following statement:

```
v_ptr = (int *) malloc (50*sizeof(int));
```

4.2.4.9 Input and Output.

There are several functions that are available in the standard C libraries for getting data from an input device and putting program results into an output device. An input can be a keyboard interface, a file, or a computer port, while an output can be a monitor display, a file, or a computer port.

Two such functions that are mostly used by C programmers are *printf* and *scanf*. The *printf* function prints a string or the value of a variable to display and *scanf* gets a value from keyboard and stores it in a specified variable. The data type that must be read or written is specified with the '%' symbol followed by pre-defined characters. For example, *%d* corresponds to variables of integer type, while *%s* corresponds to strings. Two examples for these IO functions are shown below:

```
printf ("The value of a = %d and b = %c", a_var, b_var);
scanf ("Enter a value: %d", &int_var);
```

In the above *printf* example, if the value of *a_var* is 10 and the value of *b_var* is 'p', then the *printf* function will display *The value of a = 10 and b = p* on the monitor.

In the above *scanf* example, the program shows *Enter a value:* on the monitor and waits for user to type a word. Then it stores this input word to the integer type variable *int_var*.

There are other IO functions for manipulating files. These functions include: *fopen, fread, fwrite, fseek, ftell, fclose,* etc. These functions work with a pointer of type *FILE*. The other set of file related functions are *open, read, write, seek, close,* etc. These functions work with a file handle that is defined as an integer. Functions *fprintf* and *fscanf* are two other functions for working with files. These functions work like *printf* and *scanf* functions.

4.2.4.10 Sort Program Description.

The above subsections presented an overview of the constructs of C that are most often used. At this point if we go back to the sort program that was presented at the beginning of this section in Figure 4.4, we can describe it using the above C constructs.

The sort program begins with preprocessor directives (lines 01 to 05). The first two directives are used for including standard header files (*stdio.h* and *stdlib.h*). Line 03 of this code defines a macro named *SWAP*. This macro swaps the contents of its two parameters (*s1* and *s2*). This macro is used for swapping two array members in *sort* function. On line 07, a global variable named *g_swp* is defined. This variable is used in swapping two numbers.

This program consists of three utility functions (*sort, get_array_size,* and *get_data*) and a *main()* function. The *get_array_size* function gets the input file pointer as its input and returns the number of data items that must be sorted (lines 17 to 23). The *get_data* function gets the input file pointer and a pointer to an integer array as its inputs, reads the numbers that must be sorted from its input file, stores the numbers in its input array, and returns OK (defined as *0* on line 05) if it can read all numbers from file without any error. The *sort* function performs the main sorting function in this program. This function gets a pointer to an integer array and uses a nested loop on the indexes of this array and the *SWAP* macro to sort the existing numbers in its input array (lines 11 to 13).

The *main* function uses the above functions to handle the whole process of sorting. It gets the name of input file from the first program argument and opens it (lines 36 to 40). Then it calls the *get_array_size* function to find the amount of integer numbers it must read. Then it allocates the necessary amount of memory for storing the integer numbers and calls *get_data* to store numbers in *s_arr* array variable (line 49). If all the function calls and memory allocations pass, it closes its input file and calls the *sort* function to sort the numbers in *s_arr* (line 55). From this point in the program to line 58 (where the memory allocated for *s_arr* is freed) we can use *s_arr* as an array of numbers that are sorted in the descending order. Line 59 exits the program with 0 status.

This section provided an introduction to C. After reading this section you should be able to read and understand C language programs, and be able to write simple programs like the sort program in Figure 4.4 or the calculator program in Chapter 9.

4.3 Instruction Set Architecture

As shown in Figure 4.5, a computer system is comprised of two major parts, *hardware* and *software*. The interface between these parts is called *Instruction Set Architecture* (*ISA*). ISA defines how data that is being read from a CPU memory (CPU program) and that is regarded as an instruction, is interpreted by the hardware of the CPU.

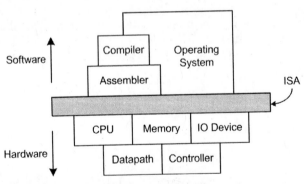

Figure 4.5 Computer System Components

4.3.4.1 Hardware. The hardware part of a computer has three major components (CPU, Memory Unit, and IO Device). Breaking down the CPU into its composing parts, shows that the CPU is built from an interconnection of datapath and controller.

Datapath consists of *functional units* and *storage elements*. A functional unit (such as an adder, a subtractor and an arithmetic-logical unit (ALU)) performs an arithmetic or logical operation. A storage element (e.g., a register or a register-file) is needed to store data. *Bussing structure* defines the way functional units and storage elements are connected. Datapath shows interconnections for the flow of data from one component to another. Controller is used to control the flow of data or the way data is processed in the datapath.

Figure 4.6 shows an example of a datapath that has two registers *R0* and *R1* and an adder/subtractor unit. This datapath is able to add (subtract) *R0* to (from) *R1* and store the result in *R0*. The controller controls the datapath to perform addition or subtraction.

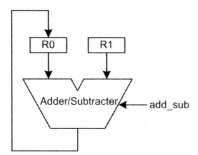

Figure 4.6 A Typical Datapath

4.3.4.2 Software. The software part of a computer consists of the Operating System, Compiler, and Assembler. The operating system provides an interface between the user and the hardware. The compiler translates the high-level language programs to assembly language programs. The assembler translates the assembly language programs to machine language programs.

4.3.4.3 Hardware/Software Interface. By specifying the format and structure of the instruction set, ISA specifies the interface between hardware and software of a processing unit. In other words, ISA provides the details of the instructions that a computer should be able to understand and execute. Each instruction specifies an operation to be performed on a set of data, called *operands*. The operands also show where the result of the instruction should be stored. The *instruction format* describes the specific fields of the instruction assigned to an operation and its operands. The *opcode* field specifies the operation, and the *operand fields* specify the required data. The way in which the operands can be delivered to an instruction is called *addressing mode*. For example, an operand may be a constant value (*immediate addressing*), contents of a register (*register addressing*), contents of a memory location (*direct addressing*), or contents of a memory location addressed by another memory location (*indirect addressing*). Figure

4.7 shows an instruction format with two operands. It is common to specify some operands explicitly in the instruction and the other operands implicitly. The implicit operands refer to the CPU registers. For example, the instruction *add 50*, may mean "add the content of the memory location *50* with *acc* and store the result in *acc*". Here, *acc* is an *implicit operand* and *50* is an *explicit operand*.

Figure 4.7 Instruction Format

4.4 SMPL-CPU Design

The previous sections introduced basic concepts of a computer system. This section shows the complete design of a simple CPU, which we refer to as *SMPL-CPU*. The purpose of this example is to show the hardware of a machine, and consequently show how hardware and software interact. Furthermore, we show assembly and machine languages of our simple example processor.

Here we describe two different implementations of *SMPL-CPU*, *Single-Cycle* and *Multi-Cycle* implementations. The presentation is incremental to show details of processor architecture.

The single-cycle implementation reads an instruction from the memory and in one clock cycle it processes it, executes it, and writes the result back into its data memory. In a multi-cycle processor, an instruction is executed in several steps, each of which takes one clock cycle. Since the design process is easier to show in a single-cycle implementation, we will only show the details of this implementation.

4.4.1 CPU Specification

The CPU design begins with the specification of the CPU, including the number of *general purpose registers*, *memory organization*, *instruction format*, and *addressing modes*. A CPU is defined according to the application it will be used for.

4.4.1.1 CPU External Busses. The *SMPL-CPU* has an 8-bit external data bus and a 6-bit address bus. The address bus connects to the memory in order to address locations that are being read from or written into. Data read from the memory are instructions and instruction operands, and data written into the memory are instruction results and temporary information. The CPU also communicates with its IO devices through its external busses. The address bus addresses a specific device or a device register, while the data bus contains data that is to be written or read from the device.

4.4.1.2 General Purpose Registers.
The *SMPL-CPU* has an 8-bit register, called *accumulator (AC)*. The *AC* register plays an important role in this CPU. All data transfers and arithmetic instructions use *AC* as an operand. In a real CPU, there may be multiple accumulators or an array of registers that is referred to as a *register-file*.

4.4.1.3 Memory Organization.
The *SMPL-CPU* is capable of addressing 64 words of memory; each word has an 8-bit width. We assume the memory read and write operations can be done synchronous with the CPU clock in one clock period. Reading from the memory is done by putting the address of the location that is being read on the address bus and issuing the memory read signal (often called read-enable or *ren*). Writing into the memory is done by assigning the right address to the address bus, putting data that is to be written on the data bus, and issuing the memory write signal (write-enable).

4.4.1.4 Instruction Format.
Each instruction of *SMPL-CPU* is eight bits and occupies a memory word. The instruction format of the *SMPL-CPU*, as shown in Figure 4.8, has an *explicit operand* (immediate data or memory location the address of which is specified in the instruction), and an *implicit operand* (*AC*). The *SMPL-CPU* has four instructions, divided into three classes of *arithmetic* (**add**), *data-transfer* (**lda**, **sta**), and *control-flow instructions* (**jmp**).

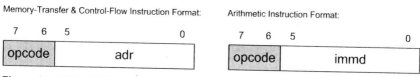

Figure 4.8 SMPL CPU Instruction Format

SMPL-CPU instructions are described below. A tabular list and summary of this instruction set is shown in Table 4.2.

- **add** *immd*: adds the *immd* data with *AC* and stores the result back in AC.

- **lda** *adr*: reads the content of the memory location addressed by *adr* and writes it into *AC*.

- **sta** *adr*: writes the content of *AC* into the memory location addressed by *adr*.

- **jmp** *adr*: jump to the memory location addressed by *adr*.

Table 4.2 SMPL-CPU Instruction Set

Opcode	Instruction	Instruction Class	Description
00	**add** *immd*	Arithmetic	AC ← AC + *immd*
01	**lda** adr	Data-Transfer	AC ← Mem [*adr*]
10	**sta** adr	Data-Transfer	Mem [*adr*] ← AC
11	**jmp** adr	Control-Flow	PC ← *adr*

4.4.1.5 Addressing Mode. The *SMPL-CPU* uses direct addressing. For an instruction that refers to the memory, the memory location is its explicit operand and *AC* is its implicit operand.

4.4.2 Single-Cycle Implementation

In a single cycle implementation the processor accesses data and instruction memories independently, reads instruction and data and performs its operation in a single clock cycle. This implementation requires more hardware, but is faster than the alternative multi-cycle implementation. Details of this implementation of our example processor are described in the sub-sections that follow. We will show the design of the datapath first and then the controller.

4.4.2.1 Datapath Design. Datapath design is an incremental process, at each increment we consider a class of instructions and build up a portion of the datapath which is required for execution of this class. Then we combine these partial datapaths to generate the complete datapath. In these steps, we decide on the control signals that control events in the datapath. In the design of the datapath, we are only concerned with how control signals affect flow of data and function of data units, and not how control signals are generated.

Step 1: Program Sequencing. Instruction execution begins with reading an instruction from the memory, called *Instruction Fetch (IF)*. So an *instruction memory* is needed to store the instructions. We also need a register to hold the address of the current instruction to be read from the *instruction memory*. This register is called *Program Counter* or *PC*. When an instruction execution is completed, the next instruction from the next memory location should be read and executed. After the completion of the current instruction, *PC* should be incremented by one to point to the next instruction in the *instruction memory*. This leads to the use of an adder for incrementing *PC*. Because the memory size is 64 (= 2^6) words, *PC* should be a 6-bit register. Figure 4.9 shows the sequencing part of the datapath.

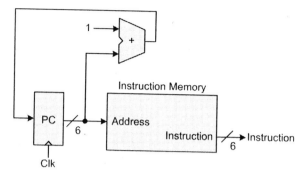

Figure 4.9 Program Sequencing Datapath

Step 2: Arithmetic Instruction Datapath. The arithmetic add instruction operates on the AC and the 6-bit immediate data that comes from bits 5 to 0 of the instruction. The result of the operation will be stored in AC. We need an *arithmetic-logical unit* (ALU), which performs the addition of the two operands of an arithmetic instruction. According to the instruction, the ALU operation will be controlled by an input, *alu_op*. The ALU is designed like other combinational circuits using the methods presented in Chapter 2. In this example the ALU is a simple adder and does not require any control signal.

Figure 4.10 shows the arithmetic instruction datapath. ALU inputs are AC and 6-bit *immd* field of the instruction. ALU output is connected to AC. Two zeros appended to the left of *immd* make an 8-bit input for the ALU. At this point in our incremental design, AC needs no control signals, because in all instructions of the type we have considered so far (arithmetic), AC is loaded with the ALU output.

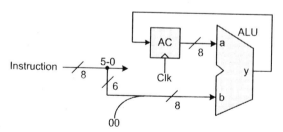

Figure 4.10 Arithmetic Instructions Datapath

Step 3: Combining the Two Previous Datapaths. Combining the previous datapaths results in the datapath of Figure 4.11. In addition to what is shown, for accessing data from the data memory, the *adr* field (bits 5 to 0) of the instruction read from the *instruction memory* must be connected to the address bus of the *data memory*. This datapath can sequence the program and execute our arithmetic instruction.

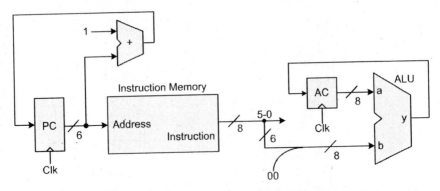

Figure 4.11 Datapath for Program Sequencing & Arithmetic Instructions

Step 4: Data-Transfer Instruction Datapath. There are two data-transfer instructions in *SMPL-CPU*, *lda* and *sta*. The *lda* instruction uses the *adr* field of the instruction to read an 8-bit data from the *data memory* and stores it in the *AC* register. The *sta* instruction writes the content of *AC* into a *data memory* location that is pointed by the *adr* field. Figure 4.12 shows the datapath that satisfies requirements of data-transfer instructions. Because *lda* reads from the *data memory* while *sta* writes into it, the *data memory* must have two control signals, *rd_mem* and *wr_mem* for control of reading from or writing into it. In data-transfer instructions, only *lda* writes into the *AC*. When executing an *sta* instruction, *AC* should be left intact. Having a register without a clock control causes data to be written into it with every clock. In order to control this clocking, the *ld_ac* (load-control, or clock enable) signal is needed for the *AC* register.

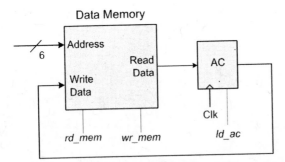

Figure 4.12 Datapath for the Data-Transfer Instructions

Step 5: Combining the Two Previous Datapaths. Combining the two datapaths, may result in multiple connections to the input of an element. For example, in *Step 3* (Figure 4.10 and Figure 4.11) the *ALU* output is connected to the *AC* input and in *Step 4* (Figure 4.12) the *data memory* output (*ReadData*) is connected to the *AC* input. To have both connections, we need a *multiplexer* (or a *bus* which is implemented using tri-state buffers) to select one of the *AC* sources. When the multiplexer select input, *ac_src* is **0** the *ALU* output is selected, and when *ac_src* is **1** the *data memory* output is selected. Figure 4.13 shows the combined datapath.

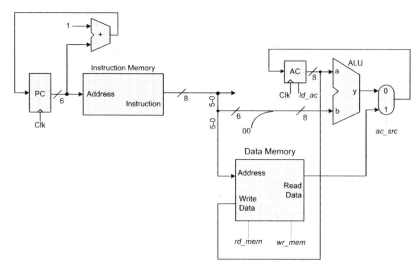

Figure 4.13 Datapath for Program Sequencing, Arithmetic and Data-Transfer

Step 6: Control-Flow Instruction Datapath. *SMPL-CPU* has an unconditional jump instruction, ***jmp***. The ***jmp*** instruction writes the *adr* field of the instruction (bits 5 to 0) into *PC*. So we need a path between the *adr* field of the instruction and the *PC* input. The required datapath is shown in Figure 4.14.

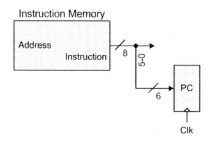

Figure 4.14 Control-Flow Instructions Datapath

Step 7: Combining the Two Previous Datapaths. Performing *Step 6* created another partial datapath for satisfying operations of our CPU. In this step we are to combine the result of the last step with the combined datapath of Figure 4.13. Considering these two partial datapaths, there are two sources for the *PC* register. One was created in *Step 1*, shown in Figure 4.9 and carried over to Figure 4.13 by *Step 5*, and the other was created in *Step 6* that is shown in Figure 4.14. As in the case of *AC*, we need a multiplexer in the combined datapath to select the appropriate source for the *PC* input. The multiplexer select input is called *pc_src*. If this control signal is **0**, incrementing *PC* is selected and if it is **1**, the address field of the instruction being fetched will be selected. Figure 4.15 shows the combined datapath. This step completes the datapath design of *SMPL-CPU*.

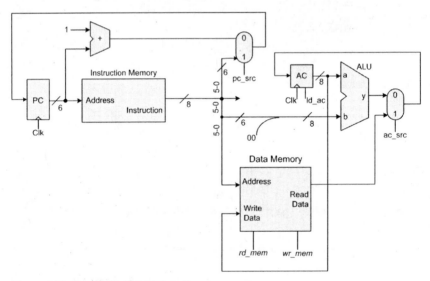

Figure 4.15 The *SMPL-CPU* Datapath

4.4.2.2 Instruction Execution. Now that we have a complete datapath, it is useful to show how a typical instruction, e. g., *add 50*, is executed in the *SMPL-CPU* datapath. On the rising edge of the clock, a new value will be written into *PC*, and *PC* points the *instruction memory* to read the instruction *add 50*. After a short delay, the memory read operation is complete and the controller starts decoding the instruction. Instruction decoding is the process of controller deciding what control signals to issue to execute the given instruction. For our add example, the controller issues appropriate control signals to control the flow of data in the datapath. When data propagation is completed in the datapath, on the next rising edge of the clock, the

ALU output that is the result of the add operation is written into *AC*. To complete the execution of the current instruction *PC+1* is written into *PC*. This new value of *PC* points to the next instruction. Because the execution of the instruction is completed in one clock cycle, the implementation is called a *single-cycle* implementation.

4.4.2.3 Controller Design.

As described before, controller issues the control signals based on the *opcode* field of the instruction. On the other hand, the *opcode* field will not change while the instruction is being executed. Therefore, the control signals will have fixed values during the execution of an instruction, and consequently the controller will be implemented as a combinational circuit.

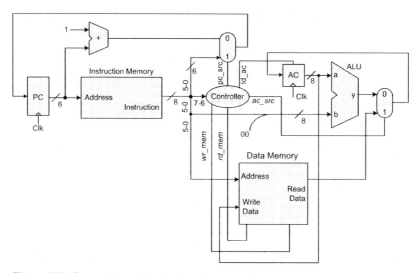

Figure 4.16 Datapath and Controller Interconnection

Figure 4.16 shows the interconnection of the datapath and controller. As shown in this figure, the controller issues all control signals. For all instructions, except *jmp*, the *pc_src* signal is **0** and that causes *PC* to be incremented. For the *jmp* instruction, the *pc_src* signal is **1** and consequently the *adr* field of the instruction from the *instruction memory* is put into the *PC* register. As shown in this figure, a *reset* primary input has been added to clear *PC*. This input is used to reset the CPU and forces to fetch the next instruction from memory location 0.

As described before, the controller of a single-cycle implementation of a system is designed as a combinational circuit. In Chapter 2 you learned how to specify a combinational logic using a truth table.

Table 4.3 Controller Truth Table

Instruction Class	Inst	opcode	rd_mem	wr_mem	ld_ac	ac src	pc src
Data Transfer	lda	00	1	0	1	1	0
	sta	01	0	1	0	X	0
Control Flow	jmp	10	0	0	0	X	1
Arithmetic	add	11	0	0	1	0	0

Table 4.3 shows the truth table of the controller. The values shown are based on activities and flow of data in the datapath. In what follows, we indicate status of control signals as they are necessary for controlling the flow of data in the datapath.

- **Arithmetic Class:**
  - ○ *ld_ac = 1,* to store the *ALU* result in *AC*.
  - ○ *ac_src = 0,* to direct *ALU* output to the *AC* input.
  - ○ *pc_src = 0,* to direct *PC+1* to the *PC* input.

- **Data-Transfer Class:**
  - ○ ***lda*** Instruction:
    - ▪ *rd_mem = 1,* to read an operand from the *data memory.*
    - ▪ *ld_acc = 1,* to store the *data memory* output in *AC*.
    - ▪ *ac_src = 1,* to direct *data memory* output to the *AC* input.
    - ▪ *pc_src = 0,* to direct *PC+1* to the *PC* input.
  - ○ ***sta*** Instruction:
    - ▪ *wr_mem = 1,* to write *AC* into the *data memory.*
    - ▪ *ld_ac = 0,* so that the value of *AC* remains unchanged.
    - ▪ *ac_src = X,* because *AC* clocking is disabled and its source is not important.
    - ▪ *pc_src = 0,* to direct *PC+1* to the *PC* input.

- **Control-Flow Class:**
  - ○ *rd_mem* and *wr_mem* are **0**, because *jmp* does not read from or write into the *data memory.*
  - ○ *ld_ac = 0,* because *AC* does not change during *jmp.*
  - ○ *ac_src = X,* because *AC* clocking is disabled and its source is not important.
  - ○ *pc_src = 1,* direct the jump address (bits 5-0 of instruction) to the *PC* input.

4.4.2.4 Verilog Description. To give a detailed view of the hardware of this machine, and also to show how its hardware interacts with its machine language, we develop the complete Verilog code of our *SMPL-CPU* by developing code for the blocks of Figure 4.16. We start with the components of the datapath, and when done, we will form the Verilog code of the datapath by instantiating and wiring these components. The controller will be described next, using a combinational circuit coding style. At the end, the description of our example will be completed by wiring datapath and controller in a top-level Verilog module. The Verilog testbench of this machine shows how machine language contents of the memory are read and processed by its hardware.

Figure 4.17 shows the datapath and controller interconnection, and the *SMPL-CPU* interface. Our top-level Verilog module corresponds to the *SMPL-CPU* block shown in this figure.

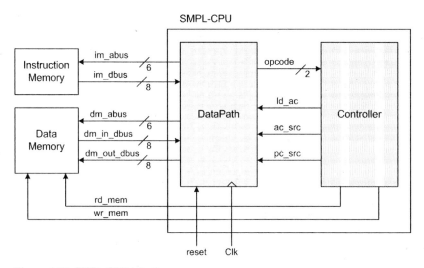

Figure 4.17 SMPL-CPU Interface

4.4.2.5 Datapath components.
Datapath components of *SMPL-CPU* could be described at the behavioral level (using **always** blocks) or at dataflow level (using **assign** statements). Verilog code for *PC*, *AC*, *ALU*, *adder* and *multiplexer* modules are shown in Figure 4.18.

The program counter (*PC*) is a simple register with a *reset* input. Accumulator (*AC*) is a simple register with *load* enable control input. This input is driven by the control signal coming from the controller through datapath ports. The *ALU* is a combinational logic that adds its two 8-bit inputs. The 6-bit *adder* is used to increment *PC*. The 6 and 8-bit *multiplexers* of Figure 4.16 are described using **assign** statements.

```
module PC (input [5:0] d_in, input reset, clk,
           output reg [5:0] d_out);
   always @(posedge clk)
      if (reset) d_out <= 0; else d_out <= d_in;
endmodule
//
module AC (input [7:0] d_in, input load, clk,
           output reg [7:0] d_out);
   always @(posedge clk)
      if (load) d_out <= d_in;
endmodule
//
module ALU (input [7:0] a, b,
            output reg [7:0] alu_out);
   always @(a or b) alu_out = a + b;
endmodule
//
module ADDER (input [5:0] a, b,
              output [5:0] adder_out);
   assign adder_out = a + b;
endmodule
//
module MUX2TO1_6B (input [5:0] i0, i1, input sel,
                   output [5:0] mux_out);
   assign mux_out = sel ? i1 : i0;
endmodule
//
module MUX2TO1_8B (input [7:0] i0, i1, input sel,
                   output [7:0] mux_out);
   assign mux_out = sel ? i1 : i0;
endmodule
```

Figure 4.18 Datapath Components of *SMPL-CPU*

4.4.2.6 Datapath Description. Figure 4.19 shows the datapath description of *SMPL-CPU*. The module name for this description is *DataPath* and it corresponds to Figure 4.15.

The inputs of the Verilog code of Figure 4.19 are control signals coming from the controller, and the external data busses coming from instruction and data memory. The outputs of this module are the *opcode*, address and data busses. The *opcode* goes out to the controller, the output address busses go to the instruction and data memory for instruction and operand fetch, and the output data bus (*dm_out_dbus*) is used for writing into a data memory location.

Following the input and output declarations, the *DataPath* module declares internal datapath busses and signals. As shown, these declarations are followed by instantiation of data components, *PC*, *AC*, *ALU*, *adder* and *multiplexers*. Interconnection of these components are done through wires and busses declared by wire net declarations. Control signals responsible for loading, clearing registers, and bus selections connect to the control inputs of *AC*, *PC* and *multiplexers*. In the last part of the *DataPath* module, several **assign** statements are used for bus assignments.

```
module DataPath (
    input reset, ld_ac, ac_src, pc_src, clk,
    output [1:0] opcode,
    output [5:0] im_abus,  // Instruction memory address bus
    input  [7:0] im_dbus,  // Instruction memory data bus

    output [5:0] dm_abus,     // Data memory address bus
    input  [7:0] dm_in_dbus,  // Data memory input data bus
    output [7:0] dm_out_dbus);// Data memory output data bus

    wire [7:0] ac_out, alu_out, mux2_out;
    wire [5:0] pc_out, adder_out, mux1_out;

    PC          pc     (mux1_out, reset, clk, pc_out);
    AC          ac     (mux2_out, ld_ac, clk, ac_out);
    ALU         alu    (ac_out, im_dbus[5:0], alu_out);
    ADDER       adder  (pc_out, 6'b000_001, adder_out);
    MUX2TO1_6B  mux1   (adder_out, im_dbus[5:0],
                        pc_src, mux1_out);
    MUX2TO1_8B  mux2   (alu_out, dm_in_dbus,
                        ac_src, mux2_out);

    assign opcode       = im_dbus[7:6];
    assign im_abus      = pc_out;
    assign dm_abus      = im_dbus [5:0];
    assign dm_out_dbus  = ac_out;
endmodule
```

Figure 4.19 Datapath Description of *SMPL-CPU*

4.4.2.7 Controller Description.

The *Controller* code for our *SMPL-CPU* example is shown in Figure 4.20. This code corresponds to the controller truth table shown in Table 4.3. The *Controller* has the *opcode* input that comes to the controller from the *DataPath* module. This part is a combinational circuit and is described using an **always** block.

```
module Controller (
    input [1:0] opcode,
    output reg rd_mem, wr_mem, ac_src, ld_ac, pc_src);

    always @(opcode) begin
        rd_mem = 1'b0;        wr_mem = 1'b0;
        ac_src = 1'b0;        ld_ac  = 1'b0;
        pc_src = 1'b0;
```

```
        case (opcode)
           2'b00: // lda  adr
              begin
                 rd_mem = 1'b1;
                 ac_src = 1'b1;
                 ld_ac  = 1'b1;
              end
           2'b01: wr_mem = 1'b1;  // sta  adr
           2'b10: pc_src = 1'b1;  // jmp  adr
           2'b11: ld_ac  = 1'b1;  // add  adr
        endcase
   end
endmodule
```

Figure 4.20 Controller Description of *SMPL-CPU*

4.4.2.8 The Complete Design.
The top-level module for our example processor is shown in Figure 4.21. In the *SMPL-CPU* module shown, *DataPath* and *Controller* modules are instantiated. Port connections of the *Controller* include its output control signals, and the *opcode* input from *DataPath*. Port connections of *DataPath* consist of address and data external busses, *opcode* output, the *reset* external input, and control signal inputs.

```
module smpl_cpu (

   input clk, reset, output rd_mem, wr_mem,
   output [5:0] im_abus, input  [7:0] im_dbus,
   output [5:0] dm_abus, input  [7:0] dm_in_dbus,
   output [7:0] dm_out_dbus);

   wire [1:0] opcode;
   wire ac_src, ld_ac, pc_src;

   DataPath dpu (reset, ld_ac, ac_src, pc_src, clk, opcode,
        im_abus, im_dbus, dm_abus, dm_in_dbus, dm_out_dbus);

   Controller cu (opcode, rd_mem, wr_mem, ac_src,
        ld_ac, pc_src);

endmodule
```

Figure 4.21 Description of *SMPL-CPU*

4.4.2.9 Testing Single-Cycle SMPL-CPU.
To develop a testbench for the *SMPL-CPU* module, we first model a simple instruction memory with a read operation, and a data memory with read and write operations. We load the instruction memory with a test program that is

written in the processor's machine language, and let the processor Verilog model run this program.

The test contents of the instruction memory and its corresponding assembly instructions are shown in Figure 4.22. The test program starts at address 0 and consists of an infinite loop by using *jmp 00* instruction located at address 03. In a loop iteration, the content of the data memory location 0 is loaded into *AC (lda 00)*, then the content of *AC* is added with the value of 2 (*add 02*), and finally the new value of *AC* is stored into the memory location 0 (*sta 00*). The initial value of the data memory location 0 is 0. The initial content of the data memory is also shown in Figure 4.22.

Figure 4.23 shows the Verilog description of the instruction and data memories. For simplicity, the contents of the instruction and data memory are specified in an **initial** block.

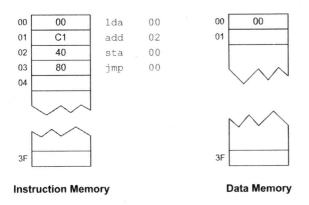

Instruction Memory　　　　　　　　**Data Memory**

Figure 4.22 Instruction and Data Memory Contents

Figure 4.24 shows the testbench for testing *SMPL-CPU*. As shown, after declarations, the processor, the instruction memory, and data memory are instantiated in this module. Following this, an **initial** block resets the machine by applying a **1** to its *reset* input. An **always** block is used to apply a waveform to the clock input.

After the reset, the CPU fetches instructions from the instruction memory starting from location 0, and executes them. With every fetch, a machine language instruction is read. The opcode of the instruction causes appropriate control signals to be issued.

```verilog
module InstMemory (input [5:0] abus,
                   output reg [7:0] dbus);
  reg [7:0] im_array [0:63];

  always @(abus) dbus = im_array [abus];

  initial begin
    im_array[0] = 8'h00;        // lda   00
    im_array[1] = 8'hC2;        // add   02
    im_array[2] = 8'h40;        // sta   00
    im_array[3] = 8'h80;        // jmp   00
  end
endmodule
//
module DataMemory (
    input rd, wr, input [5:0] abus,
    input [7:0] in_dbus, output reg [7:0] out_dbus);
  reg [7:0] dm_array [0:63];

  always @(rd or abus)
     if (rd) out_dbus = dm_array [abus];
  always @(wr or abus or in_dbus)
     if (wr) dm_array [abus] = in_dbus;

  initial dm_array[0] = 8'h00;
endmodule
```

Figure 4.23 Verilog Description of Instruction and Data Memory

```verilog
module TestSmplCPU;
  reg clk, reset;
  wire [5:0] im_abus, dm_abus;
  wire [7:0] im_dbus, dm_in_dbus, dm_out_dbus;

  smpl_cpu uut(clk, reset,rd_mem, wr_mem, im_abus,
               im_dbus, dm_abus, dm_in_dbus, dm_out_dbus);
  InstMemory IM (im_abus, im_dbus);
  DataMemory DM (rd_mem, wr_mem, dm_abus,
                 dm_out_dbus, dm_in_dbus);

  initial begin
     clk   = 1'b0;
     reset = 1'b1;
     #20 reset = 1'b0;
     #500 $finish;
  end
  always #10 clk = ~clk;
endmodule
```

Figure 4.24 *SMPL-CPU* Testbench

4.4.3 Multi-Cycle Implementation

In the single-cycle implementation of *SMPL-CPU*, we used two memory units, and two functional units (an *ALU* and an *adder*). To reduce the required hardware, we can share some of the hardware used for instruction execution. This leads to a multi-cycle implementation of *SMPL-CPU*, in which instructions are executed in a series of steps. Each step takes one clock cycle to execute.

4.4.3.1 Datapath Design. For the design of the datapath of multi-cycle version of *SMPL-CPU,* we start with the single-cycle datapath and try to use a single memory unit that stores both instructions and data. Sharing hardware adds one or more registers to store the output of the shared unit to be used in the next clock cycle.

To use a single memory, we need a common bus (that can be implemented using tri-state buffers or multiplexers) to choose between the address of the memory unit from the *PC* output (to address instructions) and bits 5 to 0 of the instruction (to address data). To store the instruction that is read from the memory, a register is used at the output of the memory unit, called *instruction register (IR)*.

The program counter (*PC*) is implemented as a counter to increment *PC* for program sequencing. Using these registers, the multi-cycle implementation of datapath is shown in Figure 4.25. The bussing shown here is appropriate for handling the necessary operations of our machine. As shown here, the datapath has an internal *dbus* bus. The external bidirectional *data_bus* drives and is driven by *dbus*. This bus connects to the input of *IR* in order to bring instruction read from the memory into this register.

IR has a load input (*ld_ir*) that is activated to cause it load from *data_bus*. Similarly, this bus connects to *AC* to bring data read from the memory into this register. The control signal for loading *AC* is *ld_ac*. This control signal is issued when the *lda* instruction is expected. *PC* has three control signals *ld_pc, inc_pc* and *clr_pc* to load, increment and clear it, respectively. The right most six bits of *IR* connect to the input of *PC* for execution of the *jmp* instruction.

For executing *sta*, *AC* is placed on the left input of *ALU* and from there to *dbus*, which eventually goes on *data_bus*. At the same time, *IR* is placed on *adr_bus* to specify the address in which *AC* data is to be stored. For this purpose, the adder unit (*ALU*) has a *pass* control input to make it pass its left input data to its output.

Execution of **add** is done by taking one of the add operands from *AC* and the other from *IR*. For this instruction, activating the *add* control input of *ALU* causes the *ALU* to perform addition.

The simple bussing structure described above facilitates execution of all four instructions of our *SMPL-CPU*. When a bus has more

than one source driving it, e.g., *IR* and *PC* driving *adr_bus*, control
signals from the controller select the source.

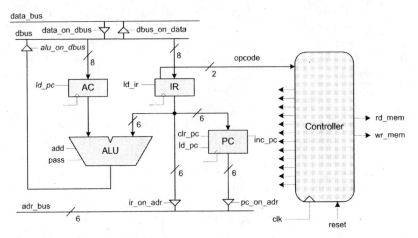

Figure 4.25 *SMPL-CPU* Multi-Cycle Datapath

4.4.3.2 Controller Design.
After the design of the datapath and fig-
uring control signals and their role in activities in the datapath, the
design of the controller becomes a simple matter. The block diagram
of this part is shown in Figure 4.26.

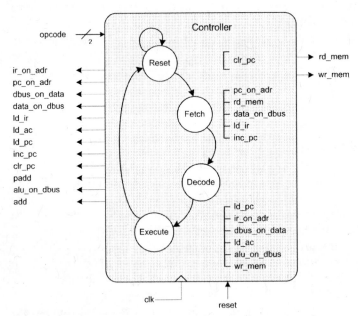

Figure 4.26 *SMPL-CPU* Multi-Cycle Controller

The controller of our simple von Neumann machine has four states, *Reset, Fetch, Decode* and *Execute*. As the machine cycles through these states, various control signals are issued. In state *Reset*, for example, the *clr_pc* control signal is issued. State *Fetch* issues *pc_on_adr*, *rd_mem*, *data_on_dbus*, *ld_ir*, and *inc_pc*, to read memory from the present *PC* location, route it to *IR*, load it into *IR*, and increment *PC* for the next memory fetch. Depending on *opcode* bits, that are the controller inputs, the *Execute* state of the controller issues control signals for execution of **lda, sta, add** and **jmp** instructions. The next section discusses details of the controller signals and their role in execution of these instructions.

4.4.3.3 Verilog Description. As before, our processor description has a datapath and a control component. In this description, we use a slightly different style of coding for the datapath. Instead of coding the individual components of datapath and instantiating them in the datapath module, our datapath module will include all the codes for its components. The controller will be described next, using a state machine coding style. At the end, the description of our small von Neumann example will be completed by wiring datapath and controller in a top-level Verilog module.

4.4.3.4 Datapath Description. Datapath components of *SMPL-CPU* are described by **always** and **assign** statements according to their functionalities described above. These descriptions are directly coded in the datapath module of our machine. Figure 4.27 shows the Verilog code of the datapath. Structure and signal names in this description are according to those shown in Figure 4.25.

The first three **always** statements in the body of the *DataPath* module are responsible for datapath registers, *AC, IR* and *PC*. These registers use signals from the controller for selecting their operations. The fourth **always** statement represents the *ALU* of our machine. All left-hand side outputs of these units are declared as **reg**.

In the last part of *DataPath*, bus assignments take place. We use bus control signals coming from the controller to drive a left-hand-side bus either with one of its sources or high-impedance. For example, *pc_on_adr* control signal either puts *PC* output (*pc_out*) or all **Z**s on *adr_bus*. The *dbus* bus is declared to connect to the external bidirectional *data_bus*. Two assignments are made to *dbus* using *alu_on_dbus* and *data_on_dbus* control signals. Placement of this intermediate bus to the external data bus of the datapath (*data_bus*) is controlled by *dbus_on_data* control signal. The last **assign** statement shown in Figure 4.27 places most significant *IR* bits on the *opcode* output of *DataPath* that goes out to the controller.

Although we have used tri-state busses, when synthesizing this circuit, we can direct our synthesis tool to use AND-OR or multiplexers to implement these busses.

```verilog
module DataPath (
    input ir_on_adr, pc_on_adr, dbus_on_data, data_on_dbus,
          ld_ir, ld_ac, ld_pc, inc_pc, clr_pc, pass, add,
          alu_on_dbus, clk,
    output [5:0] adr_bus, output [1:0] opcode,
    inout [7:0] data_bus);

    reg [7:0] ir_out, a_side, alu_out;
    reg [5:0] pc_out;
    wire [7:0] dbus;

    always @(posedge clk) if(ld_ac) a_side <= dbus;
    always @(posedge clk) if (ld_ir) ir_out <= dbus;
    always @(posedge clk) if(clr_pc) pc_out <= 6'b000_000;
        else if(ld_pc) pc_out <= ir_out;
        else if(inc_pc) pc_out <= ir_out + 1;
    always @(a_side, ir_out, pass, add)
        if (pass) alu_out = a_side;
        else if (add) alu_out = a_side + {2'b00,ir_out[5:0]};
        else alu_out = 0;

    assign adr_bus = ir_on_adr ? ir_out[5:0] : 6'bzz_zzzz;
    assign adr_bus = pc_on_adr ? pc_out : 6'bzz_zzzz;
    assign dbus = alu_on_dbus ? alu_out : 8'bzzzz_zzzz;
    assign data_bus = dbus_on_data ? dbus : 8'bzzzz_zzzz;
    assign dbus = data_on_dbus ? data_bus : 8'bzzzz_zzzz;

    assign opcode = ir_out[7:6];

endmodule
```

Figure 4.27 Datapath Description

4.4.3.5 Controller Description.
The controller code for our *SMPL-CPU* example is shown in Figure 4.28. This code corresponds to the right hand side of Figure 4.25 which is shown in more details in Figure 4.26. In addition to *clk* and *reset*, the controller has the *opcode* input that is driven by *IR* and comes to the controller from the *DataPath* module (see Figure 4.25).

The sequencing of control states is implemented by a Huffman style Verilog code. In this style, an **always** block handles assignment of values to *present_state*, and another **always** statement uses this register output as the input of a combinational logic determining *next_state*. This combinational block also sets values to control signals that are outputs of the controller.

```verilog
`define Reset 2'b00
`define Fetch 2'b01
`define Decode 2'b10
`define Execute 2'b11

module Controller (
    input reset, clk, input [1:0] op_code,
    output reg rd_mem, wr_mem, ir_on_adr, pc_on_adr,
        dbus_on_data, data_on_dbus,
        ld_ir, ld_ac, ld_pc, inc_pc, clr_pc, pass, add,
        alu_on_dbus);
    reg [1:0] present_state, next_state;

    always @( posedge clk )
        if( reset ) present_state <= `Reset;
        else present_state <= next_state;
    always @( present_state or reset ) begin : Combinational
        rd_mem=1'b0; wr_mem=1'b0; ir_on_adr=1'b0;
        pc_on_adr=1'b0; dbus_on_data=1'b0; data_on_dbus=1'b0;
        ld_ir=1'b0; ld_ac=1'b0; ld_pc=1'b0; inc_pc=1'b0;
        clr_pc=1'b0; pass=0; add=0; alu_on_dbus=1'b0;
        case ( present_state )
            `Reset : begin next_state = reset ? `Reset : `Fetch;
                clr_pc = 1;
            end
            `Fetch : begin next_state = `Decode;
                pc_on_adr = 1; rd_mem = 1; data_on_dbus = 1;
                ld_ir = 1; inc_pc = 1;
            end
            `Decode : next_state = `Execute;
            `Execute: begin next_state = `Fetch;
                case( op_code )
                    2'b00: begin
                        ir_on_adr = 1; rd_mem = 1;
                        data_on_dbus = 1; ld_ac = 1;
                    end
                    2'b01: begin
                        pass = 1; ir_on_adr = 1; alu_on_dbus = 1;
                        dbus_on_data = 1; wr_mem = 1;
                    end
                    2'b10: ld_pc = 1;
                    2'b11: begin
                        add = 1; alu_on_dbus = 1; ld_ac = 1;
                    end
                endcase
            end
            default : next_state = `Reset;
        endcase
    end
endmodule
```

Figure 4.28 Controller Description

The first **always** block of the controller code synthesizes to a register with active high *reset*, and the second one, i.e., *combinational*, synthesizes to a combinational block. This block uses *present_state* and *reset* on its sensitivity list. For synthesis purposes and to avoid output latches, all outputs of this block, that are the control signals, are set to their inactive, **0**, values. In the body of the *combinational* **always** block, a **case** statement checks *present_state* against the states of the machine ('*Reset*, '*Fetch*, '*Decode*, and '*Execute*), and activates the proper control signals.

The '*Reset* state activates *clr_pc* to clear *PC* and sets '*Fetch* as the next state of the machine. In the '*Fetch* state, *pc_on_adr*, *rd_mem*, *data_on_dbus*, *ld_ir*, and *inc_pc* become active, and '*Decode* is set to become the next state of the machine. By activating *pc_on_adr* and *rd_mem*, the *PC* output goes on the memory address and a read operation is issued. Assuming the memory responds in the same clock, contents of memory at the *PC* address will be put on *data_bus*. Issuance of *data_on_dbus* puts the contents of this bus on the internal *dbus* of *DataPath*. This bus is connected to the input of *IR* and issuance of *ld_ir* loads its contents into this register. The next state of the controller is '*Decode* that makes the new contents of *IR* available for the controller. In the '*Execute* state a newly fetched instruction in *IR* decides on control signals to issue to execute the instruction.

In the '*Execute* state, *op_code* is used in a **case** expression to decide on control signals to issue depending on the opcode of the fetched instruction. The **case** alternatives in this statement are four *op_code* values of **00**, **01**, **10** and **11** that correspond to *lda*, *sta*, *jmp* and *add* instructions.

For *lda*, *ir_on_adr*, *rd_mem*, *data_on_dbus* and *ld_ac* are issued. These control signals cause the address from *IR* to be placed on the *adr_bus* address bus, memory read to take place, and data from memory to be loaded into *AC*. Data from the memory come through *data_bus* onto *dbus* of *DataPath* by the control signal *data_on_dbus*.

The controller executes the *sta* instruction by issuing *pass*, *ir_on_adr*, *alu_on_dbus*, *dbus_on_data* and *wr_mem*. As shown in Figure 4.25, these signals take contents of *AC* to the input bus of the memory (i.e., *data_bus*), and *wr_mem* causes the writing into the memory to take place. Note that *pass* causes *AC* to pass through *ALU* unchanged.

The *jmp* instruction is executed by enabling *PC* load input, which takes the jump address from *IR* (see Figure 4.25).

The last instruction of this machine is *add*, for execution of which, *add*, *alu_on_dbus*, and *ld_ac* are issued. This instruction adds data in the upper six bits of *IR* with *AC* and loads the result into *AC*. The *add* control signal instructs *ALU* to add its two inputs; the *alu_on_dbus* puts this output on the internal datapath *dbus*; and the *ld_ac* causes *AC* to be loaded with the result of addition.

4.4.3.6 The Complete Design. The top-level module for our adding machine example is shown in Figure 4.29. In the *SMP-CPU* module shown, *DataPath* and *Controller* modules are instantiated. Port connections of the *Controller* include its output control signals, the *opcode* input from *DataPath* and the *reset* external input. Port connections of *DataPath* consist of *adr_bus* and *data_bus* external busses, *opcode* output, and control signal inputs.

```
module SMPL_CPU (
                input reset, clk,
                output [5:0] adr_bus, output rd_mem, wr_mem,
                inout [7:0] data_bus);

    wire ir_on_adr, pc_on_adr, dbus_on_data, data_on_dbus,
        ld_ir, ld_ac, ld_pc, inc_pc, clr_pc, pass, add,
        alu_on_dbus;
    wire [1:0] op_code;

    Controller cu (reset, clk, op_code, rd_mem, wr_mem,
                    ir_on_adr, pc_on_adr, dbus_on_data,
                    data_on_dbus, ld_ir, ld_ac, ld_pc, inc_pc,
                    clr_pc, pass, add, alu_on_dbus);

    DataPath dp (ir_on_adr, pc_on_adr, dbus_on_data,
                    data_on_dbus, ld_ir, ld_ac, ld_pc, inc_pc,
                    clr_pc, pass, add, alu_on_dbus, clk, adr_bus,
                    op_code, data_bus );
endmodule
```

Figure 4.29 SMPL-CPU Top-Level Module

4.4.3.7 Testing Multi-Cycle Implementation of SMPL-CPU. In the testbench for the *SMPL-CPU* module, we model a simple memory with read and write operations. The memory is file-based and we will use file I/O tasks for reading and writing from and to the memory. The testbench uses a task for converting instructions in mnemonic form from an external file to binary memory data. This testbench also performs the task of an assembler by reading mnemonics and converting them to the processor's machine language.

4.4.3.8 Testbench / Assembler Outline. The outline of the testbench that also acts as an assembler for our machine is shown in Figure 4.30. This module reads the *InstructionFile.mem* file which contains instruction mnemonics and their addresses, converts them to hex and writes them to *HexadecimalFile.mem* file. After this conversion is done, every addressed memory read or write uses this file. Because the Unit Under Test (UUT) does not have a large memory, no image of its memory is kept in the testbench as an array of **reg**, and all read and write operations are directly performed on the *Hexadecimal-File.mem* file.

As shown in Figure 4.30, after declarations and instantiation of *SMPL-CPU*, an **initial** block calls the *Convert* **task** to translate instruction mnemonics to hex, opens the *HexadecimalFile.mem* file, and sets the end of the simulation run time. The **$fopen** task opens this hex file and assigns the *HexFile* descriptor that is a declared integer to it.

```verilog
module Test_SMPL_CPU;
  reg reset=1, clk=0;
  wire [5:0] adr_bus;
  wire rd_mem, wr_mem;
  wire [7:0] data_bus;
  reg [7:0] mem_data=8'b0;
  reg control=0;
  integer HexFile, check;

  SMPL_CPU UUT (reset, clk, adr_bus,
                rd_mem, wr_mem, data_bus);

  always #10 clk = ~clk;

  initial begin
    Convert; // Read mnemonics, write machine language
    HexFile = $fopen ("HexadecimalFile.mem", "r+");
    #25 reset=1'b0;
    #405 $fclose (HexFile);
    $stop;
  end

  always @(posedge clk) begin : Memory_Read_Write
    // . . .
  end
  // . . .
  task Convert; // The assembler
    // . . .
  endtask

endmodule
```

Figure 4.30 Outline of SMPL-CPU Testbench

This **initial** block is followed by the *Memory_Read_Write* block. This block assumes 64 8-bit hex data are available in *Hexadecimal-File.mem*. For accessing this file, its descriptor *HexFile*, will be used.

Figure 4.31 shows the details of *Memory_Read_Write* **always** block. After a short delay (1 ns) after the *posedge* of *clk*, *rd_mem* and *wr_mem* are expected to be stable. At this time, if *rd_mem* is 1, data on *adr_bus* is used to set the position of the next read from *HexFile*. Since data in *HexadecimalFile.mem* are in hex (2 bytes), a total of 4 bytes that include two "end of line" bytes are used for each memory entry. Therefore, **$fseek** of Figure 4.31 positions the next reading from *4*adr_bus*. The **$fscanf** task that follows this task reads the hex data at the file position into *mem_data*. This variable is local to the testbench and is put on *data_bus* only when reading from the memory is to take place. The *control* variable is used to drive *data_bus* with *mem_data* or 8'hZZ.

```
always @ (posedge clk) begin : Memory_Read_Write
   control = 0;
   #1;
   if (rd_mem) begin
      #1;
      check = $fseek (HexFile, 4 * adr_bus, 0);
      check = $fscanf (HexFile, "%h", mem_data);
      control = 1;
   end
   if (wr_mem) begin
      #1;
      check = $fseek (HexFile, 4 * adr_bus, 0);
      $fwrite (HexFile, "%h", data_bus);
      $fflush (HexFile);
   end
end

assign data_bus = (control) ? mem_data: 8'hZZ;
```

Figure 4.31 Memory Read and Write

The next part of the **always** block of Figure 4.31 handles writing into the memory. For this purpose, after file positioning, the **$fwrite** task writes contents of *data_bus* into *HexadecimalFile.mem*. After every writing, **$fflush** writes any buffered output to this file.

4.4.3.9 Assembler. The testbench outline of Figure 4.30 shows the *Convert* **task** that is used for converting instruction mnemonics of *InstructionFile.mem* to hex data in *HexadecimalFile.mem*. Figure 4.32 shows six lines of *InstructionFile.mem* and their corresponding hex translation in memory locations 0 to 15.

```
InstructionFile              HexadecimalFile

Line:                        Location:
    1:  00 lda 0f                00:  0F
    2:  0f ::: 0f                01:  4A
    3:  01 sta 0a                02:  C1
    4:  02 add 01                03:  4B
    5:  03 sta 0b                04:  80
    6:  04 jmp 00                06:  00
                                 07:  00
                                 08:  00
                                 09:  00
                                 10:  00
                                 11:  00
                                 12:  00
                                 13:  00
                                 14:  00
                                 15:  0F
```

Figure 4.32 Instruction Mnemonics and Hex Memory Data

The *Convert* **task** reads a line of *InstructionFile.mem* that contains a memory location, instruction mnemonic and its operand. It converts this line to an opcode and its data and writes it in its specified location in *HexadecimalFile.mem*. For example, the third line of the instruction file of Figure 4.32 (*sta 0A*) is translated to *4A* and is put in location 1 of the hexadecimal file. For direct memory data, the instruction file uses the ":::" notation. *0F ::: 0F* shown in Figure 4.32 is translated to data *0F* in location 15 of the hexadecimal file.

The *Convert* **task** is shown in Figure 4.33. Initially all locations of *HexadecimalFile.mem* are initialized to "00". The *InstructionFile.mem* is opened for reading (i.e., with *r* argument), and *HexadecimalFile.mem* is opened for reading and writing (i.e., with *r+* argument). File descriptors for these two files are *InstFile* and *HexFile*.

Convert has a **while** loop that reads data from *InstFile*, converts it to hex and puts it in its corresponding location in *HexFile*. The **$fscanf** task shown in this loop reads the first two hex digits of a line of instruction into *addr*. This variable is then used for setting the write position for the *HexFile* file. File positioning is done by **$fseek**. This is followed by **$fgets** that reads the *opcode* string from the instruction file (*InstFile*). A **case** statement in *convert* translates string opcodes to their hex equivalent, and an **$fwrite task** writes this hex data into the hex file (*HexFile*) at the location set by the **$fseek task**.

If the opcode string read from *InstFile* is ":::", the hex data that follows this string will be written into *HexFile* location specified by *addr*. The last part of *Convert* flushes *HexFile* and closes both instruction and hexadecimal files.

```
task Convert; // Our solution for the assembler
  begin: block
    reg [5: 0] addr;
    reg [3 * 8: 1] opCode;
    reg [7: 0] data, writeData;
    reg JustData;
    integer i, HexFile, InstFile, check;
    HexFile = $fopen ("HexadecimalFile.mem");
    for (i = 0; i < 64; i = i + 1)
        $fwrite (HexFile, "00\n");
    $fflush (HexFile); $fclose (HexFile);
    InstFile = $fopen ("InstructionFile.mem", "r");
    HexFile = $fopen ("HexadecimalFile.mem", "r+");
    while ($fscanf(InstFile, "%h", addr)!= -1) begin
      check = $fseek (HexFile, addr * 4, 0);
      check = $fgets (opCode, InstFile);
      JustData = 0;
      case (opCode)
        "lda": writeData[7: 6] = 0;
        "sta": writeData[7: 6] = 1;
        "jmp": writeData[7: 6] = 2;
        "add": writeData[7: 6] = 3;
        ":::": begin JustData = 1;
          check = $fscanf(InstFile,"%h", writeData);
        end
        default: begin JustData = 1;
          check = $fscanf(InstFile,"%h", writeData);
        end
      endcase
      if(JustData == 0) begin
        check = $fscanf (InstFile, "%h", data);
        writeData[5: 0] = data[5: 0];
      end
      $fwrite(HexFile, "%h", writeData);
    end
    $fflush (HexFile);
    $fclose (HexFile); $fclose(InstFile);
  end
endtask
```

Figure 4.33 Converting Instructions to Hex

4.5 SAYEH Design and Test

This section shows design, description, and test of a small computer in Verilog. The CPU is SAYEH (Simple Architecture, Yet Enough Hardware) that has been designed for educational and benchmarking purposes. The design is simple, and follows the design strategy used for the processor of the previous section. This processor will be used as an embedded processor in the later chapters of this book.

4.5.1 Details of Processor Functionality

The simple CPU example discussed here has a register file that is used for data processing instructions. The CPU has a 16-bit data bus and a 16-bit address bus. The processor has 8 and 16-bit instructions. Short instructions contain shadow instructions, which effectively pack two such instructions into a 16-bit word. Figure 4.34 shows SAYEH interface signals.

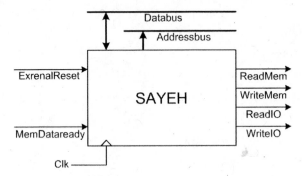

Figure 4.34 SAYEH Interface

4.5.1.1 CPU Components. SAYEH uses its register file for most of its data instructions. Addressing modes of this processor also take advantage of this structure. Because of this, the addressing hardware of SAYEH is a simple one and the register file output is used in address calculations.

SAYEH components that are used by its instructions include the standard registers such as the Program Counter, Instruction Register, the Arithmetic Logic Unit, and Status Register. In addition, this processor has a register file forming registers *R0*, *R1*, *R2* and *R3* as well as a Window Pointer that defines *R0*, *R1*, *R2* and *R3* within the register file. CPU components and a brief description of each are shown below.

- **PC:** Program Counter, 16 bits
- **R0, R1, R2, and R3:** General purpose registers part of the register file, 16 bits
- **Reg File:** The general purpose registers form a window of 4 in a register file of 8 registers
- **WP:** Window Pointer points to the register file to define *R0*, *R1*, *R2* and *R3*, 3 bits
- **IR:** Instruction Register that is loaded with a 16-bit, an 8-bit, or two 8-bit instructions, 16 bits

- **ALU:** The ALU that can AND, OR, NOT, Shift, Compare, Add, Subtract and Multiply its inputs, 16 bit operands
- **Z flag:** Becomes **1** when the ALU output is **0**
- **C flag:** Becomes **1** when the ALU has a carry output

Table 4.4 Instruction Set of SAYEH

Instruction Mnemonic and Definition		Bits 15:0	RTL notation: *comments or condition*
nop	No operation	0000-00-00	No operation
hlt	Halt	0000-00-01	Halt, fetching stops
szf	Set zero flag	0000-00-10	Z <= '1'
czf	Clr zero flag	0000-00-11	Z <= '0'
scf	Set carry flag	0000-01-00	C <= '1'
ccf	Clr carry flag	0000-01-01	C <= '0'
cwp	Clr Window pointer	0000-01-10	WP <= "000"
mvr	Move Register	0001-D-S	$R_D <= R_S$
lda	Load Addressed	0010-D-S	$R_D <= (R_S)$
sta	Store Addressed	0011-D-S	$(R_D) <= R_S$
inp	Input from port	0100-D-S	In from R_S write to R_D
oup	Output to port	0101-D-S	Out to port R_D from R_S
and	AND Registers	0110-D-S	$R_D <= R_D$ & R_S
orr	OR Registers	0111-D-S	$R_D <= R_D \mid R_S$
not	NOT Register	1000-D-S	$R_D <= \sim R_S$
shl	Shift Left	1001-D-S	$R_D <=$ sla R_S
shr	Shift Right	1010-D-S	$R_D <=$ sra R_S
add	Add Registers	1011-D-S	$R_D <= R_D + R_S + C$
sub	Subtract Registers	1100-D-S	$R_D <= R_D - R_S - C$
mul	Multiply Registers	1101-D-S	$R_D <= R_D * R_S$:*8-bit multiplication*
cmp	Compare	1110-D-S	R_D, R_S *(if equal:Z=1; if $R_D<R_S$: C=1)*
mil	Move Immd Low	1111-D-00-I	$R_{DL} <= \{8'bZ, I\}$
mih	Move Immd High	1111-D-01-I	$R_{DH} <= \{I, 8'bZ \}$
spc	Save PC	1111-D-10-I	$R_D <= PC + I$
jpa	Jump Addressed	1111-D-11-I	$PC <= R_D + I$
jpr	Jump Relative	0000-01-11-I	$PC <= PC + I$
brz	Branch if Zero	0000-10-00-I	$PC <= PC + I$:*if Z is 1*
brc	Branch if Carry	0000-10-01-I	$PC <= PC + I$:*if C is 1*
awp	Add Win pntr	0000-10-10-I	$WP <= WP + I$

4.5.1.2 SAYEH Instructions. The general format of 8-bit and 16-bit SAYEH instructions is shown in Figure 4.35. The 16-bit instructions have the *Immediate* field and the 8-bit instructions do not. The *OP-CODE* filed is a 4-bit code that specifies the type of instruction. The *Left* and *Right* fields are two bit codes selecting *R0* through *R3* for source and/or destination of an instruction. Usually, *Left* is used for destination and *Right* for source. The *Immediate* filed is used for immediate data, or if two 8-bit instructions are packed, it is used for the second instruction.

15 12	11 10	09 08	07 00
OPCODE	*Left*	*Right*	*Immediate*

Figure 4.35 SAYEH Instruction Format

Our processor has a total of 29 instructions as shown in Table 4.4. Instructions with *I* immediate field are 16-bit instructions and the rest are 8-bit instructions. Instructions that use the *Destination* and *Source* fields (designated by *D* and *S* in the table of instruction set) have an opcode that is limited to 4 bits. Instructions that do not require specification of source and destination registers use these fields as opcode extensions. In addition to *nop*, hex code 0F is used as filler for the right most 8-bits of a 16-bit word that only contains an 8-bit instruction in its 8 left-most bits.

In the instruction set, addressed locations in the memory are indicated by enclosing the address in a set of parenthesis. For these instructions, the processor issues *ReadMem* or *WriteMem* signals to the memory. When input and output instructions (*inp, oup*) are executed, SAYEH issues *ReadIO* or *WriteIO* signals to its IO devices.

4.5.2 SAYEH Datapath

The datapath of SAYEH is shown in Figure 4.36. The main components of this machine are: *Addressing Unit* that consists of *PC (Program Counter)* and *Address Logic, IR (Instruction Register), WP (Window Pointer), Register File* that consists of *Left Decoder1* and *Right Decoder2, ALU (Arithmetic Logic Unit)*, and *Flags*. As shown in Figure 4.36, these components are either hardwired or connected through three-state busses. Component inputs with multiple sources, such as the right hand side input of *ALU*, use three-state busses. Three-state busses in this structure are *Dastabus* and *OpndBus*. In this figure, signals that are in italic are control signals issued by the controller. These signals control register clocking, logic unit operations and placement of data in busses.

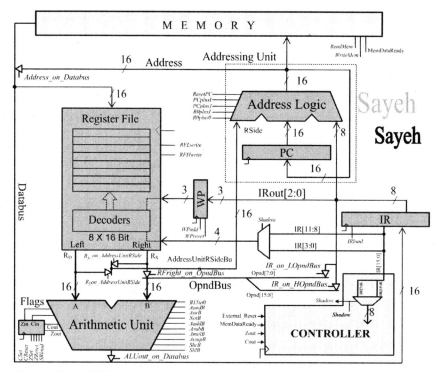

Figure 4.36 SAYEH Datapath

4.5.2.1 Datapath Components.
Figure 4.37 shows the hierarchical structure of SAYEH components. The processor has a *datapath* and a *controller*. *Datapath* components are *Addressing Unit, IR, WP, Register File, Arithmetic Unit,* and the *Flags* register. The *Addressing Unit* is further partitioned into the *PC* and *Address Logic.*

The *Addressing Logic* is a combinational circuit that is capable of adding its inputs to generate a 16-bit output that forms the address for the processor memory. *Program Counter* and *Instruction Register* are 16-bit registers. *Register File* is a two-port memory and a file of 8, 16-bit registers. The *Window Pointer* is a 3-bit register that is used as the base of the *Register File*. Specific registers for read and write (*R0, R1, R2* or *R3*) in the *Register File* are selected by its 4-bit input bus coming from the *Instruction Register*. Two bits are used to select a source register and other two bits select the destination register.

When the Window Pointer is enabled, it adds its 3-bit input to its current data. The *Flags* register is a 2-bit register that saves the flag outputs of the *Arithmetic Unit*. The *Arithmetic Unit* is a 16-bit arithmetic and logic unit that has the functions shown in Table 4.5. A 9-bit input selects the function of the ALU shown in this table. This code is provided by the processor controller.

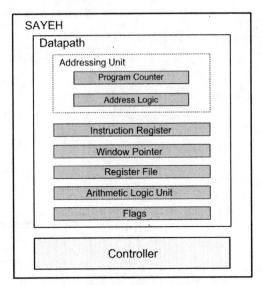

Figure 4.37 SAYEH Hierarchical Structure

Table 4.5 ALU Operations

Mnemonic	Description	Code
B15to0H	Place B on the output	1000000000
AandBH	Place A and B on the output	0100000000
AorBH	Place A or B on the output	0010000000
notBH	Place not B on the output	0001000000
shlBH	Shift B one bit to the left	0000100000
shrBH	Shift B one bit to the right	0000010000
AaddBH	Place A + B on the output	0000001000
AsubBH	Place A - B on the output	0000000100
AmulBH	Place A * B on the output	0000000010
AcmpBH	Z = 1 if A = B; C = 1 if A < B	0000000001

Controller of SAYEH has eleven states for various reset, fetch, decode, execute, and halt operations. Signals generated by the controller control logic unit operations and register clocking in the datapath.

SAYEH sequential data components and its controller are triggered on the falling edge of the main system clock. Control signals remain active after one falling edge through the next. This duration allows for propagation of signals through the busses and logic units in the datapath.

4.5.3 SAYEH Verilog Description

SAYEH is described according to the hierarchical structure of Figure 4.37. Data components are described separately, and then wired to form the datapath. Controller is described in a single Verilog module. In the complete SAYEH description, the datapath and controller are wired together.

The coding style used for the description of this processor is similar to that of the multi-cycle description of *SMPL-CPU*. The CD accompanying this book has the complete code of this processor and will not be shown here. SAYEH top-level description will be shown in order to present its testbench in the next section.

4.5.3.1 SAYEH Top-Level Description. The top-level Verilog code of SAYEH that is shown in Figure 4.38 consists of instantiation of *DataPath* and *controller* modules. In the *Sayeh* module, control signal outputs of *controller* are wired to the similarly named signals of *DataPath*. The ports of the processor are according to the block diagram of Figure 4.34.

```
module Sayeh (
    input clk, output ReadMem, WriteMem, ReadIO, WriteIO,
    inout [15: 0] Databus, output [15: 0] Addressbus,
    input ExternalReset, MemDataready);

    wire [15:0] Instruction;
    wire esetPC, PCplusI, PCplus1, RplusI, Rplus0,
        . . .
    DataPath dp (
        clk, Databus, Addressbus,
        ResetPC, PCplusI, PCplus1, RplusI, Rplus0, ...);
    controller ctrl (
        ExternalReset, clk,
        ResetPC, PCplusI, PCplus1, RplusI, Rplus0, ...);
endmodule
```

Figure 4.38 SAYEH Top Level Description

4.5.4 SAYEH Top-Level Testbench / Assembler

In the testbench of SAYEH, we instantiate the processor of Figure 4.38, and through a memory model, we apply instructions to the CPU and watch its response to these test instructions.

The testbench style is similar to that of multi-cycle version of *SMPL-CPU* that was discussed in the previous section. SAYEH testbench also has a translation program that is, of course, much larger than that of *SMPL-CPU*. This **task** is called *Convert* and functions like an assembler for SAYEH.

The testbench reads its program that is written in SAYEH assemble code from an input file (*inst.mem*), translates it to SAYEH machine language and writes a hex file that contains the machine language equivalent of the program. This file (*memFile.mem*) represents the memory image of the program and is read and written into by the SAYEH processor Verilog model.

```
0000   mil   r0 00    :r0=768    starting address in memory
0001   mih   r0 03    :
0002   lda   r1 r0    :r1=       total number of elements
0003   awp   5        :
0004   mil   r0 01    :r5=1      for adding with index each time
0005   mih   r0 00    :
0006   cwp            :
0006   add   r1 r0    :r1=       limit for final r4
0007   mvr   r2 r1    :
0008   awp   2        :
0009   sub   r0 r3    :r2=       limit for index r3
0009   cwp            :
000A   mvr   r3 r0    :r3=       outer index
000A   nop            :
000B   cwp            :
000B   cmp   r3 r2    :          is outer index == to its limit
000C   brz   19       :          branch to 0025 if zero
000D   awp   3        :
000E   add   r0 r2    :r3=r3+1   increment outer index
000E   mvr   r1 r0    :r4=r3     init inner index to outer
000F   cwp            :
0010   awp   1        :
0011   cmp   r3 r0    :          check if inner index == limit
0012   brz   10       :          branch to 0022 if zero
0013   awp   2        :
0014   lda   r3 r0    :r6=(r3)
0015   awp   1        :
0016   add   r0 r1    :r4=r4+r5  increment inner index
0016   lda   r3 r0    :r7=(r4)
0017   cmp   r2 r3    :          check if r6 is > than r7
0018   brc   07       :          branch to 001F if carry
0019   lda   r1 r0    :r5=(r4)   r5 as an temporary register
0019   sta   r0 r2    :(r4)=r6
001A   cwp            :
001B   awp   3        :
001C   sta   r0 r2    :(r3)=r5
001D   mil   r2 01    :
001E   mih   r2 00    :r5=1      for adding with index
001F   cwp            :
0020   awp   5        :
0021   jpa   r0 0E    :          jump to 000F
0022   cwp            :
0023   awp   5        :
0024   jpa   r0 0A    :          jump to 000B
0025   hlt            :
```

Figure 4.39 Sorting Program for SAYEH

An example sort program in SAYEH assembly language is shown in Figure 4.39. This is the assembly equivalent of the C program that was discussed in Section 4.2.4 (Figure 4.4). The program reads data from the CPU memory and sorts them in descending order. The number of data items to sort is in location 768 and data begins in the next memory location. Like its equivalent C program, this program uses two loops for its sorting function. When completed, the CPU is put into the halt state.

The program shown in Figure 4.39 is translated into its hexadecimal equivalent and is put in the *memFile.mem*. As discussed in the previous section, SAYEH testbench reads instructions from this file and applies them to the CPU.

4.5.5 SAYEH Hardware Realization

SAYEH CPU described in this chapter has been synthesized and programmed into a number of FPGAs and tested on Altera development boards. One implementation has been on Altera's Cyclone EP1C12 device of an Altera UP3. We used a RAM from Altera's megafunctions and configured it as a memory of 256 16-bit words. The number of logic elements used by this CPU was 874, which is 7% of the available 12060 LEs. Memory bits used was 2096, which is 2% of the available memory bits. This usage indicates that we can form a complete system with a keyboard and VGA output on a Cyclone FPGA.

4.6 Summary

In this chapter we discussed processing units, their hardware and software. These topics are generally covered in detail in two or more courses on computer architecture and computer software. However, the discussion here only focused on those issues that are needed for designing embedded systems with processors and processor software. This discussion provided the necessary background for using embedded processors for implementation of parts of a larger design, or as stand alone systems. This chapter also showed CPU hardware, and design methodology for large RT level designs. Application of processing elements in design of embedded systems will be discussed in later chapters of this book.

5 Field Programmable Devices

The need for getting designs done quickly has led to the creation and evolution of Field Programmable Devices. The idea began from Read Only Memories (ROM) that were just an organized array of gates and has evolved into System On Programmable Chips (SOPC) that use programmable devices, memories and configurable logic all on one chip. This chapter shows the evolution of basic array structures like ROMs into complex CPLD (Complex Programmable Logic Devices) and FPGAs (Field Programmable Gate Array). This topic can be viewed from different angles, like logic structure, physical design, programming technology, transistor level, software tools, and perhaps even from historic and commercial aspects. However our treatment of this subject is more at the structural level. We discuss gate level structures of ROMs, PLAs, PALs, CPLDs, and FPGAs. The material is at the level needed for understanding configuration and utilization of CPLDs and FPGAs in digital designs.

5.1 Read Only Memories

We present structure of ROMs by showing the implementation of a 3-input 4-output logic function. The circuit with the truth table shown in Figure 5.1 is to be implemented.

5.1.1 Basic ROM Structure

The simplest way to implement the circuit of Figure 5.1 is to form its minterms using AND gates and then OR the appropriate minterms for formation of the four circuit outputs. The circuit requires eight 3-input AND gates and four OR gates that can take up-to eight inputs. It is easiest to draw this structure in an array format as shown in Figure 5.2.

m:	a	b	c	w	x	y	z
0:	0	0	0	0	0	0	1
1:	0	0	1	1	1	0	0
2:	0	1	0	1	0	1	1
3:	0	1	1	1	0	0	1
4:	1	0	0	0	0	0	1
5:	1	0	1	0	1	1	1
6:	1	1	0	0	0	1	0
7:	1	1	1	0	0	0	0

Figure 5.1 A Simple Combinational Circuit

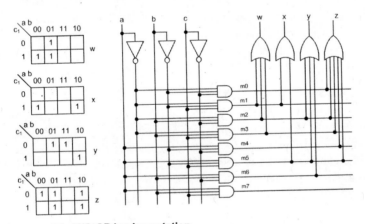

Figure 5.2 AND-OR Implementation

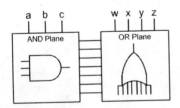

Figure 5.3 AND and OR Planes

The circuit shown has an array of AND gates and an array of OR gates, that are referred to as the AND-plane and the OR-plane. In the AND-plane all eight minterms for the three inputs, a, b, and c are generated. The OR plane uses only the minterms that are needed for the outputs of the circuit. See for example minterm 7 that is generated in the AND-plane but not used in the OR-plane. Figure 5.3 shows the block diagram of this array structure.

5.1.2 NOR Implementation

Since realization of AND and OR gates in most technologies are difficult and generally use more delays and chip area than NAND or NOR implementations, we implement our example circuit using NOR gates. Note that a NOR gate with complemented outputs is equivalent to an OR, and a NOR gate with complemented inputs is equivalent to an AND gate. Our all NOR implementation of Figure 5.4 uses NOR gates for generation of minterms and circuit outputs. To keep functionality and activity levels of inputs and outputs intact, extra inverters are used on the circuit inputs and outputs. These inverters are highlighted in Figure 5.4. Although NOR gates are used, the left plane is still called the AND-plane and the right plane is called the OR-plane.

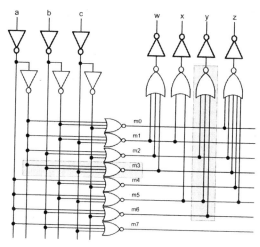

Figure 5.4 All NOR Implementation

5.1.3 Distributed Gates

Hardware implementation of the circuit of Figure 5.4 faces difficulties in routing wires and building gates with large number of inputs. This problem becomes more critical when we are using arrays with tens of inputs. Take for example, a circuit with 16 inputs, which is very usual for combinational circuits. Such a circuit has 64k (2^{16}) minterms. In the AND-plane, wires from circuit inputs must be routed to over

64,000 NOR gates. In the OR-plane, the NOR gates must be large enough for every minterm of the function (over 64,000 minterms) to reach their inputs.

Such an implementation is very slow because of long lines, and takes too much space because of the requirement of large gates. The solution to this problem is to distribute gates along array rows and columns.

In the AND-plane, instead of having a clustered NOR gate for all inputs to reach to, the NOR gate is distributed along the rows of the array. In Figure 5.4, the NOR gate that implements minterm 3 is highlighted. Distributed transistor-level logic of this NOR gate is shown in Figure 5.5. This figure also shows a symbolic representation of this structure.

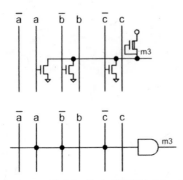

Figure 5.5 Distributed NOR of the AND-plane

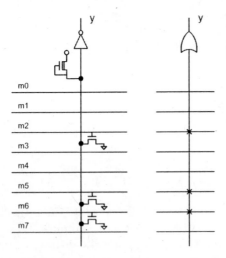

Figure 5.6 Distributed NOR Gate of Output y

Likewise, in the OR-plane, instead of having large NOR gates for the outputs of the circuit, transistors of output NOR gates are distributed along the corresponding output columns. Figure 5.6 shows the distributed NOR structure of the y output of circuit of Figure 5.4. A symbolic representation of this structure is also shown in this figure.

As shown in Figure 5.5 and Figure 5.6, distributed gates are symbolically represented by gates with single inputs. In each case, connections are made on the inputs of the gate. For the AND-plane, the inputs of the AND gate are a, b, and c forming minterm 3, and for the OR gate of Figure 5.6, the inputs of the gate are $m2$, $m5$ and $m6$. The reason for the difference in notations of connections in the AND-plane and the OR-plane (dots versus crosses) becomes clear after the discussion of the next section.

5.1.4 Array Programmability

For the a, b and c inputs, the structure shown in Figure 5.4 implements w, x, y and z functions. In this implementation, independent of our outputs, we have generated all minterms of the three inputs. For any other functions other than w, x, y and z, we would still generate the same minterms, but use them differently. Hence, the AND-plane with which the minterms are generated can be wired independent of the functions realized. On the contrary, the OR-plane can only be known when the output functions have been determined.

We can therefore generate a general purpose array logic with all minterms in its AND-plane, and capability of using any number of the minterms for any of the array outputs in its OR-plane. In other words, we want a fixed AND-plane and a programmable (or configurable) OR-plane. As shown in Figure 5.7, transistors for the implementation of minterms in the AND-plane are fixed, but in the OR-plane there are fusible transistors on every output column for every minterm of the AND-plane. For realization of a certain function on an output of this array, transistors corresponding to the used minterms are kept, and the rest are blown to eliminate contribution of the minterm to the output function.

Figure 5.7 shows configuration of the OR-plane for realizing outputs shown in Figure 5.1. Note for example that for output y, only transistors on rows $m2$, $m5$, and $m6$ are connected and the rest are fused off.

Instead of the complex transistor diagram of Figure 5.7, the notation shown in Figure 5.8 is used for representing the programmability of the configurable arrays. The dots in the AND-plane indicate permanent connections, and the crosses in the OR-plane indicate programmable or configurable connections.

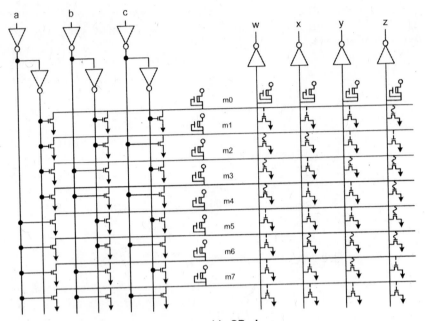

Figure 5.7 Fixed AND-plane, Programmable OR-plane

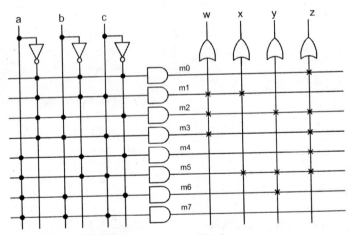

Figure 5.8 Fuse Notation for Configurable Arrays

5.1.5 Memory View

Let us look at the circuit of Figure 5.8 as a black box of three inputs and four outputs. In this circuit, if an input value between 0 and 7 is applied to the abc inputs, a 4-bit value is read on the four circuit outputs. For example $abc=011$ always reads $wxyz=1001$.

If we consider abc as the address inputs and $wxyz$ as the data read from abc designated address, then the black box corresponding to Figure 5.8 can be regarded as a memory with an address space of 8 words and data of four bits wide. In this case, the fixed AND-plane becomes the memory decoder, and the programmable OR-plane becomes the memory array (see Figure 5.9). Because this memory can only be read from and not easily written into, it is referred to as Read Only Memory or ROM.

The basic ROM is a one-time programmable logic array. Other variations of ROMs offer more flexibility in programming, but in all cases they can be read more easily than they can be written into.

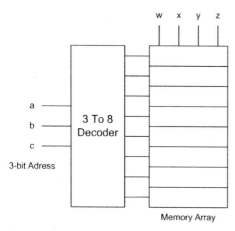

Figure 5.9 Memory View of ROM

5.1.6 ROM Variations

The acronym, ROM is generic and applies to most read only memories. What is today implied by ROM may be ROM, PROM, EPROM, EEPROM or even flash memories. These variations are discussed here.

5.1.6.1 ROM. ROM is a mask-programmable integrated circuit, and is programmed by a mask in IC manufacturing process. The use of mask-programmable ROMs is only justified when a large volume is needed. The long wait time for manufacturing such circuits makes it a less attractive choice when time-to-market is an issue.

5.1.6.2 PROM. Programmable ROM is a one-time programmable chip that, once programmed, cannot be erased or altered. In a PROM, all minterms in the AND-plane are generated, and connections of all AND-plane outputs to OR-plane gate inputs are in place. By applying a high voltage, transistors in the OR-plane that correspond to the minterms that are not needed for a certain output are burned out. Referring to Figure 5.7, a fresh PROM has all transistors in its OR-plane connected. When programmed, some will be fused out permanently. Likewise, considering the diagram of Figure 5.8, an unprogrammed PROM has X's in all wire crossings in its OR-plane.

5.1.6.3 EPROM.

An Erasable PROM is a PROM that once programmed, can be completely erased and reprogrammed. Transistors in the OR-plane of an EPROM have a normal gate and a floating gate as shown in Figure 5.10. The non-floating gate is a normal NMOS transistor gate, and the floating-gate is surrounded by insulating material that allows an accumulated charge to remain on the gate for a long time.

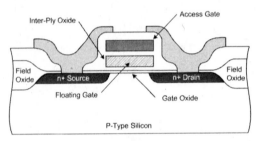

Figure 5.10 Floating Gate

When not programmed, or programmed as a '**1**', the floating gate has no extra charge on it and the transistor is controlled by the non-floating gate (access gate). To fuse-out a transistor, or program a '**0**' into a memory location, a high voltage is applied to the access gate of the transistor which causes accumulation of negative charge in the floating-gate area. This negative charge prevents logic 1 values on the access gate from turning on the transistor. The transistor, therefore, will act as an unconnected transistor for as long as the negative charge remains on its floating-gate.

To erase an EPROM it must be exposed to ultra-violate light for several minutes. In this case, the insulating materials in the floating-gates become conductive and these gates start loosing their negative charge. In this case, all transistors return to their normal mode of operation. This means that all EPROM memory contents become **1**, and ready to be reprogrammed.

Writing data into an EPROM is generally about a 1000 times slower than reading from it. This is while not considering the time needed for erasing the entire EPROM.

5.1.6.4 EEPROM. An EEPROM is an EPROM that can electrically be erased, and hence the name: Electrically Erasable Programmable ROM. Instead of using ultra-violate to remove the charge on the non-floating gate of an EPROM transistor, a voltage is applied to the opposite end of the transistor gate to remove its accumulated negative charge. An EEPROM can be erased and reprogrammed without having to remove it. This is useful for reconfiguring a design, or saving system configurations. As in EPROMs, EEPROMs are non-volatile memories. This means that they save their internal data while not powered. In order for memories to be electrically erasable, the insulating material surrounding the floating-gate must be much thinner than those of the EPROMS. This makes the number of times EEPROMs can be reprogrammed much less than that of EPROMs and in the order of 10 to 20,000. Writing into a byte of an EEPROM is about 500 times slower than reading from it.

5.1.6.5 Flash Memory. Flash memories are large EEPROMs that are partitioned into smaller fixed-size blocks that can independently be erased. Internal to a system, flash memories are used for saving system configurations. They are used in digital cameras for storing pictures. As external devices, they are used for temporary storage of data that can be rapidly retrieved.

Various forms of ROM are available in various sizes and packages. The popular 27xxx series EPROMs come in packages that are organized as byte addressable memories. For example, the 27256 EPROM has 256K bits of memory that are arranged into 32K bytes. This package is shown in Figure 5.11.

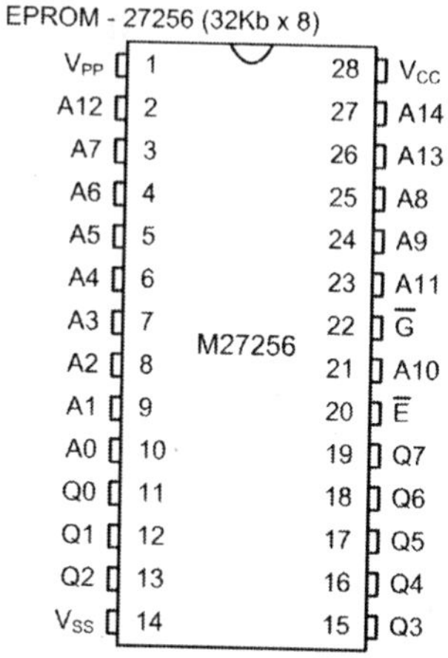

EPROM - 27256 (32Kb x 8)

		M27256		
V_{PP}	1		28	V_{CC}
A12	2		27	A14
A7	3		26	A13
A6	4		25	A8
A5	5		24	A9
A4	6		23	A11
A3	7		22	$\overline{G}$
A2	8		21	A10
A1	9		20	$\overline{E}$
A0	10		19	Q7
Q0	11		18	Q6
Q1	12		17	Q5
Q2	13		16	Q4
V_{SS}	14		15	Q3

Figure 5.11 27256 EPROM

The 27256 EPROM has a Vpp pin that is used for the supply input during read-only operations and is used for applying programming voltage during the programming phase. The 15 address lines address 256K of 8-bit data that are read on to $O7$ to $O0$ outputs. Active low CS and OE are for three-state control of the outputs and are used for cascading EPROMs and/or output bussing.

EPROMs can be cascaded for word length expansion, address space expansion or both. For example, a 1Meg 16-bit word memory can be formed by use of a four by two array of 27256s.

5.2 Programmable Logic Arrays

The price we are paying for the high degree of flexibility of ROMs is the large area occupied by the AND-plane that forms every minterm of the inputs of the ROM. PLAs (Programmable Logic Arrays) constitutes an alternative with less flexibility and less use of silicon. For this discussion we look at ROMs as logic circuits as done in the earlier parts of Section 5.1, and not the memory view of the later parts of this section.

For illustrating the PLA structure, we use the 3-input, 4-output example circuit of Figure 5.1. The AND-OR implementation of this circuit that is shown in Figure 5.2 led to the ROM structure of Figure 5.8, in which minterms generated in the AND-plane are used for function outputs in the OR-plane.

An easy step to reduce the area used by the circuit of Figure 5.8 is to implement only those minterms that are actually used. In this example, since minterm 7 is never used, the last row of the array can be completely eliminated. In large ROM structures, there will be a much larger percentage of unused minterms that can be eliminated. Furthermore, if instead of using minterms, we minimize our output functions and only implement the regained product terms we will be able to save even more rows of the logic array.

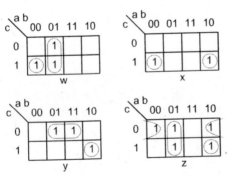

Figure 5.12 Minimizing Circuit of Figure 5.1

Figure 5.12 shows Karnaugh maps for minimization of w, x, y and z outputs of table of Figure 5.1. In this minimization sharing product terms between various outputs is particularly emphasized.

The resulting Boolean expressions for the outputs of circuit described by the tables of Figure 5.12 are shown in Figure 5.13. Common product terms in these expressions are vertically aligned.

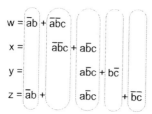

$$w = \overline{ab} + \overline{a}bc$$
$$x = \overline{a}\overline{b}c + a\overline{b}c$$
$$y = a\overline{b}c + b\overline{c}$$
$$z = \overline{ab} + a\overline{b}c + \overline{b}c$$

Figure 5.13 Minimized Boolean Expressions

Implementation of w, x, y and z functions of a, b and c inputs in an array format using minimized expressions of Figure 5.13 is shown in Figure 5.14. This array uses five rows that correspond to the product terms of the four output functions. Comparing this with Figure 5.8, we can see that we are using fewer number of rows by generating only the product terms that are needed and not every minterm.

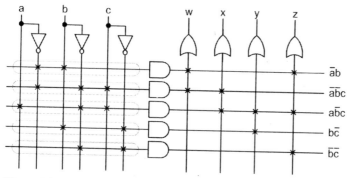

Figure 5.14 PLA Implementation

The price we are paying for the area gained in the PLA implementation of Figure 5.14 is that we now have to program both AND and OR planes. In Figure 5.14 we use X's in both planes where in Figure 5.8 dots are used in the fixed AND-plane and X's in the programmable OR-plane.

While ROM structures are used for general purpose configurable packages, PLAs are mainly used as structured array of hardware for implementing on-chip logic. A ROM is an array logic with fixed AND-plane and programmable OR-plane, and PLA is an array with programmable AND-plane and programmable OR-plane.

A configurable array logic that sits between a PLA and a ROM is one with a programmable AND-plane and a fixed OR-plane. This logic structure was first introduced by Monilitic Memories Inc. (MMI) in the late 1970s and because of its similarity to PLA was retuned to PAL or Programmable Array Logic.

The rationale behind PALs is that outputs of a large logic function generally use a limited number of product terms and the capability of being able to use all product terms for all function outputs is in most cases not utilized. Fixing the number of product terms for the circuit outputs significantly improves the speed of PALs.

5.2.1 PAL Logic Structure

In order to illustrate the logical organization of PALs, we go back to our 3-input, 4-output example of Figure 5.1. Figure 5.15 shows PAL implementation of this circuit. This circuit uses w, x, y and z expressions shown in Figure 5.13. Recall that these expressions are minimal realizations for the outputs of our example circuit and are resulted from the k-maps of Figure 5.12.

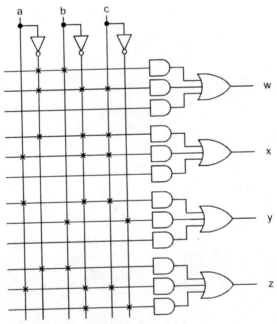

Figure 5.15 PAL Implementation

The PAL structure of Figure 5.15 has a programmable AND-plane and a fixed OR-plane. Product terms are formed in the AND-plane and three such terms are used as OR gate inputs in the OR-plane. This structure allows a maximum of three product terms per output.

Implementing expressions of Figure 5.13 is done by programming fuses of the AND-plane of the PAL. The z output uses all three available product terms and all other outputs use only two.

5.2.2 Product Term Expansion

The limitation on the number of product terms per output in a PAL device can be overcome by providing feedbacks from PAL outputs back into the AND-plane. These feedbacks are used in the AND-plane just like regular inputs of the PAL. Such a feedback allows ORing a subset of product terms of a function to be fed back into the array to further be ORed with the remaining product terms of the function.

Consider for example, PAL implementation of expression w shown below:

$$w = \overline{a} \cdot b \cdot \overline{c} + \overline{a} \cdot \overline{b} \cdot c + a \cdot \overline{b} + a \cdot b \cdot c$$

Let us assume that this function is to be implemented in a 3-input PAL with three product terms per output and with outputs feeding back into the AND-plane, as shown in Figure 5.16.

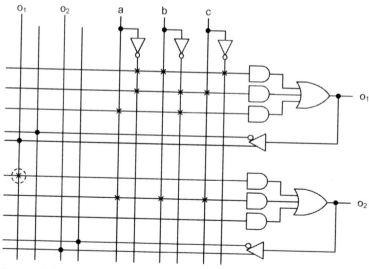

Figure 5.16 A PAL with Product Term Expandability

The partial PAL shown in this figure allows any of its outputs to be used as a circuit primary output or as a partial sum-of-products to be completed by ORing more product terms to it. For implementation of expression w, the first three product terms are generated on the o_1. The structure shown does not allow the last product term $(a \cdot b \cdot c)$ to be ORed on the same output. Therefore, the feedback from this output is used as an input into the next group of product terms. The circled X connection in this figure causes o_1 to be used as an input into the o_2 group. The last product terms $(a.b.c)$ is generated in the AND-plane driving the o_2 output and is ORed with o_1 using the OR-gate of the o_2 output. Expression w is generated on o_2. Note that the feedback of o_2 back into the AND-plane does exist, but not utilized.

5.2.3 Three-State Outputs

A further improvement to the original PAL structure of Figure 5.15 is done by adding three-state controls to its outputs as shown in the partial structure of Figure 5.17.

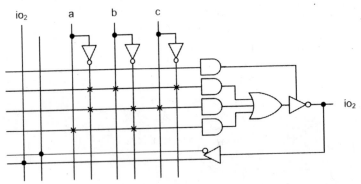

Figure 5.17 PAL Structure with Three Output Control

In addition to the feedback from the output, this structure has two more advantages. First, the pin used as output or partial sum-of-products terms can also be used as input by turning off the three-state gate that drives it. Note that the lines used for feeding back outputs into the AND-plane in Figure 5.16, become connections from the io_2 input into the AND-plane. The second advantage of this structure is that when $io2$ is used as output it becomes a three-state output that is controlled by a programmable product term.

Instead of using a three-state inverting buffer, an XOR gate with three-state output and a fusable input (see Figure 5.18) provides output polarity control when the bi-directional io_2 port is used as output

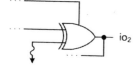

Figure 5.18 Output Inversion Control

5.2.4 Registered Outputs

A major advantage of PALs over PLAs and ROMs is the capability of incorporating registers into the logic structure. Where registers can only be added to the latter two structures on their inputs and outputs, registers added to PAL arrays become more integrated in the input and output of the PAL logic.

As an example structure, consider the registered output of Figure 5.19. The input/output shown can be used as a registered output with three-state, as a two-state output, as a registered feedback into the logic array, or as an input into the AND-plane.

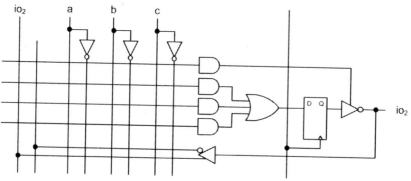

Figure 5.19 Output Inversion Control

A further enhancement to this structure provides logic for bypassing the output flip-flop when its corresponding I/O pin is being used as output. This way, PAL outputs can be programmed as registered or combinational pins.

Other enhancements to the register option include the use of asynchronous control signals for the flip-flop, direct feedback from the flip-flop into the array, and providing a programmable logic function for the flip-flop output and the feedback line.

5.2.5 Commercial Parts

PAL is a trademark of American Micro Devices Inc. More generically, these devices are referred to as PLDs or programmable logic devices.

A variation of the original PAL or a PLD that is somewhat different from a PAL is GAL (Generic Array Logic). The inventor of GAL is the Lattice Semiconductor Inc. GALs are electrically erasable; otherwise have a similar logical structure to PALs. By ability to bypass output flip-flops, GALs can be configured as combinational or sequential circuits. To familiarize readers with some actual parts, we discuss one of Altera's PLD devices.

5.2.5.1 Altera Classic EPLD Family.

Altera Corporation's line of PLDs is its Classic EPLD Family. These devices are EPROM based and have 300 to 900 usable gates depending on the specific part. These parts come in 24 to 68 pin packages and are available in dual in-line package (DIP), plastic J-lead chip carrier (PLCC), pin-grid array (PGA), and small-outline integrated circuit (SOIC) packages. The group of product terms that are ORed together are referred to as a Macrocell, and the number of Macrocells varies between 16 and 48 depending on the device. Each Macrocell has a programmable register that can be programmed as a D, T, JK and SR flip-flop with individual clear and clock controls.

These devices are fabricated on CMOS technology and are TTL compatible. They can be used with other manufacturers PAL and GAL parts. The EP1810 is the largest of these devices that has 900 usable gates, 48 Macrocells, and a maximum of 64 I/O pins. Pin-to-pin logic delay of this part is 20 ns and it can operate with a maximum frequency of 50 MHz. The architecture of this and other Altera's Classic EPLDs includes Macrocells, programmable registers, output enable or clock select, and a feedback select. The brief description of this device that follows is useful for understanding architecture of this category of programmable devices.

Macrocells. Classic macrocells, shown in Figure 5.20, can be individually configured for both sequential and combinatorial logic operation. Eight product terms form a programmable-AND array that feeds an OR gate for combinatorial logic implementation. An additional product term is used for asynchronous clear control of the internal register; another product term implements either an output enable or a logic-array-generated clock. Inputs to the programmable-AND array come from both the true and complement signals of the dedicated inputs, feedbacks from I/O pins that are configured as inputs, and feedbacks from macrocell outputs. Signals from dedicated inputs are globally routed and can feed the inputs of all device macrocells. The feedback multiplexer controls the routing of feedback signals from macrocells and from I/O pins.

The eight product terms of the programmable-AND array feed the 8-input OR gate, which then feeds one input to an XOR gate. The other input to the XOR gate is connected to a programmable bit that allows the array output to be inverted. This gate is used to implement either active-high or active-low logic, or De Morgan's inversion to reduce the number of product terms needed to implement a function.

Programmable Registers. To implement registered functions, each macrocell register can be individually programmed for D, T, JK, or SR operation. If necessary, the register can be bypassed for combinatorial operation. Registers have an individual asynchronous clear function that is controlled by a dedicated product term. These registers are cleared automatically during power-up. In addition, macrocell registers can be individually clocked by either a global clock or any input or feedback path to the AND array. Altera's proprietary programmable I/O architecture allows the designer to program output and feedback paths for combinatorial or registered operation in both active-high and active-low modes.

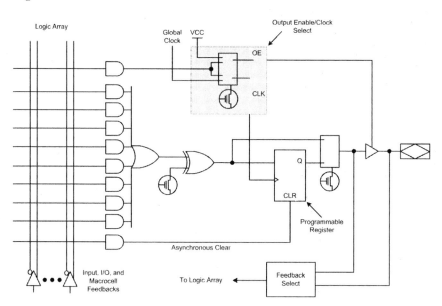

Figure 5.20 Altera's Classic Mecrocell

Output Enable / Clock Select. The box shown in the upper part of Figure 5.20 allows two modes of operations for output and clocking of a Classic macrocell. The mode is controlled by a single programmable bit that can be individually configured for each macrocell. Mode selection logic allows a product term, global clock, and VCC to rive flip-flop clock and output-enable inputs.

In Mode 0, the tri-state output buffer is controlled by a single product term, and the macrocell flip-flop is clocked by its global clock input signal. In Mode 1, the output enable buffer is always enabled, and the macrocell register can be triggered by an array clock signal generated by a product term. This mode allows registers to be individually clocked by any signal on the AND array. This product-term-controlled clock configuration also supports gated clock structures.

Feedback Select. Each macrocell in a Classic device provides feedback selection that is controlled by the feedback multiplexer. This feedback selction logic, shown at the bottom of Figure 5.20 allows the designer to feed either the macrocell output or the I/O pin input associated with the macrocell back into the AND array. The macrocell output can be either the Q output of the programmable register or the combinatorial output of the macrocell.

5.3 Complex Programmable Logic Devices

The next step up in the evolution and complexity of programmable devices is the CPLD, or Complex PLD. Extending PLDs by making their AND-plane larger and having more macrocells in order to be able to implement larger and more complex logic circuits would face difficulties in speed and chip area utilization. Therefore, instead of simply making these structures larger, CPLDs are created that consist of multiple PLDs with programmable wiring channels between the PLDs. Figure 5.21 shows the general block diagram of a CPLD.

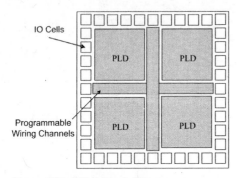

Figure 5.21 CPLD Block Diagram

The approach taken by different manufacturers for implementation of their CPLDs are different. As a typical CPLD we discuss Altera's EPM7128S that is a member of this manufacturer's MAX 7000 Programmable Device Family.

5.3.1 Altera's MAX 7000S CPLD

A member of Altera's MAX 7000 S-series is the EPM7128S CPLD. This is an EEPROM-based programmable logic device with in-system programmability feature through its JTAG interface. Logic densities for the MAX family of CPLDs range from 600 to 5,000 usable gates and the EPM7128S is a mid-rage CPLD in this family with 2,500 usable gates. Note that these figures are 2 to 4 times larger than those of the PLDs from Altera.

The EPM7128s is available in plastic J-lead chip carrier (PLCC), ceramic pin-grid array (PGA), plastic quad flat pack (PQFP), power quad flat pack (RQFP), and 1.0-mm thin quad flat pack (TQFP) packages. The maximum frequency of operation of this part is 147.1 MHz, and it has a propagation delay of 6 ns. This part can operate with 3.3 V or 5.0 V.

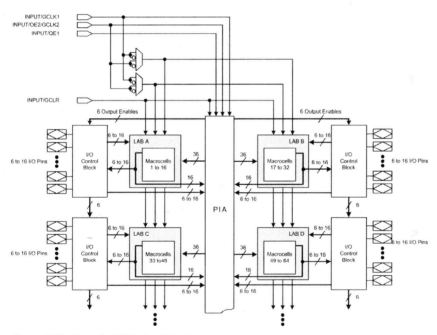

Figure 5.22 Altera's CPLD Architecture

This CPLD has 8 PLDs that are referred to as Logic Array Blocks (LABs). Each LAB has 16 macrocells, making the total number of its macrocells 128. The LABs are linked by a wiring channel that is referred to as the Programmable Interconnect Array (PIA). The macrocells include hardware for expanding product terms by linking several macrocells. The overall architecture of this part is shown in Figure 5.22. In what follows, blocks shown in this figure will be briefly described.

5.3.1.1 Logic Array Blocks. The EPM7128S has 8 LABs (4 shown in Figure 5.22) that are linked by the PIA global wiring channel. In general, a LAB has the same structure as a PLD described in the previous section. Multiple LABs are linked together via the PIA global bus that is fed by all dedicated inputs, I/O pins, and macrocells. Signals included in a LAB are 36 signals from the PIA that are used for general logic inputs, global controls that are used for secondary register functions, and direct input paths from I/O pins to the registers. Macrocells constitute substructures of LABs.

5.3.1.2 Programmable Interconnect Array. Logic is routed between LABs via the programmable interconnect array (PIA) shown in the center of Figure 5.22. This global bus is a programmable path that connects any signal source to any destination on the device. All MAX 7000 dedicated inputs, I/O pins, and macrocell outputs feed the PIA, which makes the signals available throughout the entire device. Only the signals required by each LAB are actually routed from the PIA into the LAB.

5.3.1.3 I/O Control Blocks. The I/O control block, shown on the sides of Figure 5.22, allows each I/O pin to be individually configured for input, output, or bidirectional operation. All I/O pins have a tri-state buffer that is individually controlled by one of the global output enable signals or directly connected to ground or VCC.

5.4 Field Programmable Gate Arrays

A more advanced programmable logic than the CPLD is the Field Programmable Gate Array (FPGA). An FPGA is more flexible than CPLD, allows more complex logic implementations, and can be used for implementation of digital circuits that use equivalent of several Million logic gates.

An FPGA is like a CPLD except that its logic blocks that are linked by wiring channels are much smaller than those of a CPLD and there are far more such logic blocks than there are in a CPLD. FPGA logic blocks consist of smaller logic elements. A logic element has only one flip-flop that is individually configured and controlled. Logic complexity of a logic element is only about 10 to 20 equivalent gates. A further enhancement in the structure of FPGAs is the addition of memory blocks that can be configured as a general purpose RAM. Figure 5.23 shows the general structure of an FPGA.

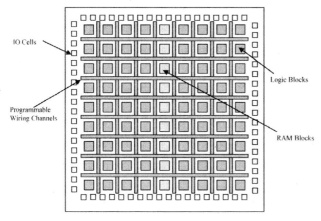

Figure 5.23 FPGA General Structure

As shown in Figure 5.23, an FPGA is an array of many logic blocks that are linked by horizontal and vertical wiring channels. FPGA RAM blocks can also be used for logic implementation or they can be configured to form memories of various word sizes and address space. Linking of logic blocks with the I/O cells and with the memories are done through wiring channels. Within logic blocks, smaller logic elements are linked by local wires.

FPGAs from different manufacturers vary in routing mechanisms, logic blocks, memories and I/O pin capabilities. As a typical FPGA, we will discuss Altera's EPF10K70 that is a member of this manufacturer's FLEX 10K Embedded Programmable Logic Device Family.

5.4.1 Altera's FLEX 10K FPGA

A member of Altera's FLEX 10K family is the EPF10K70 FPGA. This is a SRAM-based FPGA that can be programmed through its JTAG interface. This interface can also be used for FPGAs logic boundary-scan test. Typical gates of this family of FPGAs range from 10,000 to 250,000. This family has up to 40,960 RAM bits that can be used without reducing logic capacity.

Altera's FLEX 10K devices are based on reconfigurable CMOS SRAM elements, the Flexible Logic Element MatriX (FLEX) architecture is geared for implementation of common gate array functions. These devices are reconfigurable and can be configured on the board for the specific functionality required. At system power-up, they are configured with data stored in an Altera serial configuration device or provided by a system controller. Altera offers the EPC1, EPC2, EPC16, and EPC1441 configuration devices, which configure FLEX

10K devices via a serial data stream. Configuration data can also be downloaded from system RAM or from Altera's BitBlaster serial download cable, ByteBlaster parallel download cable, or USBBlaster USB port download cable. After a FLEX 10K device has been configured, it can be reconfigured in-circuit by resetting the device and loading new data. Reconfiguration requires less than 320 ms. FLEX 10K devices contain an interface that permits microprocessors to configure FLEX 10K devices serially or in parallel, and synchronously or asynchronously. The interface also enables microprocessors to treat a FLEX 10K device as memory and configure the device by writing to a virtual memory location.

The EPF10K70 has a total of 70,000 typical gates that include logic and RAM. There are a total of 118,000 system gates. The entire array contains 468 Logic Array Blocks (LABs) that are arranged in 52 columns and 9 rows. The LABs are the "Logic Blocks" shown in Figure 5.23. Each LAB has 8 Logic Elements (LEs), making the total number of its LEs 3,744. In the middle of the FPGA chip, a column of 9 Embedded Array Blocks (EABs), each of which has 2,048 bits, form the 18,432 RAM bits of this FPGA. The EPF10K70 has 358 user I/O pins.

5.4.1.1 FLEX 10K Blocks. The block diagram of a FLEX 10K is shown in Figure 5.24. Each group of LEs is combined into an LAB; LABs are arranged into rows and columns. Each row also contains a single EAB. The LABs and EABs are interconnected by the FastTrack Interconnect. IOEs are located at the end of each row and column of the FastTrack Interconnect.

FLEX 10K devices provide six dedicated inputs that drive the flip-flops' control inputs to ensure the efficient distribution of high-speed, low-skew (less than 1.5 ns) control signals. These signals use dedicated routing channels that provide shorter delays and lower skews than the FastTrack Interconnect. Four of the dedicated inputs drive four global signals. These four global signals can also be driven by internal logic, providing an ideal solution for a clock divider or an internally generated asynchronous clear signal that clears many registers in the device.

Signal interconnections within FLEX 10K devices and to and from device pins are provided by the FastTrack Interconnect, a series of fast, continuous row and column channels that run the entire length and width of the device.

Each I/O pin is fed by an I/O element (IOE) located at the end of each row and column of the FastTrack Interconnect. Each IOE contains a bidirectional I/O buffer and a flip-flop that can be used as either an output or input register to feed input, output, or bidirectional signals. When used with a dedicated clock pin, these registers provide exceptional performance. As inputs, they provide setup times as low as 1.6 ns and hold times of 0 ns; as outputs, these registers provide clock-to-output times as low as 5.3 ns. IOEs provide a variety of features, such as JTAG BST support, slew-rate control, tri-state buffers, and open-drain outputs.

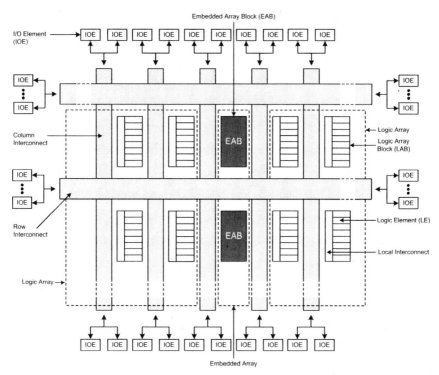

Figure 5.24 FLEX 10K Block Diagram

5.4.1.2 Embedded Array Block.

Each device contains an embedded array to implement memory and specialized logic functions, and a logic array to implement general logic. The embedded array consists of a series of EABs. When implementing memory functions, each EAB provides 2,048 bits, which can be used to create RAM, ROM, dual-port RAM, or first-in first-out (FIFO) functions. When implementing logic, each EAB can contribute 100 to 600 gates towards complex logic functions, such as multipliers, microcontrollers, state machines, and DSP functions. EABs can be used independently, or multiple EABs can be combined to implement larger functions.

Logic functions are implemented by programming the EAB with a read-only pattern during configuration, creating a large look-up table. With tables, combinatorial functions are implemented by looking up the results, rather than by computing them. This implementation of combinatorial functions can be faster than using algorithms implemented in general logic, a performance advantage that is further enhanced by the fast access times of EABs. The large capacity of EABs enables designers to implement complex functions in one logic level. For example, a single EAB can implement a 4×4 multiplier with eight inputs and eight outputs.

EABs can be used to implement synchronous RAM that generates its own WE signal and is self-timed with respect to the global clock. A circuit using the EAB's self-timed RAM need only meet the setup and hold time specifications of the global clock. When used as RAM, each EAB can be configured in any of the following sizes: 256×8, 512×4, 1,024×2, or 2,048×1. Larger blocks of RAM are created by combining multiple EABs.

5.4.1.3 Logic Array Block. Referring to Figure 5.24, the logic array of FLEX 10K consists of logic array blocks (LABs). Each LAB contains eight LEs and a local interconnect. An LE consists of a 4-input look-up table (LUT), a programmable flip-flop, and dedicated signal paths for carry and cascade functions. Each LAB represents about 96 usable gates of logic.

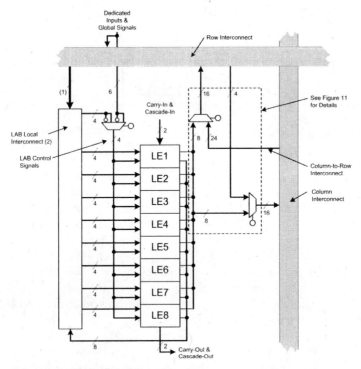

Figure 5.25 FLEX 10K LAB Architecture

Each LAB (see Figure 5.25) provides four control signals with programmable inversion that can be used in all eight LEs. Two of these signals can be used as clocks; the other two can be used for clear/preset control. The LAB clocks can be driven by the dedicated clock input pins, global signals, I/O signals, or internal signals via the LAB local interconnect. The LAB preset and clear control signals can be driven by the global signals, I/O signals, or internal signals via the LAB local interconnect. The global control signals are typically used for global clock, clear, or preset signals because they provide asynchronous control with very low skew across the device. If logic is required on a control signal, it can be generated in one or more LEs in any LAB and driven into the local interconnect of the target LAB. In addition, the global control signals can be generated from LE outputs.

Logic Element. The LE eight of which are contained in a LAB, as shown in Figure 5.25, is the smallest unit of logic in the FLEX 10K architecture. Each LE contains a four-input LUT, which is a function generator that can compute any function of four variables. In addition, each LE contains a programmable flip-flop with a synchronous enable, a carry chain, and a cascade chain. Each LE drives both the local and the FastTrack Interconnect. See Figure 5.26.

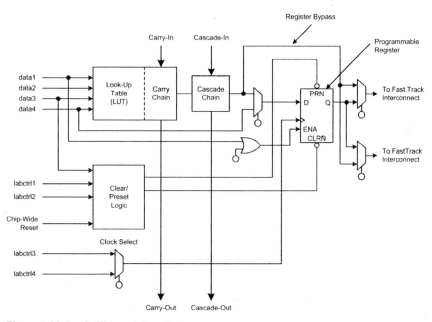

Figure 5.26 Logic Element Structure

The programmable flip-flop in the LE can be configured for D, T, JK, or SR operation. The clock, clear, and preset control signals on the flip-flop can be driven by global signals, general-purpose I/O pins, or any internal logic. For combinatorial functions, the flip-flop is bypassed and the output of the LUT drives the output of the LE.

The LE has two outputs that drive the interconnect; one drives the local interconnect and the other drives either the row or column FastTrack Interconnect. The two outputs can be controlled independently. For example, the LUT can drive one output while the register drives the other output. This feature, called register packing, can improve LE utilization because the register and the LUT can be used for unrelated functions.

The FLEX 10K architecture provides two types of dedicated high-speed data paths that connect adjacent LEs without using local interconnect paths: carry chains and cascade chains. The carry chain supports high-speed counters and adders; the cascade chain implements wide-input functions with minimum delay. Carry and cascade chains connect all LEs in an LAB and all LABs in the same row. Intensive use of carry and cascade chains can reduce routing flexibility. Therefore, the use of these chains should be limited to speed-critical portions of a design.

5.4.1.4 FastTrack Interconnect.

In the FLEX 10K architecture, connections between LEs and device I/O pins are provided by the FastTrack Interconnect. This is a series of continuous horizontal and vertical routing channels that traverse the device. This global routing structure provides predictable performance, even in complex designs.

5.4.1.5 I/O Element.

An I/O element (IOE) of FLEX 10K (see the top-level architecture of Figure 5.24) contains a bidirectional I/O buffer and a register that can be used either as an input register for external data that requires a fast setup time, or as an output register for data that requires fast clock-to-output performance. In some cases, using an LE register for an input register will result in a faster setup time than using an IOE register. IOEs can be used as input, output, or bidirectional pins. For bidirectional registered I/O implementation, the output register should be in the IOE, and the data input and output enable register should be LE registers placed adjacent to the bidirectional pin.

When an IOE connected to a row (as shown in Figure 5.27), is used as an input signal it can drive two separate row channels. The signal is accessible by all LEs within that row. When such an IOE is used as an output, the signal is driven by a multiplexer that selects a signal from the row channels. Up to eight IOEs connect to each side of each row channel.

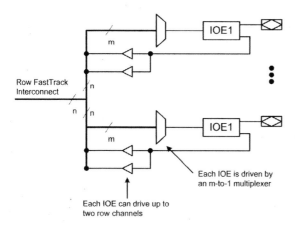

Row FastTrack
Interconnect

Each IOE is driven by
an m-to-1 multiplexer

Each IOE can drive up to
two row channels

Figure 5.27 FLEX 10K Row-to-IOE Connections

Connections of columns to IOEs are similar to those of rows, as shown in Figure 5.27. When an IOE connected to a column is used as an input, it can drive up to two separate column channels. When an IOE is used as an output, the signal is driven by a multiplexer that selects a signal from the column channels. Two IOEs connect to each side of the column channels. Each IOE can be driven by column channels via a multiplexer. The set of column channels that each IOE can access is different for each IOE.

In this section we have shown FPGA structures by using Altera's EPF10K70 that is a member of the FLEX 10K family as an example. The focus of the above discussion was on the description of the main components of this programmable device. Some detailed logic structures and many of the timing and logical configuration details have been eliminated. The *"FLEX 10K Embedded Programmable Logic Device Family"* datasheet is a detailed document about this and other FLEX 10K members. Interested readers are encouraged to study this document for advanced features and details of logical configurations of this FPGA family.

5.4.2 Altera's Cyclone FPGA

The Cyclone field programmable gate array family is based on a 1.5-V, 0.13-μm, all-layer copper SRAM process, with densities up to 20,060 logic elements (LEs) and up to 288 Kbits of RAM. With features like phase-locked-loops (PLLs) for clocking and a dedicated double data rate (DDR) interface to meet DDR SDRAM and fast cycle RAM (FCRAM) memory requirements, Cyclone devices are a cost-effective solution for data-path applications. The Cyclone device family offers a range of logic-elements, memory bits, phase-lock-loops, and IO pins. Table 5.1 shows Cyclone devices and their features.

Cyclone devices are available in quad flat pack (QFP) and space-saving FineLine BGA packages.

Table 5.1 Cyclone Device Features

Feature	EP1C3	EP1C4	EP1C6	EP1C12	EP1C20
LEs	2,910	4,000	5,980	12,060	20,060
M4K RAM blocks (128 × 36 bits)	13	17	20	52	64
Total RAM bits	59,904	78,336	92,160	239,616	294,912
PLLs	1	2	2	2	2
Maximum user I/O pins *(1)*	104	301	185	249	301

Cyclone devices contain a two-dimensional row- and column-based architecture to implement custom logic. Column and row interconnects of varying speeds provide signal interconnects between LABs and embedded memory blocks.

The logic array consists of LABs, with 10 LEs in each LAB. An LE is a small unit of logic providing efficient implementation of user logic functions. LABs are grouped into rows and columns across the device. Various Cyclone devices have between 2,910 and 20,060 LEs.

M4K RAM blocks are true dual-port memory blocks with 4K bits of memory plus parity (4,608 bits). These blocks provide dedicated true dual-port, simple dual-port, or single-port memory up to 36-bits wide at up to 250 MHz. These blocks are grouped into columns across the device in between certain LABs. Cyclone devices offer between 60 and 288 Kbits of embedded RAM.

Each Cyclone device I/O pin is fed by an I/O element (IOE) located at the ends of LAB rows and columns around the periphery of the device. I/O pins support various single-ended and differential I/O standards, such as the 66- and 33-MHz, 64- and 32-bit PCI standard and the LVDS I/O standard at up to 640 Mbps. Each IOE contains a bidirectional I/O buffer and three registers for registering input, output, and output-enable signals. Dual-purpose DQS, DQ, and DM pins along with delay chains (used to phase-align DDR signals) provide interface support with external memory devices such as DDR SDRAM, and FCRAM devices at up to 133 MHz (266 Mbps).

Cyclone devices provide a global clock network and up to two PLLs. The global clock network consists of eight global clock lines that drive throughout the entire device. The global clock network can provide clocks for all resources within the device, such as IOEs, LEs, and memory blocks. The global clock lines can also be used for control signals. Cyclone PLLs provide general-purpose clocking with clock multiplication and phase shifting as well as external outputs for high-speed differential I/O support.

Figure 5.28 shows the general outline of a typical Cyclone device (EP1C12). The number of M4K RAM blocks, PLLs, rows, and columns vary per device, and are shown in Table 5.2.

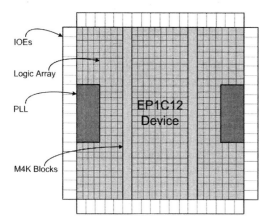

Figure 5.28 General Outline of a Cyclone Device

Table 5.2 Cyclone Device Resources

Device	M4K RAM		PLLs	LAB Columns	LAB Rows
	Columns	Blocks			
EP1C3	1	13	1	24	13
EP1C4	1	17	2	26	17
EP1C6	1	20	2	32	20
EP1C12	2	52	2	48	26
EP1C20	2	64	2	64	32

5.4.2.1 Logic Array Blocks.

A Cyclone LAB structure and its surrounding environment are shown in Figure 5.29. Each LAB consists of 10 LEs, LE carry chains, LAB control signals, a local interconnect, look-up table (LUT) chain, and register chain connection lines. The local interconnect transfers signals between LEs in the same LAB. LUT chain connections transfer the output of one LE's LUT to the adjacent LE for fast sequential LUT connections within the same LAB. Register chain connections transfer the output of one LE's register to the adjacent LE's register within an LAB. The Quartus® II Compiler places associated logic within an LAB or adjacent LABs, allowing the use of local, LUT chain, and register chain connections for performance and area efficiency.

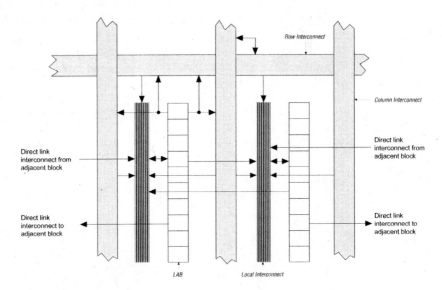

Figure 5.29 Cyclone LAB Structure

LAB Interconnects. The LAB local interconnect can drive LEs within the same LAB. The LAB local interconnect is driven by column and row interconnects and LE outputs within the same LAB. Neighboring LABs, PLLs, and M4K RAM blocks from the left and right can also drive an LAB's local interconnect through the direct link connection. The direct link connection feature minimizes the use of row and column interconnects, providing higher performance and flexibility. Each LE can drive 30 other LEs through fast local and direct link interconnects.

LAB Control Signals. Each LAB contains dedicated logic for driving control signals to its LEs. The control signals include two clocks, two clock enables, two asynchronous clears, synchronous clear, asynchronous preset/load, synchronous load, and add/subtract control signals. This gives a maximum of 10 control signals at a time.

Each LAB can use two clocks and two clock enable signals. Each LAB's clock and clock enable signals are linked. For example, any LE in a particular LAB using the *labclk1* signal will also use *labclkena1*. If the LAB uses both the rising and falling edges of a clock, it also uses both LAB-wide clock signals. De-asserting the clock enable signal will turn off the LAB-wide clock.

Each LAB can use two asynchronous clear signals and an asynchronous load/preset signal. The asynchronous load acts as a preset when the asynchronous load data input is tied high.

With the LAB-wide *addnsub* control signal, a single LE can implement a one-bit adder and subtractor. This saves LE resources and improves performance for logic functions such as DSP correlators and signed multipliers that alternate between addition and subtraction depending on data.

The LAB row clocks [5..0] and LAB local interconnect generate the LAB wide control signals. The MultiTrack interconnect's inherent low skew allows clock and control signal distribution in addition to data. Figure 5.30 shows the LAB control signal generation circuit.

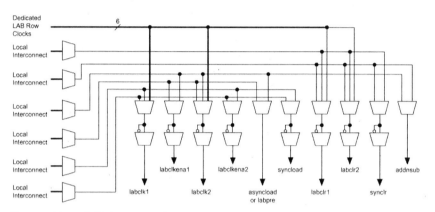

Figure 5.30 LAB-Wide Control Signals

5.4.2.2 Logic Elements. The smallest unit of logic in the Cyclone architecture, the LE (Figure 5.31), is compact and provides advanced features with efficient logic utilization. Each LE contains a four-input LUT, which is a function generator that can implement any function of four variables. In addition, each LE contains a programmable register and carry chain with carry select capability. A single LE also supports dynamic single bit addition or subtraction mode selectable by an LAB-wide control signal. Each LE drives all types of interconnects: local, row, column, LUT chain, register chain, and direct link interconnects.

Each LE's programmable register can be configured for D, T, JK, or SR operation. Each register has data, true asynchronous load data, clock, clock enable, clear, and asynchronous load/preset inputs. Global signals, general-purpose I/O pins, or any internal logic can drive the register's clock and clear control signals. Either general-purpose I/O pins or internal logic can drive the clock enable, preset, asynchronous load, and asynchronous data. The asynchronous load data input comes from the *data3* input of the LE. For combinatorial functions, the LUT output bypasses the register and drives directly to the LE outputs.

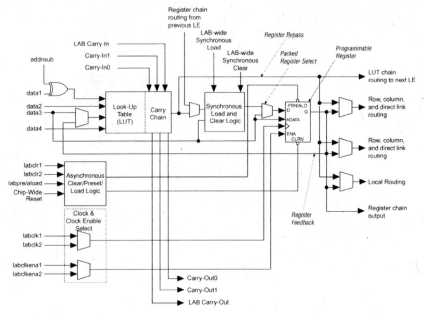

Figure 5.31 Cyclone LE

Each LE has three outputs that drive the local, row, and column routing resources. The LUT or register output can drive these three outputs independently. Two LE outputs drive column or row and direct link routing connections and one drives local interconnect resources. This allows the LUT to drive one output while the register drives another output. This feature, called register packing, improves device utilization because the device can use the register and the LUT for unrelated functions. Another special packing mode allows the register output to feed back into the LUT of the same LE so that the register is packed with its own fan-out LUT. This provides another mechanism for improved fitting. The LE can also drive out registered and unregistered versions of the LUT output.

LUT Chain & Register Chain. In addition to the three general routing outputs, the LEs within an LAB have LUT chain and register chain outputs. LUT chain connections allow LUTs within the same LAB to cascade together for wide input functions. Register chain outputs allow registers within the same LAB to cascade together. The register chain output allows an LAB to use LUTs for a single combinatorial function and the registers to be used for an unrelated shift register implementation. These resources speed up connections between LABs while saving local interconnect resources.

addnsub Signal. The LE's dynamic adder/subtractor feature saves logic resources by using one set of LEs to implement both an adder and a subtractor. This feature is controlled by the LAB-wide control signal addnsub. The addnsub signal sets the LAB to perform either A + B or A – B. The LUT computes addition; subtraction is computed by adding the two's complement of the intended subtractor. The LAB-wide signal converts to two's complement by inverting the B bits within the LAB and setting carry-in = 1 to add one to the least significant bit (LSB). The LSB of an adder/subtractor must be placed in the first LE of the LAB, where the LAB-wide *addnsub* signal automatically sets the carry-in to 1.

LE Operating Modes. The Cyclone LE can operate in normal or dynamic arithmetic operating modes. Each mode uses LE resources differently. In each mode, eight available inputs to the LE – the four data inputs from the LAB local interconnect, *carry-in0* and *carry-in1* from the previous LE, the LAB carry-in from the previous carry-chain LAB, and the register chain connection – are directed to different destinations to implement the desired logic function. LAB-wide signals provide clock, asynchronous clear, asynchronous preset/load, synchronous clear, synchronous load, and clock enable control for the register. These LAB-wide signals are available in all LE modes. The addnsub control signal is allowed in arithmetic mode.

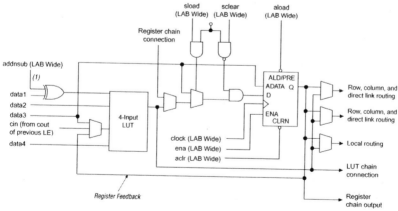

Figure 5.32 LE in Normal Mode

The normal mode is suitable for general logic applications and combinatorial functions. In normal mode, four data inputs from the LAB local interconnect are inputs to a four-input LUT (see Figure

5.32). Each LE can use LUT chain connections to drive its combinatorial output directly to the next LE in the LAB. Asynchronous load data for the register comes from the *data3* input of the LE. LEs in normal mode support packed registers. The *addnsub* signal in Figure 5.32 is only allowed in the normal mode if the LE is at the end of an adder/subtractor chain.

The dynamic arithmetic mode is ideal for implementing adders, counters, accumulators, wide parity functions, comparators, and other iterative logic functions. An LE in dynamic arithmetic mode uses four 2-input LUTs configurable as a dynamic adder/subtractor. The first two 2-input LUTs compute two summations based on a possible carry-in of 1 or 0; the other two LUTs generate carry outputs for the two chains of the carry select circuitry. As shown in Figure 5.33, the LAB carry-in signal selects either the *carry-in0* or *carry-in1* chain. The selected chain's logic level in turn determines which parallel sum is generated as a combinatorial or registered output. For example, when implementing an adder, the sum output is the selection of two possible calculated sums:

data1 + data2 + carry-in0

or

data1 + data2 + carry-in1

The other two LUTs use the *data1* and *data2* signals to generate two possible carry-out signals - one for a carry of **1** and the other for a carry of **0**. The *carry-in0* signal acts as the carry select for the *carry-out0* output and *carry-in1* acts as the carry select for the *carry-out1* output. LEs in arithmetic mode can drive out registered and unregistered versions of the LUT output.

The dynamic arithmetic mode also offers clock enable, counter enable, synchronous up/down control, synchronous clear, synchronous load, and dynamic adder/subtractor options. The LAB local interconnect data inputs generate the counter enable and synchronous up/down control signals. The synchronous clear and synchronous load options are LAB-wide signals that affect all registers in the LAB. The *addnsub* LAB-wide signal controls whether the LE acts as an adder or subtractor. This signal is tied to the carry input for the first LE of a carry chain only.

Carry-Select Chain. The carry-select chain provides a very fast carry-select function between LEs in dynamic arithmetic mode. The carry-select chain uses the redundant carry calculation to increase the speed of carry functions. The LE is configured to calculate outputs for a possible carry-in of **0** and carry-in of **1** in parallel. The *carry-in0* and *carry-in1* signals from a lower order bit feed forward into the higher-order bit via the parallel carry chain and feed into both the LUT and the next portion of the carry chain. Carry-select chains can begin in any LE within an LAB.

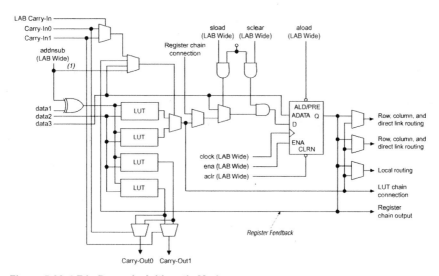

Figure 5.33 LE in Dynamic Arithmetic Mode

The speed advantage of the carry-select chain is in the parallel precomputation of carry chains. Since the LAB carry-in selects the precomputed carry chain, not every LE is in the critical path. Only the propagation delays between LAB carry-in generation (LE 5 and LE 10) are now part of the critical path. This feature allows the Cyclone architecture to implement high-speed counters, adders, multipliers, parity functions, and comparators of arbitrary width.

Figure 5.34 shows the carry-select circuitry in an LAB for a 10-bit adder. One portion of the LUT generates the sum of two bits using the input signals and the appropriate carry-in bit; the sum is routed to the output of the LE. The register can be bypassed for simple adders or used for accumulator functions. Another portion of the LUT generates carryout bits. An LAB-wide carry-in bit selects which chain is used for the addition of given inputs. The carry-in signal for each chain, *carry-in0* or *carry-in1*, selects the carry-out to carry forward to the carry-in signal of the next-higher-order bit. The final carry-out signal is routed to an LE, where it is fed to local, row, or column interconnects.

Clear & Preset Logic Control. LAB-wide signals control the logic for the register's clear and preset signals. The LE directly supports an asynchronous clear and preset function. The register preset is achieved through the asynchronous load of a logic high. The direct asynchronous preset does not require a NOT gate push-back technique. Cyclone devices support simultaneous preset/ asynchronous load and clear signals. An asynchronous clear signal takes precedence if both signals are asserted simultaneously. Each LAB supports up to two clears and one preset signal.

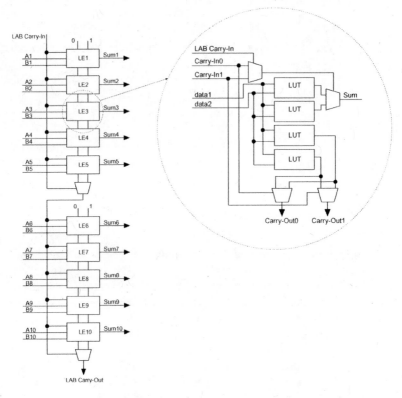

Figure 5.34 Carry Select Chain

In addition to the clear and preset ports, Cyclone devices provide a chip-wide reset pin (*DEV_CLRn*) that resets all registers in the device. An option set before compiling a design in the Quartus II software controls this pin. This chip-wide reset overrides all other control signals.

5.4.2.3 MultiTrack Interconnect. In the Cyclone architecture, connections between LEs, M4K memory blocks, and device I/O pins are provided by the MultiTrack interconnect structure. This structure consists of continuous, performance-optimized routing lines of different speeds used for inter- and intra-design block connectivity. The Quartus II Compiler automatically places critical design paths on faster interconnects to improve design performance.

The MultiTrack interconnect, shown in Figure 5.35, consists of row and column interconnects that span fixed distances. Dedicated row interconnects route signals to and from LABs, PLLs, and M4K memory blocks within the same row. These row resources include *direct link* interconnects between LABs and adjacent blocks, and *R4* interconnects traversing four blocks to the right or left.

The direct link interconnect allows an LAB or M4K memory block to drive into the local interconnect of its immediate left and right neighbors. Only one side of a PLL block interfaces with direct link and row interconnects. The direct link interconnect provides fast communication between adjacent LABs and/or blocks without using row interconnect resources.

The R4 row interconnects span four LABs, or two LABs and one M4K RAM block. These resources are used for fast row connections in a four-LAB region. Every LAB has its own set of R4 interconnects to drive either left or right. R4 interconnects can drive and be driven by M4K memory blocks, PLLs, and row IOEs. For LAB interfacing, a primary LAB or LAB neighbor can drive a given R4 interconnect. For R4 interconnects that drive to the right, the primary LAB and right neighbor can drive on to the interconnect. For R4 interconnects that drive to the left, the primary LAB and its left neighbor can drive on to the interconnect. R4 interconnects can drive other R4 interconnects to extend the range of LABs they can drive. R4 interconnects can also drive C4 column interconnects for vertical connections from one row to another.

The column interconnect operates similarly to the row interconnect. Each column of LABs is served by a dedicated column interconnect, which vertically routes signals to and from LABs, M4K memory blocks, and row and column IOEs. These column resources include LUT chain interconnects within an LAB, register chain interconnects within an LAB, and C4 interconnects traversing a distance of four blocks in an up and down direction.

Every LAB has its own set of C4 interconnects to drive either up or down. The C4 column interconnects can drive and be driven by all types of architecture blocks, including PLLs, M4K memory blocks, and column and row IOEs. For LAB interconnection, a primary LAB or its LAB neighbor can drive a given C4 interconnect. C4 interconnects can drive each other to extend their range as well as drive row interconnects for column-to-column connections.

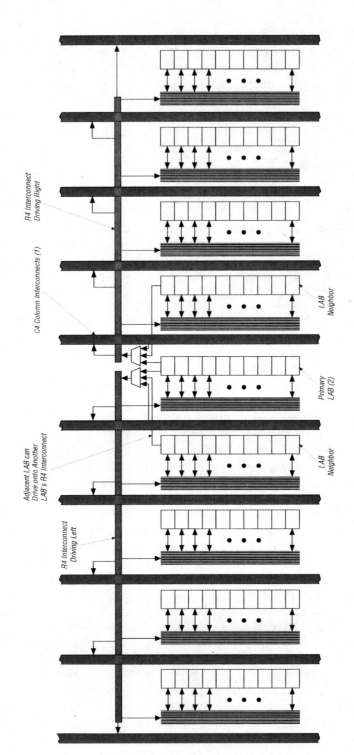

Figure 5.35 Row R4 Interconnect Connections (C4 Is Similar, but Vertical)

All embedded blocks communicate with the logic array similar to LAB-to-LAB interfaces. Each block (i.e., M4K memory or PLL) connects to row and column interconnects and has local interconnect regions driven by row and column interconnects. These blocks also have direct link interconnects for fast connections to and from a neighboring LAB.

5.4.2.4 Embedded Memory.

The Cyclone embedded memory consists of columns of M4K memory blocks. EP1C3 and EP1C6 devices have one column of M4K blocks, while EP1C12 and EP1C20 devices have two columns. Each M4K block can implement various types of memory with or without parity, including true dual-port, simple dual-port, and single-port RAM, ROM, and FIFO buffers. The M4K blocks support the following features:

- 4,608 RAM bits
- 250 MHz performance
- True dual-port memory
- Simple dual-port memory
- Single-port memory
- Byte enable
- Parity bits
- Shift register
- FIFO buffer
- ROM
- Mixed clock mode

Memory Modes. The M4K memory blocks include input registers that synchronize writes and output registers to pipeline designs and improve system performance. M4K blocks offer a true dual-port mode to support any combination of two-port operations: two reads, two writes, or one read and one write at two different clock frequencies.

In addition to true dual-port memory, the M4K memory blocks support simple dual-port and single-port RAM. Simple dual-port memory supports a simultaneous read and write. Single-port memory supports non-simultaneous reads and writes.

The memory blocks also enable mixed-width data ports for reading and writing to the RAM ports in dual-port RAM configuration. For example, the memory block can be written in ×1 mode at port A and read out in ×16 mode from port B.

The Cyclone memory can be configured as a fully synchronous RAM by registering both the input and output signals to the M4K RAM block. All M4K memory block inputs are registered, providing synchronous write cycles. In synchronous operation, the memory block generates its own self-timed strobe write enable (*wren*) signal derived from a global clock. In contrast, a circuit using asynchronous RAM must generate the RAM *wren* signal while ensuring its data and address signals meet setup and hold time specifications relative to the *wren* signal. The output registers can be bypassed. Pseudo-asynchronous reading is possible in the simple dual-port mode of M4K blocks by clocking the read enable and read address registers on the negative clock edge and bypassing the output registers.

When configured as RAM or ROM, an initialization file can be used to pre-load the memory contents.

Two single-port memory blocks can be implemented in a single M4K block as long as each of the two independent block sizes is equal to or less than half of the M4K block size.

Parity Bit Support. The M4K blocks support a parity bit for each byte. The parity bit, along with internal LE logic, can implement parity checking for error detection to ensure data integrity.

Shift Register Support. You can configure M4K memory blocks to implement shift registers for DSP applications such as pseudo-random number generators, multichannel filtering, auto-correlation, and cross-correlation functions. For these and other DSP applications that require local data storage, instead of using a large number of LEs, using embedded memory as a shift register block saves logic cell and routing resources and provides a more efficient implementation with the dedicated circuitry.

The size of a $w \times m \times n$ shift register is determined by the input data width (w), the length of the taps (m), and the number of taps (n). The size of a $w \times m \times n$ shift register must be less than or equal to the maximum number of memory bits in the M4K block (4,608 bits). The total number of shift register outputs (number of taps $n \times$ width w) must be less than the maximum data width of the M4K RAM block ($\times 36$). To create larger shift registers, multiple memory blocks are cascaded together.

Data is written into each address location at the falling edge of the clock and read from the address at the rising edge of the clock. The shift register mode logic automatically controls the positive and negative edge clocking to shift the data in one clock cycle.

Memory Configuration Sizes. The memory address depths and output widths can be configured as 4,096 × 1, 2,048 × 2, 1,024 × 4, 512 × 8 (or 512 × 9 bits), 256 × 16 (or 256 × 18 bits), and 128 × 32 (or 128 × 36 bits). The 128 × 32- or 36-bit configuration is not available in the true dual-port mode. Mixed-width configurations are also possible, allowing different read and write widths. When the M4K RAM block is configured as a shift register block, you can create a shift register up to 4,608 bits ($w \times m \times n$).

Byte Enables. M4K blocks support byte writes when the write port has a data width of 16, 18, 32, or 36 bits. The byte enables allow the input data to be masked so the device can write to specific bytes. The unwritten bytes retain the previous written value.

Control Signals & M4K Interface. The M4K blocks allow for different clocks on their inputs and outputs. Either of the two clocks feeding the block can clock M4K block registers (*renew (read-enable-not-write-enable)*, address, byte enable, *datain*, and output registers). Only the output register can be bypassed. The six *labclk* signals or local interconnects can drive the control signals for the A and B ports of the M4K block. LEs can also control the *clock_a*, *clock_b*, *renwe_a*, *renwe_b*, *clr_a*, *clr_b*, *clocken_a*, and *clocken_b* signals.

Independent Clock Mode. The M4K memory blocks implement independent clock mode for true dual-port memory. In this mode, a separate clock is available for each port (ports A and B). Clock A controls all registers on the port A side, while clock B controls all registers on the port B side. Each port, A and B, also supports independent clock enables and asynchronous clear signals for port A and B registers.

Input/Output Clock Mode. Input/output clock mode can be implemented for both the true and simple dual-port memory modes. On each of the two ports, A or B, one clock controls all registers for inputs into the memory block: data input, *wren*, and address. The other clock controls the block's data output registers. Each memory block port, A or B, also supports independent clock enables and asynchronous clear signals for input and output registers.

Read/Write Clock Mode. The M4K memory blocks implement read/write clock mode for simple dual-port memory. You can use up to two clocks in this mode. The write clock controls the block's data inputs, *wraddress*, and *wren*. The read clock controls the data output, *rdaddress*, and *rden*. The memory blocks support independent clock enables for each clock and asynchronous clear signals for the read- and write-side registers.

Single-Port Mode. The M4K memory blocks also support single-port mode, used when simultaneous reads and writes are not required. A single M4K memory block can support up to two single-port mode RAM blocks if each RAM block is less than or equal to 2K bits in size.

5.4.2.5 Global Clock Network & Phase-Locked Loops.
Cyclone devices provide a global clock network and up to two PLLs for a complete clock management solution.

Global Clock Network. The global clock network of a Cyclone device is shown in Figure 5.36. As shown, There are four dedicated clock pins (*CLK[3..0]*, two pins on the left side and two pins on the right side of a Cyclone device) that drive the global clock network. PLL outputs, logic array, and dual-purpose clock (*DPCLK[7..0]*) pins can also drive the global clock network.

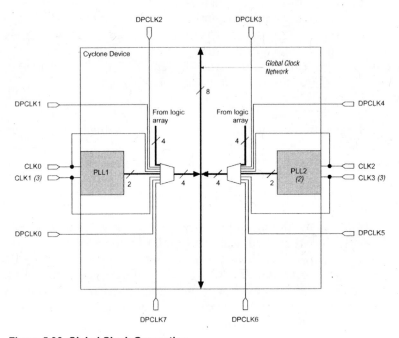

Figure 5.36 Global Clock Generation

Dual-Purpose Clock Pins. Each Cyclone device except the EP1C3 device has eight dual-purpose clock pins, *DPCLK[7..0]* (two on each I/O bank). EP1C3 devices have five *DPCLK* pins in the 100-pin TQFP package. These dual-purpose pins can connect to the global clock network (see Figure 5.36) for high-fanout control signals such as clocks, asynchronous clears, presets, and clock enables, or protocol control signals such as *TRDY* and *IRDY* for PCI, or DQS signals for external memory interfaces.

PLLs. Cyclone PLLs provide general-purpose clocking with clock multiplication and phase shifting as well as outputs for differential I/O support. Cyclone devices contain two PLLs, except for the EP1C3 device, which contains one PLL.

5.4.2.6 I/O Structure. IOEs support many features, including differential and single-ended I/O standards, 3.3-V, 64- and 32-bit, 66- and 33-MHz PCI compliance, JTAG boundary-scan test support, and tristate buffers.

Cyclone device IOEs contain a bidirectional I/O buffer and three registers for complete embedded bidirectional single data rate transfer. Figure 5.37 shows the Cyclone IOE structure. The IOE contains one input register, one output register, and one output enable register. You can use the input registers for fast setup times and output registers for fast clock-to-output times. Additionally, you can use the output enable (OE) register for fast clock-to-output enable timing. IOEs can be used as input, output, or bidirectional pins.

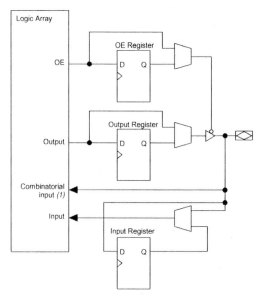

Figure 5.37 Cyclone IOE Structure

The IOEs are located in I/O blocks around the periphery of the Cyclone device. There are up to three IOEs per row I/O block and up to three IOEs per column I/O block (column I/O blocks span two columns). The row I/O blocks drive row, column, or direct link interconnects. The column I/O blocks drive column interconnects. Figure 5.38 shows how a row I/O block connects to the logic array; a similar arrangement is used for connection of a column I/O block to the logic array.

The Cyclone device IOE includes programmable delays to ensure zero hold times, minimize setup times, or increase clock to output times.

A path in which a pin directly drives a register may require a programmable delay to ensure zero hold time, whereas a path in which a pin drives a register through combinatorial logic may not require the delay. Programmable delays decrease input-pin-to-logic-array and IOE input register delays. The Quartus II Compiler can program these delays to automatically minimize setup time while providing a zero hold time. Programmable delays can increase the register-to-pin delays for output registers.

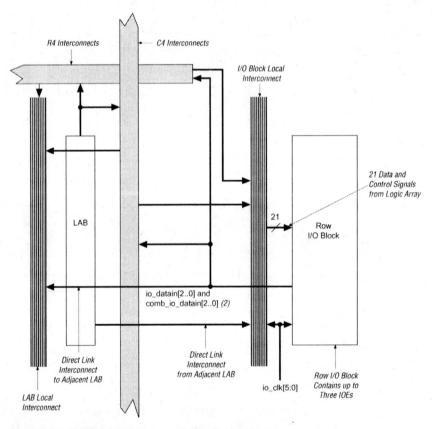

Figure 5.38 Row I/O Block Connection to the Interconnect

External RAM Interfacing. Cyclone devices support DDR SDRAM and FCRAM interfaces at up to 133 MHz through dedicated circuitry.

DDR SDRAM & FCRAM. Cyclone devices have dedicated circuitry for interfacing with DDR SDRAM. All I/O banks support DDR SDRAM and FCRAM I/O pins. However, the configuration input pins in bank 1 must operate at 2.5 V because the SSTL-2 V_{CCIO} level is 2.5 V. Additionally, the configuration output pins ($nSTATUS$ and $CONF\_DONE$) and all the JTAG pins in I/O bank 3 must operate at 2.5 V because the V_{CCIO} level of SSTL-2 is 2.5 V. I/O banks 1, 2, 3, and 4 support DQS signals with DQ bus modes of × 8.

5.4.2.7 Power Sequencing & Hot Socketing.

Because Cyclone devices can be used in a mixed-voltage environment, they have been designed specifically to tolerate any possible power-up sequence. Therefore, the V_{CCIO} and V_{CCINT} power supplies may be powered in any order.

Signals can be driven into Cyclone devices before and during power up without damaging the device. In addition, Cyclone devices do not drive out during power up. Once operating conditions are reached and the device is configured, Cyclone devices operate as specified by the user.

5.5 Summary

In an evolutionary fashion, this chapter showed how a simple idea like the ROM have evolved into FPGA programmable chips that can be used for implementation of complete systems that include several processors, memories and even some analog parts. The first part of this chapter discussed generic structures of programmable devices, and in the second part, when describing more complex programmable devices, Altera devices were used as examples. We focused on the structures and tried to avoid very specific manufacturer's details. This introduction familiarizes readers with the general concepts of the programmable devices and enables them to better understand specific manufacturer's datasheets.

6 Tools for Design and Prototyping

In a hardware-only design process, a hierarchical structure of the hardware is obtained and simulation and synthesis tools are used for verifying it and obtaining hardware for it. Often, in an FPGA based design, a development board is used for actual verification of the hardware that is being designed.

This chapter introduces tools and utilities for design and implementation of hardware. We will discuss a typical hardware design flow and use a small hierarchical example for its illustration. Application of tools for design specification, simulation, synthesis, and device programming will be shown using this example. We start by showing HDL simulation using Altera-Mentor ModelSim simulation environment. We will then show how an HDL synthesis is performed with Altera's Quartus II FPGA design environment. Post-synthesis simulation of an HDL design will be illustrated using ModelSim. After being introduced to an HDL design flow, we will show how a complete design that consists of HDL parts, gates, library parts, and other predefined components is specified, simulated and synthesized. For this purpose a complete flow from design entry to hardware translation will be shown using Quartus II. In the final part of this chapter Altera's UP3 and DE2 development boards will be discussed and use of Quartus II for programming Cyclone / Cyclone II of these boards will be illustrated.

6.1 Hardware Design Flow

This section uses a simple example that is composed of a datapath and a controller. Components of this hierarchical design are implemented using Verilog, gate instantiations, predefined Altera library functions (Megablocks), and a combination of all these. In the implementation of this system we will show how various components of a system are specified, simulated, synthesized, and wired together into a complete system. Utilization of design tools used for this implementation will be illustrated.

The design example we are using is the serial adder of Chapter 2, also shown here in Figure 6.1. For complements, some of the material of this chapter will be repeated here.

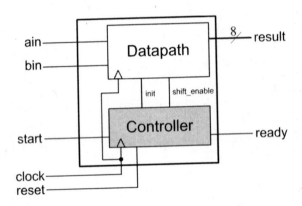

Figure 6.1 Datapath and Control of Serial-Adder example

The serial adder has two serial data inputs *ain* and *bin*, and a control input *start*. As shown in Figure 6.1, the circuit has an eight bit *result* output and a *ready* signal. After a complete pulse on *start*, operand data bits start showing up on *ain* and *bin* with every clock with least significant bits coming in first. In eight clock pulses as input data come into the circuit, they are added and the result becomes ready on *result*. At this time the *ready* signal becomes **1** and it remains **1** until a **1** is detected on the *start* input. While the circuit is performing its data collection and addition, pulses on *start* are ignored.

As shown in Figure 6.1, this circuit has a datapath and a controller. The datapath collects data, adds them, and shifts the result into a shift-register. The controller waits for *start*, controls shifting of data into the shift-register, and issues *ready* when the addition operation is complete. In what follows, the details of the two parts of this design will be discussed.

6.1.1 Datapath of Serial Adder

In the datapath of the serial adder a full-adder adds data coming in on *ain* and *bin*. With each addition, the sum is shifted into a shift-register. As data are added, the full adder carry is saved in a flip-flop to be used for the addition of the next set of data coming on *ain* and *bin*. This flip-flop must be reset before a new 8-bit addition starts.

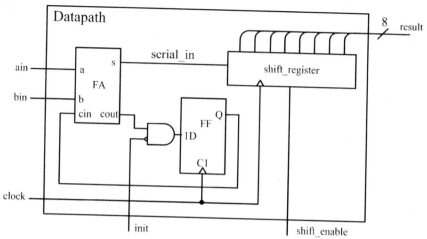

Figure 6.2 Serial Adder Datapath

Figure 6.2 shows the details of hardware of the datapath. FA is a simple Full-Adder that will be implemented with a Altera Megablock. The flip-flop shown is a rising-edge trigger D-type flip-flop, for the implementation of which a primitive component will be used. The AND gate at the input of the flip-flop provides it with a synchronous reset. This input connects to the *init* input that comes from the controller. This flip-flop saves a carry output from a lower order bit for the addition of the next upper-order bit.

The shift-register of the datapath is an 8-bit shift-register with an enable input. We will implement this part of the datapath with a Verilog module description.

6.1.2 Serial Adder Controller

Figure 6.3 shows the controller of our serial adder. On the one side there is a state machine that waits for *start* and issues *count_enable* and *shift_enable* and *ready*. The state machine waits for the *complete* signal to be issued by the counter before it returns to its initial state that waits for another pulse on *start*. The outputs of this state machine are *ready*, *init*, *count_enable*, and *shift_enable*. The *init* and *shift_enable* outputs go out to the datapath to control initialization and shift activities. This state machine assumes that after a pulse on *start*, it will not become 1 again until *ready* is issued. We will use a Verilog module for the implementation of this state machine.

On the other side of the controller is a counter that counts when *count_enable* is issued. Eight clock pulses after *init* resets this counter to **0**, and while *count_enable* is active, the counter reaches its **111** state and issues the *complete* signal. When this signal is issued, the controller disables *count_enable*, which causes the counter to hold its last state. The counter part of the controller will be implemented with a predefined configurable Mega function. That is part of our Quartus II design environment.

Parallel with *count_enable*, the controller also issues *shift_enable* that goes out to the datapath. While this signal is active, add results from the full-adder (FA) are shifted into the datapath shift-register. Note that after eight shifts, because the shift-register is disabled, the output remains on the circuit *result* output.

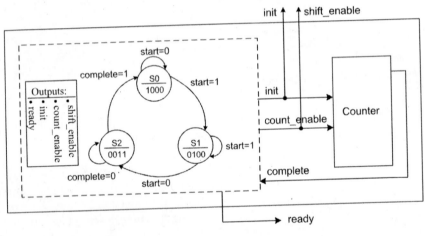

Figure 6.3 Serial Adder Controller

The complete controller of the serial adder is formed by wiring the counter and the state machine together. Using the controller, the complete serial adder is formed by wiring this (Figure 6.3) and the circuit of Figure 6.2 together to form the block diagram of Figure 6.1.

The sections that follow show the implementation of this design and programming the Cyclone FPGA of a UP3 and Cyclone II of the DE2 development board.

In the text of this material we often need to show menus selected and menu items that need to be clicked. For this purpose we use the following notation:

{Originating Window}: OriginatingMenu ⇨ MenuItem1 ⇨ . . . [Button1] ⇨ "Comment On What To Do" ⇨ [Button2]

Interpretation of the above menu selections is clear by their contents. If a certain tab needed to be selected in a certain window, the name of the window would be followed by the tab name separated by a dash, e.g., {CertainWindow – TabName}. If the first menu is originated by a mouse right-click the following format will be used.

(RightClick) ⇨ MenuItem1 ⇨ MenuItem2 ⇨ . . .

When a window name or menu item is used in text, Arial font will be used for its name. An example is: A New Window.

6.2 HDL Simulation and Synthesis

For the situations that a part of a design cannot be found in a predefined or a user library, or it cannot easily be built by discrete components, developing an HDL model is usually the best alternative. Most parts in the controller and datapath of our serial adder example are standard parts. The most likely candidate for HDL implementation is the state machine part of the controller.

We will show how the Verilog code of this component is simulated with ModelSim, synthesized with Quartus II, and post synthesis simulation is done in the ModelSim environment.

Figure 6.4 shows the Verilog code of the controller state machine shown on the left hand side of Figure 6.3. This circuit is described by use of three **always** blocks that handle state transitions, output values, and register clocking respectively. This description is synthesizable; the first two **always** blocks follow synthesizable combinational rules of Chapter 3, and the last **always** block implements a two-bit register with synchronous reset.

```verilog
module controller_SM (input reset, start, complete,
      output reg init, shift_enable, count_anable, ready);

   localparam [1:0] S0 = 2'b00, S1 = 2'b01, S2 = 2'b11;
   reg [1:0] p_state, n_state;

   always @( p_state, start, complete) begin: transition
      n_state = S0;
      case ( p_state )
         S0: if(~start) n_state = S0; else n_state = S1;
         S1: if( start) n_state = S1; else n_state = S2;
         // we don't expect start=1 before ready is issued
         S2: if(~complete) n_state = S2;
            else n_state = S0;
         default: n_state = S0;
      endcase
   end
```

```
    always @( p_state, start, complete) begin: Outputing
        {ready, init, count_enable, shift_enable} = 4'b0000;
    case ( p_state )
        S0: {ready, init, count_enable, shift_enable} =
            4'b1000;
        S1: {ready, init, count_enable, shift_enable} =
            4'b0100;
        S2: {ready, init, count_enable, shift_enable} =
            4'b0011;
        default:
            {ready, init, count_enable, shift_enable} =
            4'b0000;
    endcase
    end
    always @( posedge clk ) begin: sequential
        if (reset) p_state <= S0; else p_state <= n_state;
    end
endmodule
```

Figure 6.4 Controller State Machine Verilog Code

6.2.1 Pre-Synthesis Simulation

This section shows steps involved in simulating the *controller_SM* module of Figure 6.4 in ModelSim. We show how a project is created, an existing module is compiled, a testbench is developed, simulation is performed and waveforms are displayed.

After starting ModelSim its main window and its default toolbar appears. Definition of a simulation project starts here.

6.2.1.1 Creating a Project. Creation of a new ModelSim project starts from the File menu and continues as shown here:

{Main}: File ⇨ New ⇨ Project . . .

Figure 6.5 shows the project creation window, where project name and its corresponding directory are specified. We use *controller* for the project name and browse to our chosen directory of *Chapter6* designs. We use *work* for our library which is the default library name.

Project creation in ModelSim continues with the software wanting us to add existing files to the newly created project. This is done in the windows shown in Figure 6.6. File *controller_SM.v* (Figure 6.4) from *Chapter6* directory is added to the *controller* project. When this is done, *controller_SM.v* will be added to our workspace. The following displays the workspace window.

{Main}: View ⇨ Workspace

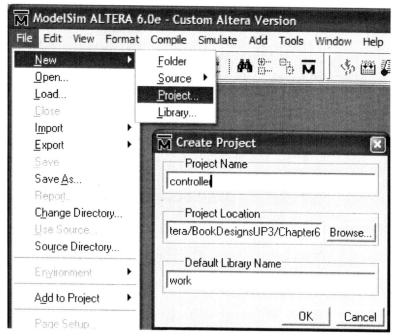

Figure 6.5 Project Creation

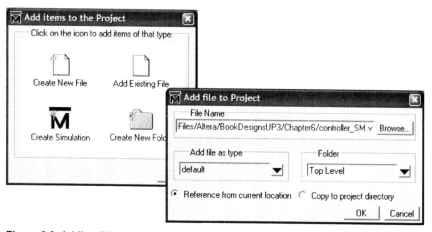

Figure 6.6 Adding Files to Project

6.2.1.2 Creating a Verilog Testbench.
A testbench is just another Verilog file that we will enter and add to our project. For this purpose follow menu items shown below:

{Main}: File ⇨ New ⇨ Source ⇨ Verilog

This causes a text window to open in which we can enter our Verilog testbench. Alternatively, we could use our own text editor to create our testbench and then add it to the project.

Figure 6.7 shows the text window of ModelSim with our testbench code typed in it. This testbench instantiates *controller_SM* of Figure 6.4 and applies data to its inputs. We name this file *controller_SM_Tester.v* and save it in *Chapter6* directory.

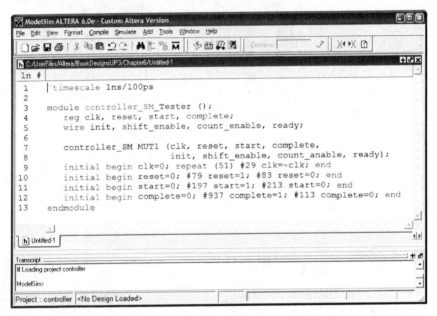

Figure 6.7 Entering a Design File

A file created as specified above must be added to our current project before it can be compiled and simulated. Follow the menu items shown below to open the Add File to Project window.

{Main}: File ⇨ Add to Project

In the window that opens browse to select *controller_SM_Tester.v*, and click OK to complete adding this file to the *controller* project.

After completion of this phase, the Project tab of the Workspace window shows *controller_SM.v* and *controller_SM_Tester.v* files added to our currently active project, i.e., *controller*. Figure 6.8 shows the new workspace and menu items that need to be selected for displaying it.

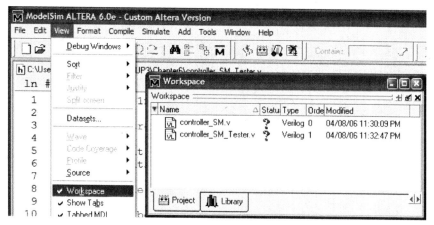

Figure 6.8 Workspace showing Files in Project

6.2.1.3 Compiling Design Files.
Before simulation can take place, all design file must be compiled. Question marks are used for the status of design files in the workspace of Figure 6.8. This shows that these files are not compiled.

We can perform compilation by selecting a file in the workspace, right-clicking it and selecting Compile in the menu that appears. Alternatively, from the main window, the following menu items perform compilation of all project files.

{Main}: Compile ⇨ Compile All

After a successful compilation, question marks in the workspace will be replaced by check marks. If an error occurs, an X mark...

6.2.1.4 Starting the Simulation.
The top-level design unit in our controller project is *controller_SM_Tester*. For simulation of this design, start in the main window and select the following menu items:

{Main}: Simulate ⇨ Start Simulation

When this is done, the Start Simulation window shown in Figure 6.9 opens. In this window under the Design tab, open the *work* library and click the top-level design unit, *controller_SM_Tester*.

Alternatively, simulation could be started from the workspace window by selection of the following menu items:

{Workspace-Library}: (RightClick) *Controller_SM_Tester* ⇨ Simulate

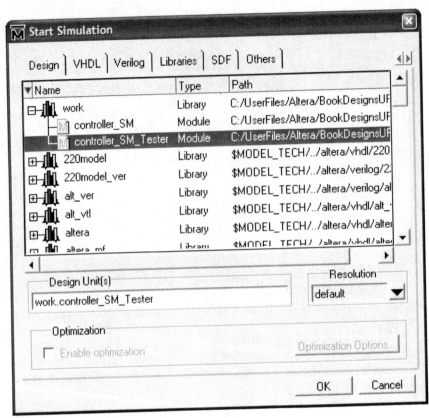

Figure 6.9 Starting Simulation

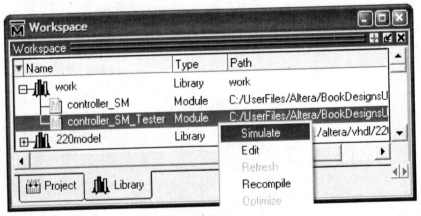

Figure 6.10 Starting Simulation from Workspace

6.2.1.5 Setting up Waveform.
After the start of simulation, an objects window opens, listing signals of the top-level testbench. Signals shown can be selected for waveform and display. If this window does not automatically open, perform the following:

{Main}: View ⇨ DebugWindows ⇨ Objects

Initially, signals shown in the Objects windows are of the top-level design components. To select signals in lower levels of hierarchy in the design, select the Sim tab in the Workspace window and select the component for which signals are to be displayed. Once selected, the corresponding signals will appear in the Objects window. Figure 6.11 shows selection of *mut1* instance and listing of its corresponding signals in the Objects window.

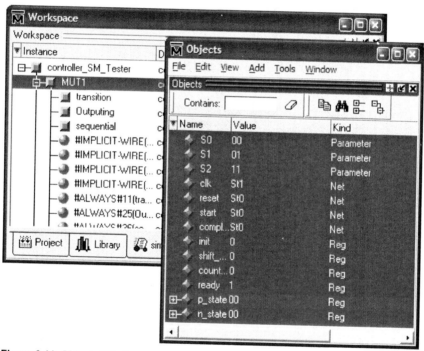

Figure 6.11 **Objects Window**

Signals from the Objects window must be selected and added to the Wave window for displaying them. For this purpose, in the Objects window, perform the following:

{Objects}: (Right Click) ⇨ Add to Wave ⇨ Signals in Region

This will open the Wave window and add signals in the Objects window to it. The next step is to simulate the design for the values given to it by its testbench.

6.2.1.6 Running the Simulation. With the Wave window open and selected signals on its left, running simulation will show waveforms on all the selected signals. To run the simulation, follow menu items listed here,

{Main}: Simulate ⇨ Run ⇨ Run-All

This will run the testbench until no more events occur in the design. We can afford running the simulation opened-ended, because our testbench limits placement of data on input signals of the design. See in Figure 6.7 that all the procedural blocks terminate at some point in time.

Simulation result of our *controller_SM* is shown in Figure 6.12. This figure shows *clk, reset, start, complete, init, shift_enable, count_enable,* and *ready* from the testbench, and the *p_state* (present state) from *MUT1* instance of the *controller_SM* module.

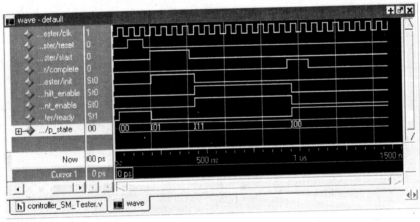

Figure 6.12 Simulation Results of *Controller_SM*

6.2.2 Module Synthesis

The next step in design of a module after a successful simulation is synthesis and generation of a netlist. For this purpose we will use Altera's Quartus II design software. In this section we will only take advantage of this program's synthesis capabilities and its other utilities such as graphical design entry, library utilization, and post-synthesis simulation will not be discussed. Such features will be illustrated in a later section when we use Quartus II for a complete design implementation, i.e., our serial adder.

This section shows a project definition in Quartus II, synthesis of the *controller_SM* module, and generation of post synthesis Verilog code. Figure 6.13 shows the main window of Altera's Quartus II when it first opens.

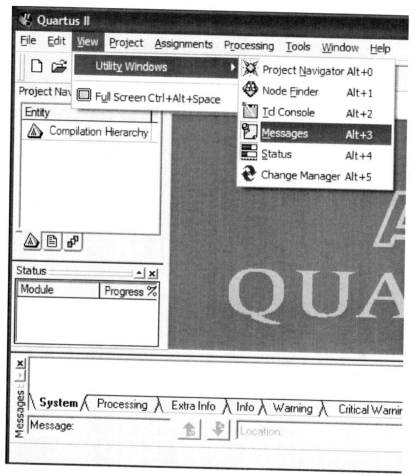

Figure 6.13 Quartus II Main Window

Shown here are utility windows: Project Navigator, Status, and Messages. Utility windows can be added or removed by:

{Main}: View ⇨ Utility Windows ⇨ The Desired Window

6.2.2.1 Synthesis Project.
We will define a new project for synthesizing the *controller_SM* module. For this purpose, in the main Quartus II window, select the following menu items:

{Main}: File ⇨ New Project Wizard

This opens the project wizard that uses five pages for defining the details of the project being created. As shown in Figure 6.14, the first page asks for project directory, project name and design file, for which we use *Chapter6*, *controller_SM*, and *controller_SM* respectively. This coincides with the ModelSim project for simulation of *controller_SM* which causes the same simulated design file to become available for synthesis.

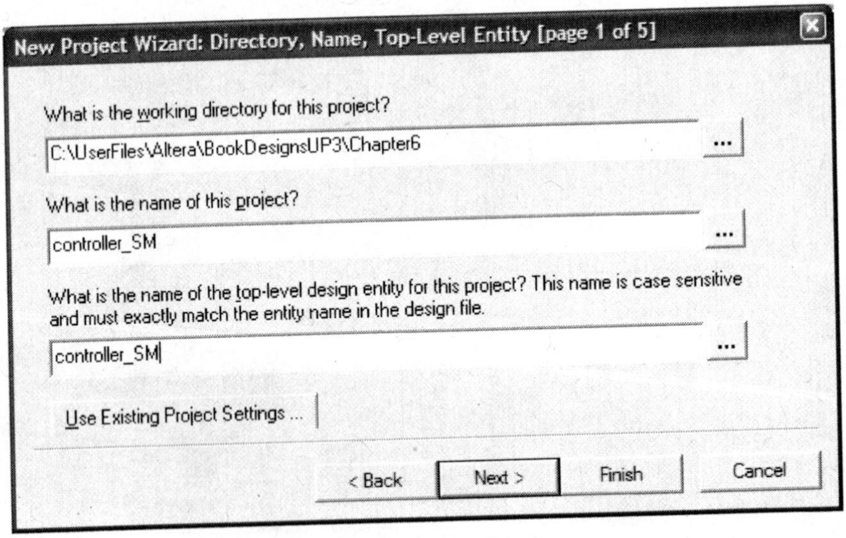

Figure 6.14 Project Directory, Name and Design File

Page 2 of project definition flow asks for files to be included in the design. Since our *controller_SM* is self contained and does not depend on any other file we will skip this step.

Page 3 of project definition asks for FPGA family and the specific device to synthesize to. As shown in Figure 6.15 we use *Cyclone* EPIC12Q240C8 that is the device on Altera's UP3 development board. For the DE2 board we would use *Cyclone II* - EPC35F672C6.

The next page of project definition (page 4) is for EDA page settings. In this page (Figure 6.16), we will specify *ModelSim(Verilog)* for our EDA simulation tool. This way, the Quartus II synthesis generates a Verilog output file (.*vo*) that can be simulated in ModelSim. This is for post-synthesis simulation in ModelSim.

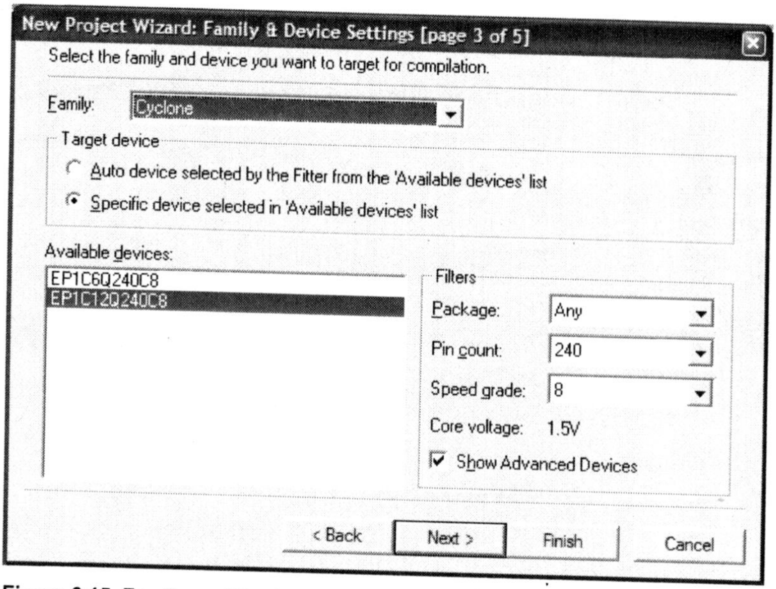

Figure 6.15 Family and Device Setting

The last page of a Quartus II project definition sequence (Figure 6.17) shows the project summary. Now that *controller_SM* project is defined, the existing *controller_SM* module in the *controller_SM.v* file of the *Chapter6* directory becomes our top-level design entity. The following list of commands opens this file.

{Main}: File ⇨ Open ⇨ {Open}: *controller_SM.v* ⇨ [Open]

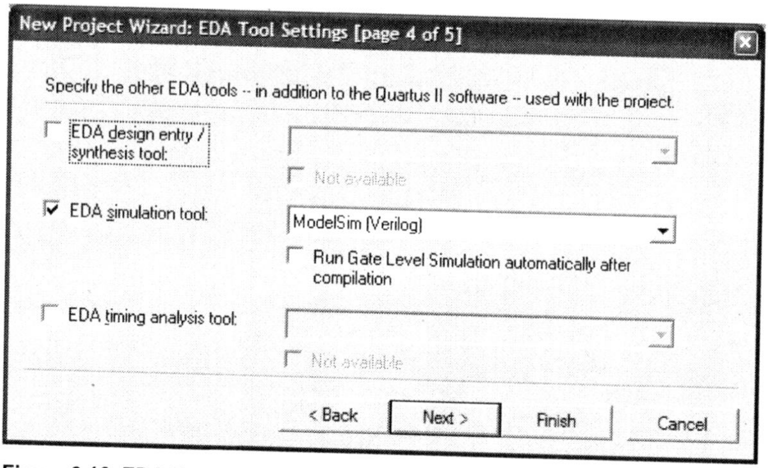

Figure 6.16 EDA Tool Setting

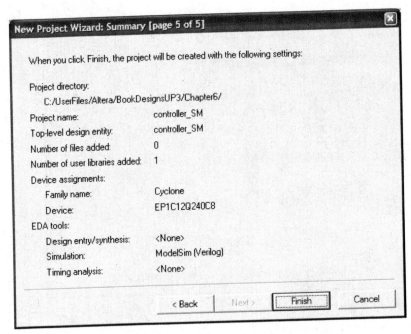

Figure 6.17 Project Summary

Figure 6.18 shows the main Quartus II window and procedure for opening the design file.

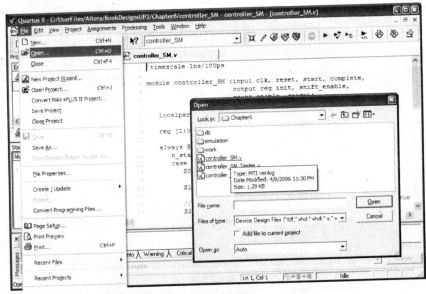

Figure 6.18 Opening Top-Level Design Entity File

6.2.2.2 Design Compilation.

Compiling *controller_SM* results in synthesis, timing file generation, FPGA placement and routing, and generation of the device programming file. For this compilation perform:

{Main}: Processing ⇨ Start Compilation

After a successful compilation, appropriate files are created and a compilation report, as shown in Figure 6.19 will be given. Files that are important in the design flow being discussed in this section are: *controller_SM.vo* and *controller_SM.sdo*. The former is post-synthesis Verilog output file using appropriate components of our target FPGA. The latter file is an SDF (Standard Delay Format) file that contains the timing of the netlist of the *controller_SM.vo* file.

6.2.2.3 Creating Symbol.

The *controller_SM* component that is now synthesized will be used in an upper level Quartus II design. For this to be possible, we have to create a symbol for this design. In order to do this, open the design file (*controller_SM.v*), and follow the sequence of menu items shown below:

{Main}: File ⇨ Create/Update ⇨ Create Symbol Files for Current File

The symbol created here will be used in Section 6.3 when we present the complete design of our serial adder.

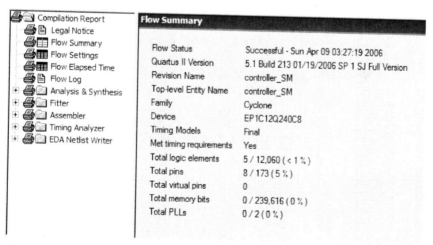

Figure 6.19 Compilation and Synthesis Report

6.2.3 Post-Synthesis Simulation

Files created in the compilation process discussed above are used for post-synthesis simulation of *controller_SM* in ModelSim. For this purpose we will copy *controller_SM.vo* and *controller_SM.sdo* from *Chapter6/simulation/modelsim* directory to our design directory, *Chapter6*, and start ModelSim again for post-synthesis simulation.

For post-synthesis simulation, we will use the *controller* Model-Sim project that we created for our pre-synthesis project. We will simulate both models simultaneously, and see the differences between pre- and post- synthesis models, i.e., *controller-SM.v* and *controller_SM.vo*. Tasks for performing this simulation are project setup, new testbench, simulation and waveform display.

6.2.3.1 Project Setup. As mentioned, we use the same project we used for pre-synthesis simulation. In order to do this, the newly created design, *controller_SM.vo* must be added to the *controller* project, which is done by starting in the main window of ModelSim and following menu items:

{Main}: File ⇨ Add to Project ⇨ Existing File ⇨
      {Add File to Project}: [Browse...] ⇨
      {Select Files to Add to Project}: *controller_SM.vo* ⇨ [Open]

The above results in addition of *controller_SM.vo* to the *controller* project. Figure 6.20 shows the new workspace.

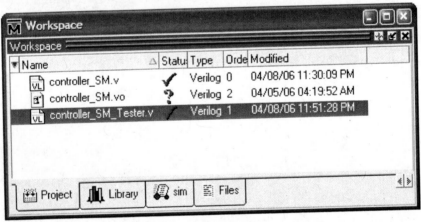

Figure 6.20 Workspace for Post-Synthesis Simulation

6.2.3.2 Testbench For Post-Synthesis Simulation. In order to be able to simulate *controller_SM.v* and *controller_SM.vo* with the same testbench, we have to use different module names. Since the post-synthesis description of our design generated by Quartus II uses the same name as the original module, we have to edit this code to use a different name. Figure 6.21 shows the header part of *controller_SM.vo* to use *controller_SM_PS* for module name.

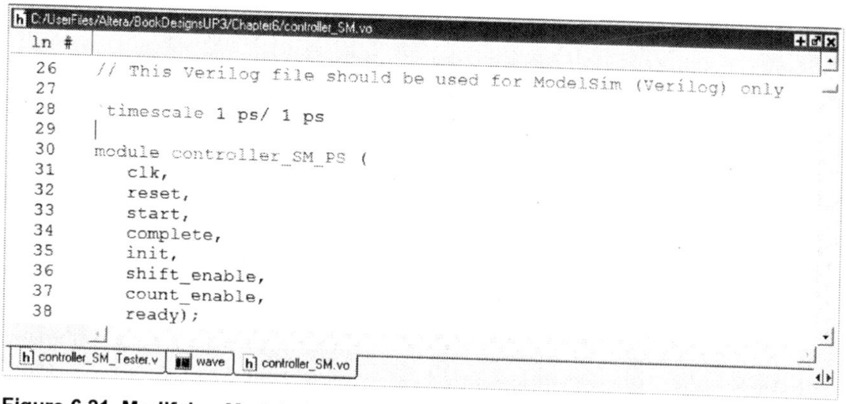

```
ln #
26    // This Verilog file should be used for ModelSim (Verilog) only
27
28    `timescale 1 ps/ 1 ps
29    |
30    module controller_SM_PS (
31        clk,
32        reset,
33        start,
34        complete,
35        init,
36        shift_enable,
37        count_enable,
38        ready);
```

controller_SM_Tester.v | ∎∎ wave | controller_SM.vo

Figure 6.21 Modifying Module Name of Post-Synthesis Description

We also need to change our testbench to instantiate both pre- and post-synthesis modules. The code of the new testbench is shown in Figure 6.22. As shown, *MUT1* is the instantiation of the original *controller_SM* and *MUT2* is its post-synthesis description. *MUT1* and *MUT2* use the same input signals, but use output names that are appended by "1" or "2" for *MUT1* or *MUT2*.

```
module controller_SM_Tester ();
    reg clk, reset, start, complete;
    wire init1, shift_enable1, count_enable1, ready1;
    wire init2, shift_enable2, count_enable2, ready2;
    controller_SM MUT1 (clk, reset, start, complete,
                init1, shift_enable1, count_enable1, ready1);
    controller_SM_PS MUT2 (clk, reset, start, complete,
                init2, shift_enable2, count_enable2, ready2);
    initial begin clk=0; repeat (51) #29 clk=~clk; end
    initial begin reset=0; #79 reset=1; #83 reset=0; end
    initial begin start=0; #197 start=1; #213 start=0; end
    initial begin
        complete=0; #937 complete=1; #113 complete=0;
    end
endmodule
```

Figure 6.22 Testbench Instantiating Pre- and Post-Synthesis Description

6.2.3.3 Simulation of Pre- and Post-Synthesis Descriptions. The
controller project workspace includes *controller_SM.v, control-*

ler_SM_Tester.v, and *controller_SM.vo*. Perform the following to compile these files:

{Main}: Compile ⇨ Compile All

For starting the simulation, the same procedure discussed in Section 6.2.1 must be followed, except that the Cyclone FPGA library must be added to our library search path. As before, activate the Start Simulation window by:

{Main}: Simulate ⇨ Start Simulation

In this window first click on *controller_SM_Tester* under the Design tab. Before clicking OK, go to Libraries tab in this window and add the path to the *Cyclone* (or Cyclone II for DE2) library of *ModelSim* to the search libraries. The default installation of Altera version of ModelSim puts the *Cyclone* library in the directory that is shown below. This directory needs to be added to the list of search Libraries.

C:/Modeltech-ae/altera/verilog/cyclone or /cycloneii

Before clicking OK in the Start Simulation window, make sure the SDF file (*controller_SM.sdo*) is in the original design directory, i.e., *Chapter6*. Simulation starts by clicking OK in the start simulation window.

6.2.3.4 Waveform Display. To prepare for running simulation and displaying the results, bring up the Objects window, select signals, and add them to the Wave window. Perform the following to run the simulation.

{Main}: Simulate ⇨ Run ⇨ Run -All

The waveform of Figure 6.23 shows pre- and post-synthesis results. The first four signals shown are common inputs. The second four signals are outputs of the pre-synthesis description, and the last four signals are *init2*, *shift_enable2*, *count_enable2*, and *ready2* outputs of the post-synthesis description.

As shown in this waveform, there is a difference of about 6.7 ns between *init1* and *init2*. This is due to the fact that in *init2* all internal FPGA delays are considered, whereas, *init1* is the output of our original controller description in which no delay values were specified.

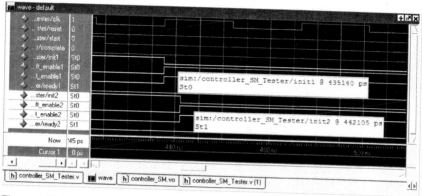

Figure 6.23 Comparing Pre- and Post-Synthesis Results

This section showed design and implementation of an HDL module. This may be a complete design, or, in our case, part of a larger design. The next section shows the use of Quartus II for design and implementation of a circuit composed of various forms of components. The *controller_SM* module is one component of this design.

6.3 Mixed-Level Design with Quartus II

Section 6.2 showed the design of a component using its Verilog code. Although many designs can be completely done this way, block diagram specification and using predefined components, instead of describing every component of a design, has its own advantages.

For small gate level components, there is too much overhead in writing a Verilog code and instantiating with other components. In addition, many RTL components such as ALUs, counter, register files, and FIFOs, are available as Mega blocks or Mega functions in most FPGA design environments and can easily be instantiated and configured to proper functionality and size. An advantage of using such functions is that they are optimized for specific target FPGAs and their implementations take advantage of especial FPGA features they are used for. Furthermore, use of predefined components, instead of defining our own, has the advantage that such components have already been tested and debugged.

Another advantage of using tools for block diagram specification is the graphical user interface that they provide. Such a graphical interface provides ways of selecting, instantiating, configuring and wiring parts from other designs, as well as using parts from the library of Mega-blocks, primitives, and parts that are specific to the design being done. Often these environments provide facilities for high level test and debug of designs.

In what follows we show implementation of a complete design that consists of various parts at various levels of abstraction and of various design libraries. The example we are using is the serial adder of Section 6.1. Although this is a simple design, but it has most characteristics of a large RTL design and it is a good candidate for showing a complete design flow.

As discussed in Section 6.1, our serial adder has a datapath and a controller. As shown in Figure 6.2, the datapath of this design has four components: a full-adder, a rising-edge D-Type flip-flop, an inverted-input AND gate and an 8-bit shift-register with clock enable input. As shown in Figure 6.3, the controller of our design has a state machine part (FSM) and a counter. Various forms of component specifications will be used for these parts:

- Full-adder: *lpm_add_sub* configurable Megafunction from Library of Parameterized Modules (LPM).

- D-Type flip-flop: *dff* primitive from library of primitives.

- Inverted-input AND: *NOT* and *AND2* primitives from library of primitives.

- 8-bit shift-register: Verilog code entered directly in Quartus II environment.

- Controller FSM: Existing Verilog code developed in Section 6.2.

- Controller counter: *lpm_counter* primitive from Quartus II LPM.

In the sections that follow we will use Quartus II and discuss the use of this software for our design. We will refer to Quartus II menus, windows and toolbars. In most cases we show complete paths of menus and toolbars to a certain Quartus II program. On the other hand, this software has a standard toolbar with shortcuts to commonly used programs and settings, shown in Figure 6.24. Instead of following a long list of menu and window selections, a program can be started directly by clicking its icon on the standard toolbar.

In case this toolbar is not active, it can be activated by starting from the main Quartus II window and following menu items shown below:

{Main}: Tools ⇨ Customize ... ⇨ {Customize}–Toolbars ⇨ "Check Standard Quartus II" ⇨ [OK]

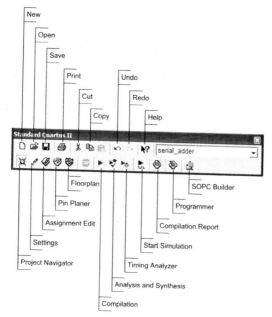

Figure 6.24 Quartus Standard Toolbar

6.3.1 Project Specification

The design and implementation of our serial adder will be done in Quartus II and programmed into a Cyclone EP1C12 or Cyclone II EP2C35. The Quartus II *serial_adder* project is created for this purpose. The method used for creating this project is similar to that described in Section 6.2 for the *controller* project.

We create the serial-adder project in the same directory as that of the *controller*, i.e., *Chapter6*. When this project is being created, Quartus II finds the *controller* project already in this directory and asks if you want to use a different directory. Answer No to use the same directory. Figure 6.25 shows Quartus II windows and the serial-adder project summary.

6.3.2 Block Diagram Design File

The top-level structure of our serial adder design is a block diagram. Components of the serial adder are placed in this block diagram and wired together.

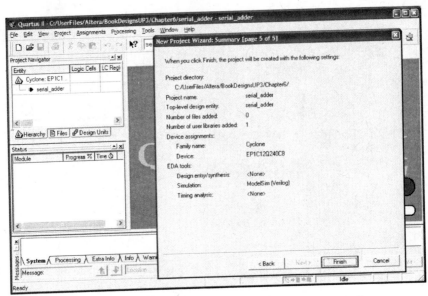

Figure 6.25 Serial-adder Project Definition

The top-level block diagram of our design must be named the same as the project it is in. To define a block diagram design file, start in the main Quartus II window, in the File menu select New..., and under the Device Design File tab of the New window select Block Diagram/Schematic File. This procedure is shown in Figure 6.26 and summarized below.

{Main}: File ⇨ New... ⇨ {New}-Device Design Files ⇨ Block Diagram/Schematic File ⇨ [OK]

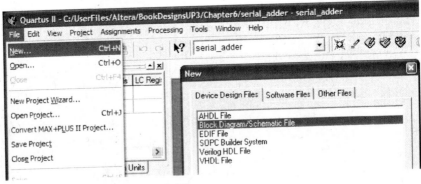

Figure 6.26 A New Block Diagram File

After following this procedure, a window opens in which *our serial_adder* components can be placed and their wirings specified.

6.3.3 Creating and Inserting Design Components

Various methods available in Quartus II for creating new components and using existing ones will be used for the components of *serial_adder*. We will start with the datapath components and then do the controller.

6.3.3.1 Inserting Primitive Components.
The procedure for inserting a new component for our schematic file is to select it from a library, and place it in the block diagram window. Primitive gates and components of our design, like the flip-flop, AND, and NOT of Figure 6.27, are the easiest to handle.

The D-type flip-flop (Figure 6.27) is a primitive of Quartus II, i.e., it exists as a component in Quartus II library of *primitives*. To place this component in the **Block Diagram** window, double click anywhere on this window for the **symbol** window to open. In this window under altera/quartus/libraries, go to primitives, and under storage select *dff*. The *dff* symbol is now selected (see Figure 6.27) and will be placed in the **Block Diagram** window when **OK** is pressed.

Repeat the same procedure for AND and NOT gates (these are under logic in the library of primitives). A two-input AND gate is called *and2* and an inverter is called *not* in the *logic* library.

6.3.3.2 Inserting IO Pins.
We follow the same procedure discussed above for placement of input/output pins. In the Quartus libraries, the *primitives* branch has a branch called *pin*. This library has *input* and *output* pins that are selected and placed in our schematic. Figure 6.28 shows our schematic file after placement of primitives discussed above. To assign names to input and output pins, double-click on the pin symbol and enter the name of the pin in the window that opens. Alternatively, the same **Properties** window opens by selecting the pin and doing a right-click on it.

Also shown in Figure 6.28 is the Block Diagram tool bar. This tool bar appears on the left side of the block diagram window. Some important tools here are: **Symbol Tool** (the AND symbol), **Orthogonal Node Tool** (the 90 this line), and the **Orthogonal Bus Tool**. Activating the **Symbol** window for selection of a library component can be done by selecting the **Symbol Tool** instead of double clicking the schematic area.

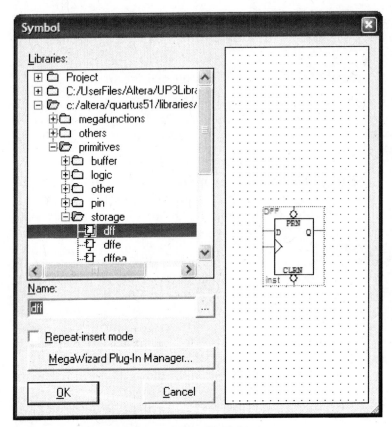

Figure 6.27 Selecting a Component for Block Diagram

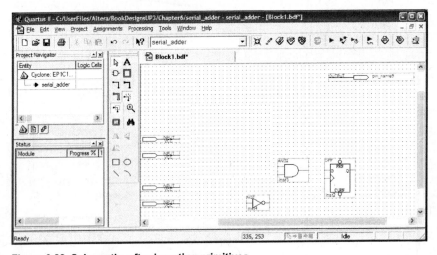

Figure 6.28 Schematic, after inserting primitives

6.3.3.3 Inserting configurable components.

In addition to basic primitives, the *quartus* library contains components that can be configured for size and functionality to be inserted in schematic files. These components are in the *megafunctions* library of Quartus II.

For the full adder of the datapath of our serial adder (Figure 6.2) and for the counter part of its controller (Figure 6.3), Altera megafunction will be used.

As with other components, inserting a configurable component in our schematic begins by double-clicking anywhere on the Block Diagram window. Instead of a double-click, pressing the right mouse button and selecting Insert and then Symbol... in the menu that opens, also opens the symbol window (Figure 6.29).

(Rightclick) ⇨ Insert ⇨ Symbol... ⇨ {Symbol Window}

The full adder of Figure 6.2 can be generated by configuring an adder-subtractor *megafunction*. In order to do this, in the Symbol window select megafunctions and under its arithmetic library branch select lpm_add_sub. This LPM component can be configured as an n-bit complex pipeline adder-subtractor, or a simple full-adder. After pressing OK in the Symbol window of Figure 6.29, the wizard for configuring this component begins working. Configuration will be done in a series of pages, of which the window of Figure 6.29 is considered the first page.

A modified version of Page 2 of the Megawizard is shown in Figure 6.30. In this page we specify a name for our component and the HDL that will be generated for it. As shown, we are using Verilog for the output file and *FA* for the name of the component. The original Page 2 looks different that what is shown here. We have altered by eliminating some of the comments on this page.

Page 3 to 6 of Megawizard, parts of which are shown in Figure 6.31, allow users to specify number of bits, data types carry input and output, and pipeline structure for *lpm_add_sub*. Our specifications for defining a full adder are shown in this figure.

The last page of Megawizard shows files that are created and will be written in our design directory. After completion of this phase of our design, the *FA* symbol will be added to the schematic window of our design. Components defined thus far are shown in Figure 6.32.

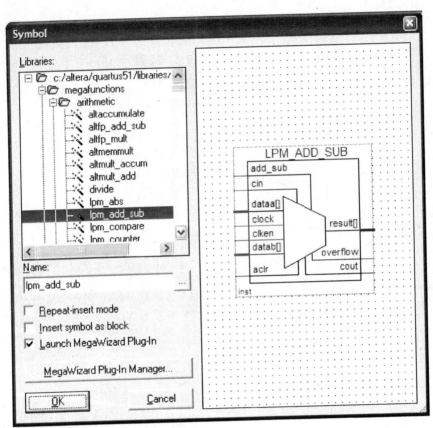

Figure 6.29 Configuring *lpm_add_sub* for FA

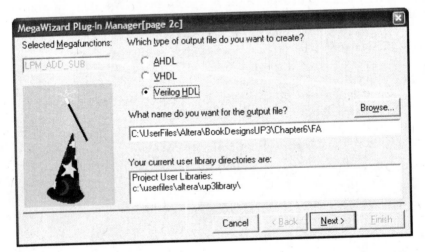

Figure 6.30 Page 2 of Megawizard

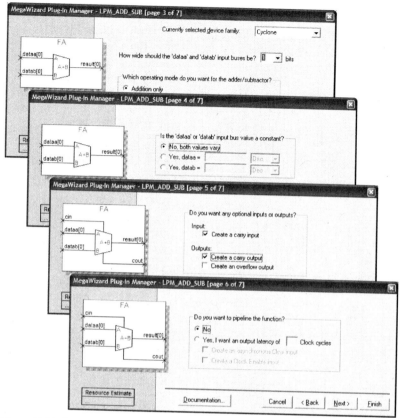

Figure 6.31 Configuring *lpm_add_sub*

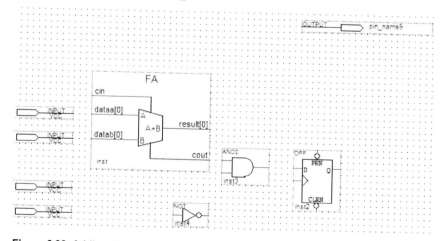

Figure 6.32 Adding FA to our Schematic

6.3.3.4 Inserting an HDL components. Next we will show how a component that we are to write Verilog code for is added to our schematic. For this demonstration, in the Quartus II environment, we will describe the *shift_register* shown in Figure 6.33 in Verilog, create a symbol for it, and add its symbol to our schematic.

We begin the definition of our *shift_register* by opening a new Verilog file. This is similar to opening a schematic file as shown in Figure 6.26, except that Verilog HDL File must be selected in the New window. The procedure is shown below:

{Main}: File ⇨ New... ⇨ {New}-Device Design Files ⇨ Verilog HDL

After execution of these menus, a new file opens in which we will enter the Verilog code of our shift register. When completed, this file should be saved using the name of the module for the file name. Saving the *shift_register* module and its Verilog window are shown in Figure 6.33.

Before the above saved file can be used in our schematic, a symbol must be created for it. Menu items shown below must be executed for creating a symbol for our shift register.

{Main}: File ⇨ Create/update ⇨ Create Symbol Files for Current File

Note in the above procedure that the symbol will be created for the current file. This means that while this procedure is being done, the *shift_register.v* Verilog window must be active.

With the above procedures, Quartus II generates a default symbol for the *shift_register.v* file. Quartus II allows us to use its symbol editor to modify a symbol and perhaps use some of our artistic abilities for generation a more attractive symbol. To modify the shift register symbol (file name is *shift_register.bsf*), start in the main Quartus window, and from File menu click Open. Open *shift_register.bsf* and edit this symbol using the basic drawing tools that this editor provides. The new *shift_register* symbol in the symbol editor is shown in Figure 6.34.

After creating a symbol for our Verilog file, the corresponding component can be inserted in our schematic just like any of the existing Quartus II components. For the *shift_register*, first activate the schematic file, double-click anywhere on this window and in the list of libraries select Project, and there, select *shift_register*. With this procedure, the symbol of Figure 6.34 will be added to the Block-diagram of our serial adder.

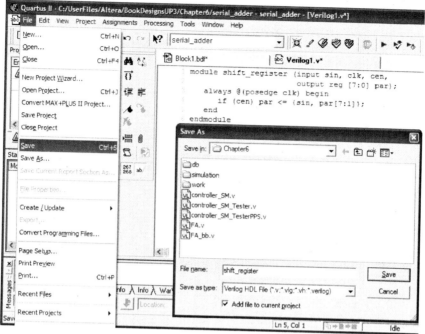

Figure 6.33 Saving a Verilog File

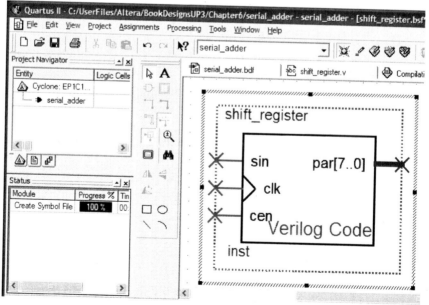

Figure 6.34 Editing Shift Register Symbol

6.3.3.5 Controller Components.
As shown in Figure 6.3, the controller of our serial adder has a state machine and a counter.

As discussed, the state machine part of the controller (*controller_SM*) was created, simulated, and synthesized in Section 6.2. In this section we also showed how a symbol was created for this component. The procedure for inserting *controller_SM* symbol in our serial adder schematic is the same as that discussed in Section 6.3.3.4 for inserting *shift_register*.

The counter part of our controller is built by selecting *lpm_counter* from the library of Quartus II Megafunctions, and configuring it as a 3-bit up-counter with an enable input, a carry output, and a synchronous clear input. The procedure for defining this component is similar to that discussed in Section 6.3.3.3 for creating and inserting *FA*. Figure 6.35 shows Megawizard pages for configuring the controller counter (*controller_CN*).

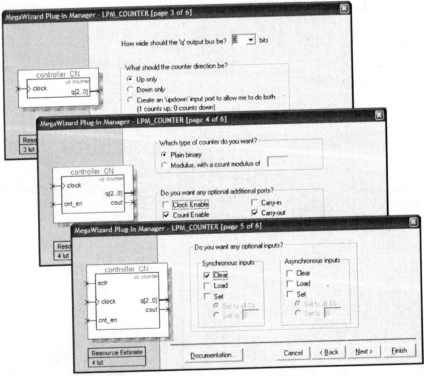

Figure 6.35 Configuring *lpm_counter* for controller count

6.3.4 Wiring Design Components

The Orthogonal Node Tool from the Block Diagram tool bar selects the tool for wiring components of the schematic window. When selected, using the mouse, wires can be drawn on the schematic of our design for connecting one node to another. The Orthogonal Bus Tool is used for wiring busses. For example connection from the 8-bit output of the *shift_register* to the output pin is done by the Bus Tool.

Figure 6.36 shows the complete schematic of *serial_adder* after completion of all wirings. Note here that the 8-bit *par* output of the *shift_register* component is wired to the 8-bit *result* which is a primary output of the serial adder. Naming the *result* output pin, is done in its Properties window. In this window its dimensions are defined as [7..0]. Quartus II uses two dots for index range.

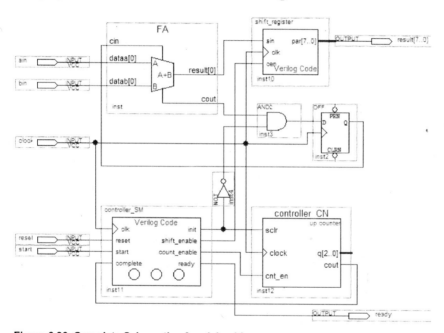

Figure 6.36 Complete Schematic of *serial_adder*

6.3.5 Design Compilation

The complete design of Figure 6.36 must be compiled for design errors and incompatibilities. For this purpose, follow menu items shown below:

{Main}: Processing ⇨ Start Compilation

Note that this process is no different than that discussed in Section 6.2.2.

6.3.6 Design Simulation

The next step in design after design entry is its simulation. Steps involved in simulation are waveform definition, and simulation run.

6.3.6.1 Waveform Definition.
To define our waveforms we create a new Vector Waveform file. For this purpose start from the File menu of the main Quartus window and follow menu items shown below:

{Main}: file ⇨ New... ⇨ {New}–Other Files ⇨ Vector Waveform File ⇨ [OK]

This is similar to opening a new schematic or HDL file as shown in Figure 6.26. The difference is that the waveform file is part of the Other Files tab. A waveform file has the *.vwf* extension.

Input ports of our design must be brought into the newly created waveform file before values can be assigned to them. For this purpose right-click in the Name area of the open waveform window and select Insert Node or Bus... in the menu that shows up. This will cause a new Insert Node or Bus window to open. In this window (shown in Figure 6.37), we can enter the input names for which waveforms are to be defined.

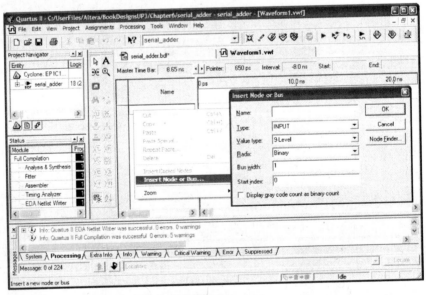

Figure 6.37 Waveform and Node Definitions

Individually naming each input can be avoided by having Quartus II find the input nodes for us. The Node Finder utility shown in Figure 6.37 can only be used if our design has been successfully compiled.

The Node Finder window opens by clicking the corresponding button in Figure 6.37. In this window (shown in Figure 6.38) click on the List button to see a list of available nodes. Note the "pins: all" selection.

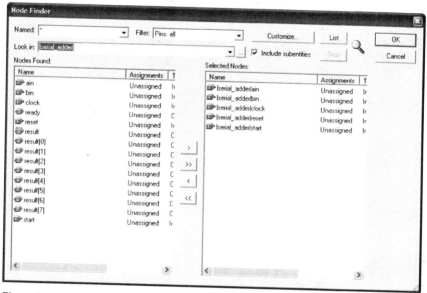

Figure 6.38 Node Finder Window

Select nodes from the left part of this window and add to the right. After clicking on OK, this window closes and the selected signals will be added to the waveform window. Another OK should be clicked.

Next is to set the end time of the simulation. For this purpose follows the procedure shown below:

{Main}: File ⇨ End Time ... ⇨ {End Time}

In the End Time window that appears, specify the end time and its units. We use 5.0 us for the simulation of our serial adder.

Figure 6.39 shows menus and windows used for definition of waveforms for the input nodes of our circuit. This is done by selecting the input and defining values for it from the waveform toolbar, or by right-clicking the input and assigning values to it using windows that open.

One way of assigning values to a waveform is by selecting it and using the mouse for highlighting certain time segments of it. The highlighter time segment can be assigned a value by clicking the value on the waveform toolbar.

Assignment of values using value menus is shown in Figure 6.39. In what is shown we are giving our *clock* signal a period of 57 ns with a duty cycle of 50%. This periodic signal continues until the simulation end time that is set to 5 us.

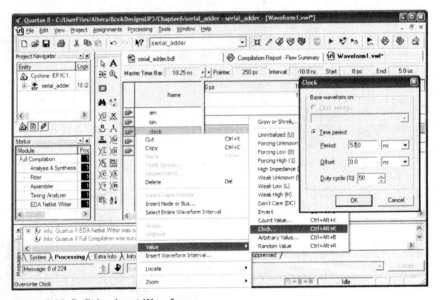

Figure 6.39 Defining Input Waveforms

6.3.6.2 Simulation Run. Simulation run begins by selecting Start Simulation of the Processing menu or by clicking the start simulation button on the Quartus standard toolbar. Figure 6.40 shows simulation result of our *serial_adder* design.

The result shown here verifies correct operation of our adder circuit. In order to use this circuit in another design or save it as a library component, a symbol must be created for it. The procedure for generation of a symbol is the same as that discussed in conjunction of Verilog files. A symbol generated as such can be used in upper level designs.

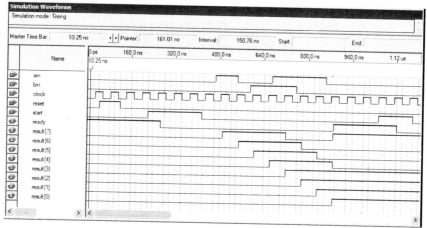

Figure 6.40 Simulation Run Results

6.3.7 Synthesis Results

Compiling a design as discussed in Section 6.3.5 synthesizes the design and generates various reports and diagrams. This section provides an overview of the information that is generated by Quartus II. Figure 6.41 shows the synthesis flow summary.

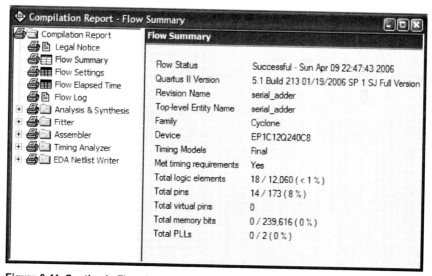

Figure 6.41 Synthesis Flow Summary

6.3.7.1 Flow Information. Quartus synthesis generates various report and log files showing the details of the synthesis process. This information includes compiler settings, device settings, elapsed time, resources used and legal notices.

6.3.7.2 Timing Information. An important part of the information provided by a synthesis tool is the timing information. This includes setup and hold time of registers, maximum clock frequency, and worst case delays. The maximum clock frequency of our *serial_adder* is reported to be 275.03 MHZ on the Cyclone device. This is shown in the timing analyzer section of the compilation report.

6.3.7.3 Hardware Information. Hardware information generated by the compiler includes diagrams showing FPGA utilized areas, interconnections, and logic diagrams. For viewing these diagrams go to Quartus Tools menu and select the appropriate viewing tool.

In what Quartus II refers to as timing Closure Floor Plan, logic blocks of the target FPGA (in our case Cyclone or Cyclone II) used for a design and input and output blocks are shown. Timings between various nodes of the implementation can be looked up by activating the corresponding timing tools. Figure 6.42 shows part of the floor plan of our design. The serial adder uses 18 of the 12060 logic elements of a Cyclone-EP1C12 FPGA. Figure 6.42 also shows some of the pins used for this design.

By right-clicking and locating a logic element in the resource property editor, the detailed view of the logic element appears. Figure 6.43 shows three of the LEs of *serial_adder*. The first LE is a flip-flop of the controller state machine of *serial_adder* with *reset, start* and data inputs going to the LUT of this LE. The second LE is one of the counter bits of the controller. This cell uses the carry input and output of the LUT for implementing the repetitive counter structure. Note here that the LUT block is used as two 3-input LUTs. The third LE is a combinational one and only its LUT is used. This LE is part of the logic of the controller state machine. The LUTs shown here are illustrative of some of many ways that LUTs can be formed in an Altera FPGA.

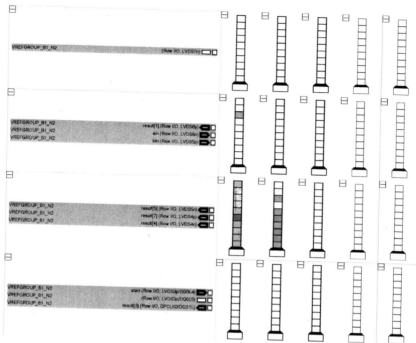

Figure 6.42 Timing closure floor plan

In what Quartus II refers to as Technology Map, mapping of FPGA resources to components of our design is shown. For every functional part of a design Logic Elements that are used are shown in this hierarchical view. Figure 6.44 shows the technology map of the *serial_adder* design. Blocks shown with darker shades, are FPGA LEs. LEs of other structures can be seen by double-clicking on them.

Gate level details of a synthesized design are available in its RTL view. In a hierarchical fashion, this view shows the complete gate level diagram of a design and their interconnections.

Figure 6.45 shows the RTL view of the serial adder. At the top-level, only the flip-flop used for saving the carry and its resetting AND gate are shown. To see gate details of other blocks, the corresponding block must be double-clicked.

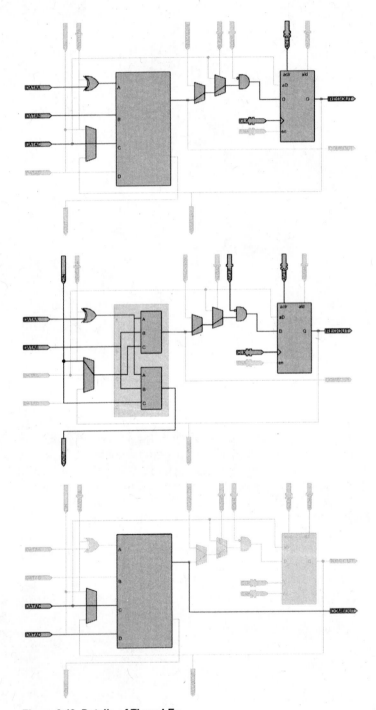

Figure 6.43 Details of Three LEs

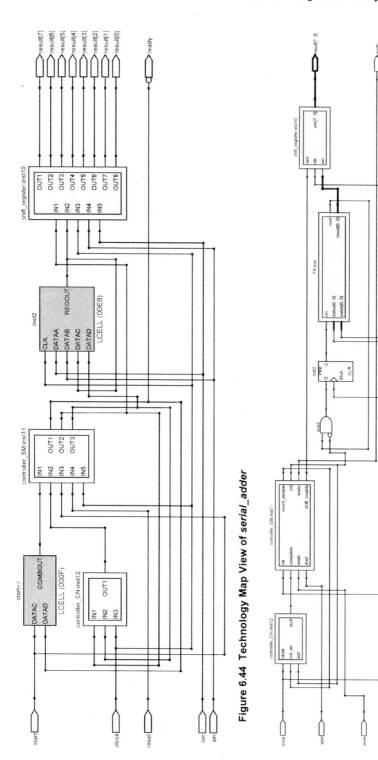

Figure 6.44 Technology Map View of *serial_adder*

Figure 6.45 RTL View of *serial_adder*

6.4 Design Prototyping

The next phase in completing a design after successful simulation, synthesis and verification of the design is design prototyping. In this phase an actual FPGA is programmed in a setup very similar to the setup the design will be used in. The FPGA is tested with actual hardware interfaces and its operation is examined.

We usually use development boards for design prototyping. A development board is a test board for a particular FPGA. In the center of such a board is the FPGA chip that the board is to be used for. A development board has standard interfaces and memories for designers to use for testing their designs.

Hardware for programming the FPGA of a development board is provided on this board. For designing with Cyclone, Altera provides several development boards. The UP3 (University Program 3) board is one such board. Another development board that will be discussed here is Altera's DE2 (or in general DE series). This board uses a Cyclone II EP2C35F672C6. The sections that follow give an overview of these development boards.

6.4.1 UP3 Board Specification

In the center of UP3 is a Cyclone EP1C12 FPGA. The board has a JTAG programming connector for direct programming of the FPGA, and a connector for AS (Active Serial) programming of its EPCS4 serial configuration device.

The board has four LEDs, four push buttons and four DIP switches that can be used for application of data or monitoring simple flags of design implemented on a Cyclone. In addition, UP3 has a PS/2 connector, a 25-pin printer port, a serial D-type connector, a standard VGA connector and a USB. Connections of all these devices (LEDs and switches included) to the Cyclone device are fixed, i.e., no user wirings are needed.

A 16×2 character LCD module is also part of UP3. This module connects to the connector on the side of the board that has permanent connections to the FPGA ports. Other features of UP3 include on board SRAM, FLASH and SDRAM, multiple clocks, an expansion header, an IDE interface connector, and I2C bus.

The front of the UP3 board is shown in Figure 6.46. The reference manual of UP3 education kit is included in the CD that accompanies this book. This is a concise document describing all details of the board and utilization of its components. In what follows, we will describe some of the main features of the UP3 board. The presentation is sufficient for the designs that are presented in chapters that follow.

Figure 6.46 UP3 Development Board

6.4.1.1 UP3 Programming.

Programming the Cyclone chip of UP3 is done by Altera's USB Blaster, or Byte Blaster. A USB Blaster connects to the USB port of your computer, while the Byte Blaster connects to the printer port. The Quartus II software must be setup to use the appropriate programming device.

The USB Blaster or Byte Blaster has a 10-pin connector to connect to UP3. This connector connects to JP11 for AS programming mode or to JP12 for JTAG programming. These connectors are on the left side of the front of the board as shown in Figure 6.46.

Active serial (AS) configuration of Cyclone is carried out through the serial device EPCS4. To program in AS mode, the download cable from a Blaster device must be connected to the 10-pin JP11 connector. With this connection, when programming is performed the EPCS4 device that is a serial FLASH memory is programmed. Upon power up, configuration data from EPCS4 is serially moved into the Cyclone FPGA. EPCS4 is an 8-pin chip labeled U15 and is located in the middle right on the back of UP3.

JTAG configuration directly downloads configuration data into Cyclone. The 10-pin header JP12 (to the right of JP11 of AS mode) is used for JTAG programming. JTAG programming configures the Cyclone memory and its data will be lost when power is removed from the board.

6.4.1.2 Basic IO Devices. Simple basic IO devices of UP3 are LEDs, Push Buttons, and DIP switches. There are four of each of these devices on UP3. These devices are located on the lower left part of the front of the UP3 board. In addition, a push button for resetting UP3 is also located in this same general area of the board.

Push buttons are connected to Cyclone IO pins through pull-up resistors. They are regarded as **0** when pressed and **1** when released. DIP switches are connected to Cyclone IO pins through pull-up resistors. These switches are considered active low, i.e., **0** when ON. The LEDs of UP3 are active high and are connected to Cyclone IO pins with current limiting resistors. Table 6.1 shows Cyclone pin connection of the basic devices discussed above.

Table 6.1 Basic Devices Pin Map (UP3)

	Push Buttons				DIP Switches				LEDs			
UP3 Labels	SW 4	SW 5	SW 6	SW 7	1	2	3	4	D6	D5	D4	D3
Cyclone Pins	48	49	57	62	58	59	60	61	56	55	54	53

6.4.1.3 The LCD Display. The LCD display of UP3 mounts on the U1 connector on the lower edge of the board. The display connects to the Cyclone pins as shown in Table 6.2.

Table 6.2 LCD FPGA Connections (UP3)

LCD Function	U1 Pin	Cyclone Pin
RS (Register Select)	4	108
R/W (Read / Write)	5	73
E (Enable Signal)	6	50
DB0	7	94
DB1	8	133
DB2	9	98
DB3	10	100
DB4	11	128
DB5	12	104
DB6	13	106
DB7	14	113

The LCD display has two rows of 16 characters. Standard ASCII characters are its default, however its user programmed RAM can be programmed for any desired character set. Before the LCD display can be written into for display, it has to be initialized and programmed. Programming is done by holding *RS* and *R/W* inputs low and applying instructions to *D7* through *D0* and issuing the *E* signal. Some of the key instructions of the LCD display and their execution time are shown in Table 6.3.

Table 6.3 LCD Instructions (UP3)

D[7:0]	Execute Time	Description
0000_0100	40 μs	Turn Off Display
0000_0001	1.64 ms	Clear Display
0000_001–	1.64 ms	Curser Home
0000_1111	40 μs	Display on, Curser on, Blink
0011_00––	40 μs	8 bit data, 1 line, 5×7 matrix
0011_10––	40 μs	8 bit data, 2 line, 5×7 matrix

After initializing the LCD (executing the first four instructions of Table 6.3), 8-bit ASCII data are written into it by setting RS, R/W to **1** and **0** respectively, placing ASCII data on D[7:0], and issuing the *E* signal and allowing 40 us. Each time a data byte is written into the LCD, the cursor moves one place to the right.

Moving the cursor back to the home position requires execution of the third instruction of Table 6.3.

6.4.1.4 PS/2 Keyboard and Mouse Connector. UP3 has a 6-pin mini-din type PS/2 connector for connecting a mouse or a keyboard. Figure 6.47 shows a PS/2 connector and its pin specifications.

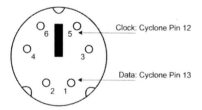

Figure 6.47 Mini-din PS/2 Connector (UP3)

A PS/2 connector has pins for bi-direction clock and data. These pins are wired to the Cyclone pins as shown in Table 6.4.

Table 6.4 PS/2 Pin Connections (UP3)

PS/2 Signal	PS/2 Pin	Cyclone Pin
Data	1	13
Clock	5	12

6.4.1.5 VGA Standard Connector. The standard VGA connector of
UP3 (Figure 6.48) is of the standard 15-pin D-type connector. This
connector has five active pins with permanent connections to the Cyclone IO pins.

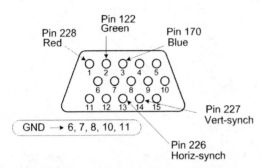

Figure 6.48 Standard VGA Connector (UP3)

Pins 1, 2 and 3 are for RGB color control, and 13 and 14 are for
horizontal and vertical synchronization. Connections of the VGA connector to Cyclone pins are shown in Table 6.5.

Table 6.5 VGA Pin Connections (UP3)

VGA Signal	VGA Pin	Cyclone Pin
Red	1	228
Green	2	122
Blue	3	170
Horizontal Synch	13	227
Vertical Synch	14	226

6.4.1.6 FLASH Memory Device. U8 that is located below the Cyclone FPGA on the UP3 board (see Figure 6.46) is a 16-Meg bit
FLASH. This device can be used as a 2-Meg 8-bit or a 1-Meg 16-bit
memory

To perform read and write operations, active low CE_n and
OE_n inputs of FLASH (Fed by Cyclone IO pins 117 and 118) must be
0. The WE_n must be **1** for reading and **0** for writing. The Ry/By_n
pin (pin 15 of FLASH, 126 of Cyclone) indicates if the FLASH is ready
for read and write operations, and if it has completed its last operation. Read and write operations can only be initiated when this
FLASH output is **1**.

The *Word/Byte* FLASH input must be **1** for configuring it as a
1M 16-bit wide words, and **0** for 2M 8-bit wide words.

Table 6.6 FLASH Cyclone Pin Connections (UP3)

FLASH Signal	FLASH Pin				Cyclone Pin			
A [20]	45				114			
A [19:16]	9	16	17	48	78	125	82	77
A [15:12]	1	2	3	4	76	75	74	68
A [11:08]	5	6	7	8	67	66	65	64
A [07:04]	18	19	20	21	63	83	84	85
A [03:00]	22	23	24	25	86	87	88	93
DQ [15:12]	45	43	41	39	114	107	105	127
DQ [11:08]	36	34	32	30	101	99	132	95
DQ [07:04]	44	42	40	38	113	106	104	128
DQ [03:00]	35	33	31	29	100	98	133	94
Write Enable (WE_n)	11				79			
Ready/Busy (RY/BY_n)	15				126			
Chip Enable (CE_n)	26				117			
Output Enable (OE_n)	28				118			
Word/Byte Makes 114 DQ[15] or A[20]	47				115			

6.4.1.7 SRAM Device. The U7 chip on the lower left side of front of UP3 is the on board SRAM of this board. This is a 64K word (65536 by 16 bits) asynchronous SRAM. The access time of this memory is about 10 ns. Table 6.7 shows pin descriptions and numbers of U7.

Writing is done by applying appropriate address and data, holding CE_n and OE_n low and applying a **0** to WE_n. For reading from the memory WE_n must be inactive (**1**). LB_n and UB_n enable lower and higher bytes, respectively.

Table 6.7 SRAM-Cyclone Pin Connections (UP3)

SRAM Signal	SRAM Pin				Cyclone Pin			
Addr [15:12]	1	2	3	4	76	75	74	68
Addr [11:08]	5	18	19	20	67	66	65	64
Addr [07:04]	21	24	25	26	63	83	84	85
Addr [03:00]	27	42	43	44	86	87	88	93
DQ [15:12]	38	37	36	35	114	107	105	127
DQ [11:08]	32	31	30	29	101	99	132	95
DQ [07:04]	16	15	14	13	113	106	104	128
DQ [03:00]	10	9	8	7	100	98	133	94
Write Enable (WE_n)	17				79			
Low Byte (LB_n)	39				77			
Upper Byte (UB_n)	40				82			
Chip Enable (CE_n)	68				116			
Output Enable (OE_n)	41				118			

6.4.1.8 SDRAM Memory Device.

The U6 chip of UP3 is an 8 Meg Byte Synchronous Dynamic RAM. This memory is located on the lower left side of the back of the UP3 board. The memory is organized as a 1Meg * 16-bit * 4-bank memory. Memory read and write operations are clocked with the rising edge the clock (*clk*) input. Pin connections of this device are shown in Table 6.8.

Table 6.8 SDRAM-Cyclone Pin Connections (UP3)

SDRAM Signal	SDRAM Pin				Cyclone Pin			
AD [11:08]	35	22	34	33	67	66	65	64
AD [07:04]	32	31	30	29	63	83	84	85
AD [03:00]	26	25	24	23	86	87	88	93
Bank [1:0]	21	20			74	68		
CAS_n	17				75			
RAS_n	18				76			
DQ [15:12]	53	51	50	48	114	107	105	127
DQ [11:08]	47	45	44	42	101	99	132	95
DQ [07:04]	13	11	10	8	113	106	104	128
DQ [03:00]	7	5	4	2	100	98	133	94
Write Enable (WE_n)	16				79			
Lower Byte (LDQM)	15				77			
Upper Byte (UB_n)	39				82			
Chip Enable (CE_n)	19				119			
Clock Enable (CKE)	37				115			
Clock (CLK)	38				11			

6.4.1.9 UP3 Clock Circuitry.

The UP3 board has a master clock chip that provides various clocks for the FPGA use. On chip clocks can be configured for various frequencies by placing appropriate jumpers on JP3 and J7 headers. These headers are located next to each other on the lower part, in the middle of the UP3 board. Clock inputs of the FPGA are pins 29, 152 and 153. Table 6.9 shows jumper settings for clock inputs of Cyclone.

Table 6.9 UP3 Clocks for Cyclone

Clock	Frequency (MHz)	Pins Shorted	Cyclone Pin
USBCLK	48	JP3: 4 and 3	29
REF0CLK	14.318	JP3: 4 and 6	29
PCICLK_E	33.33	JP3: 8 and 7	152
REF0CLK	14.318	JP3: 8 and 6	152
CPU Clock	66	J7: 1 and 2	153
CPU Clock	100	J7: 2 and 3	153

Several other clocks are available that are not connected to the FPGA pins. There is also a user clock header that is connected to pin 38 of Cyclone. Clocks shown in Table 6.9 are sufficient for most designs, and the above discussion is complete for such uses.

6.4.1.10 Other UP3 Interface Devices. Other devices or connectors not discussed above are expansion prototype connector, IDE, I2C Bus, and USB. Furthermore, the board has power and resetting circuitry that are not included in the above discussion.

These ports and devices are discussed in the UP3-1C12 Reference Manual that is included in the CD that accompanies this book.

6.4.2 DE2 Board Specification

This section discusses the DE2 board, its programming and its utilities. The DE2 development board offers many more hardware utilities than UP3. In addition, this board has a control program with a control panel for setting it up, initializing it, and loading its FLASH and other memories. This section provides a quick overview of DE2; the user's manual of this board that is provided on the CD that accompanies this book is a comprehensive document and provides detailed information on using and configuring the board. The brief introduction that follows is sufficient for the projects discussed in this book.

The DE2 board (Figure 6.49) uses Altera's Cyclone II 2C35 FPGA chip. It also has an EPCS16 for programming the FPGA. The board has a JTAG programming connector for direct programming of the FPGA, and a mode for AS (Active Serial) programming of its EPCS16 serial configuration device. The following hardware that can be used for testing various embedded designs is provided on the DE2 board:

- 4 pushbutton switches
- 18 toggle switches
- 18 red user LEDs and 9 green user LEDs
- 50-MHz oscillator and 27-MHz oscillator for clock sources
- A 2-line, 16 character LCD display
- 512-Kbyte SRAM, 8-Mbyte SDRAM, and 4-Mbyte FLASH
- SD Card socket
- 24-bit audio CODEC with line-in, line-out, and microphone jacks
- VGA DAC (10-bit triple DACs) with VGA-out connector
- TV Decoder (NTSC/PAL) and TV-in connector
- 10/100 Ethernet Controller with a connector
- USB Host/Slave Controller with type A and B connectors
- RS-232 transceiver and 9-pin connector
- PS/2 mouse/keyboard connector

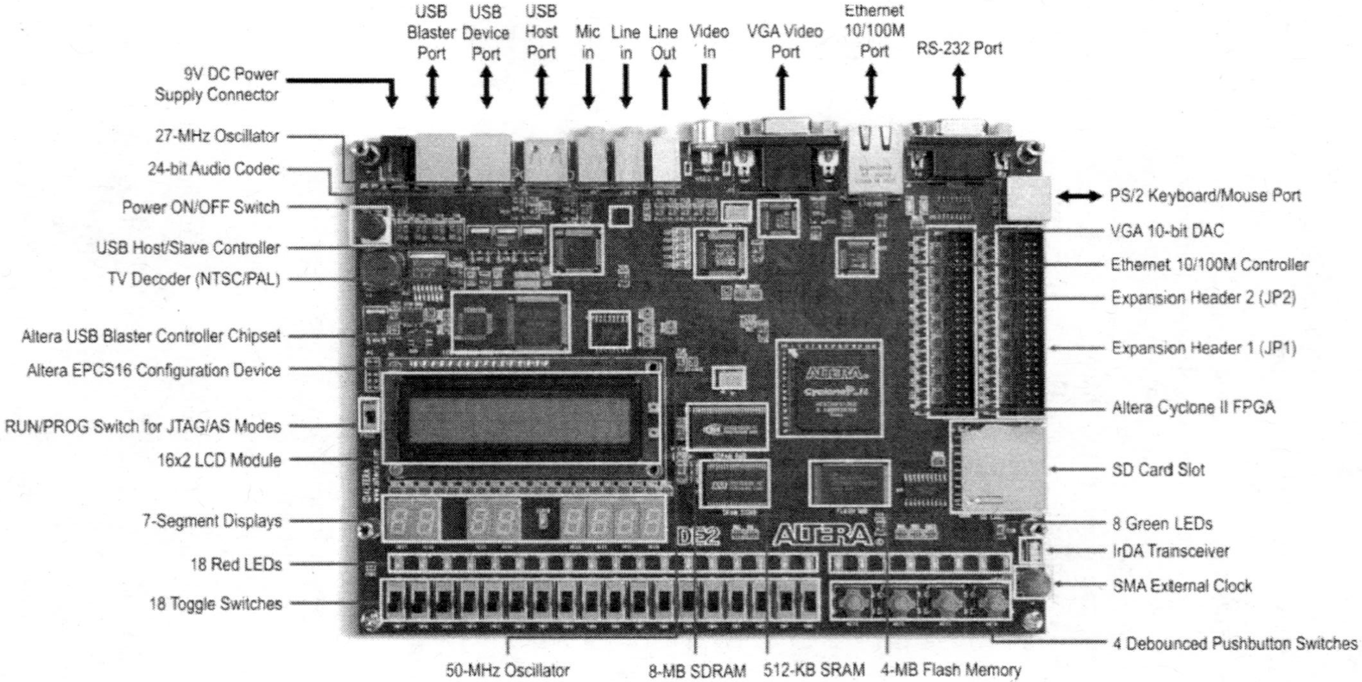

Figure 6.49 DE2 Development Board

6.4.2.1 DE2 Programming.
Programming the Cyclone II chip of DE2 is done through its built-in USB Blaster. A USB cable connects the programming USB connector of DE2 (upper left side of the board) to a USB port of the computer you are using for programming it. The Quartus II software must be setup to use this programming device. A toggle switch on the left side of the board sets the board for JTAG (RUN switch position) or AS (PROG switch position).

Active serial (AS) configuration of Cyclone II is carried out through the serial device EPCS16. To program in AS mode, the programming USB cable must be connected and the programming switch must be in PROG position. With this setup, when programming is performed the EPCS16 device that is a serial FLASH memory is programmed. Upon power up, configuration data from EPCS16 is serially moved into the Cyclone II. JTAG configuration directly downloads configuration data into Cyclone II. JTAG programming configures the Cyclone II memory and its data will be lost when power is removed from the board.

6.4.2.2 Basic IO Devices.
Simple basic IO devices of DE2 are Toggle switches, Push Buttons, LEDs, and SSD displays. There are seventeen switches, four push buttons, a total of twenty six LEDs, and eight SSD or HEX display units. These devices are located on the lower side the front of the DE2 board.

Toggle switches are connected directly to Cyclone IO. The switches provide logic **0** when in the Down position. Table 6.10 shows mapping of toggle switches to DE2 Cyclone II pins.

Table 6.10 Toggle Switches Pin Map (DE2)

DE2 ToggleSwitches				Cyclone II Pins			
SW	3	2	1	0	AE14	P25	N26	N25
SW	7	6	5	4	C13	AC13	AD13	AF14
SW	11	10	9	8	P1	N1	A13	B13
SW	15	14	13	12	U4	U3	T7	P2
SW	17	16			V2	V1		

There are four debounced push buttons (keys) on the DE2 board. These keys are directly connected to Cyclone II pins. Normally Key 0 through Key 3 provide logic **1**, and they become **0** when pressed. Table 6.11 shows connections of these keys to the Cyclone II FPGA.

Table 6.11 Push Button Pin Map (DE2)

DE2 Push Buttons				Cyclone II Pins			
KEY	3	2	1	0	W26	P23	N23	G26

DE2 provides seventeen red LEDs and nine green ones. The LEDs are connected to Cyclone II pins through pull-up resistors. Driving an LED to logic value **1** turns it on. Table 6.12 shown Cyclone II pin connections to LEDs.

Table 6.12 Red and Green LED Pin Map (DE2)

DE2 LEDs					Cyclone II Pins			
Red LEDs	3	2	1	0	AC22	AB21	AF23	AE23
	7	6	5	4	AC21	AD21	AD23	AD22
	11	10	9	8	AC14	AA13	Y13	AA14
	15	14	13	12	AE13	AF13	AE15	AD15
	17	16			AD12	AE12		
Green LEDs	3	2	1	0	V18	W19	AF22	AE22
	7	6	5	4	Y18	AA20	U17	U18
	8				Y12			

As shown in Figure 6.49, there are eight seven-segment displays on the DE2 board. The segments of these display units are like the LEDs and are directly connected to Cyclone II pins. Table 6.13 shows Cyclone II pin connections to the seven-segment displays.

Table 6.13 SSD (HEX Display) Pin Map (DE2)

DE2 HEX (SSDs)		Cyclone II Pins						
HEX0	Seg 6 to 0	V13	V14	AE11	AD11	AC12	AB12	AF10
HEX1	Seg 6 to 0	AB24	AA23	AA24	Y22	W21	V21	V20
HEX2	Seg 6 to 0	Y24	AB25	AB26	AC26	AC25	V22	AB23
HEX3	Seg 6 to 0	W24	U22	Y25	Y26	AA26	AA25	Y23
HEX4	Seg 6 to 0	T3	R6	R7	T4	U2	U1	U9
HEX5	Seg 6 to 0	R3	R4	R5	T9	P7	P6	T2
HEX6	Seg 6 to 0	M4	M5	M3	M2	P3	P4	R2
HEX7	Seg 6 to 0	N9	P9	L7	L6	L9	L2	L3

6.4.2.3 The LCD Display. The LCD display of DE2 is located above the SSDs on the left side of the board. The display connects to the Cyclone II pins as shown in Table 6.14.

The LCD display has two rows of 16 characters. Standard ASCII characters are its default, however its user programmed RAM can be programmed for any desired character set. Before the LCD display can be written into for display, it has to be initialized and programmed. Programming is done by holding RS and R/W inputs low and applying instructions to $D7$ through $D0$ and issuing the E signal.

Some of the key instructions of the LCD display and their execution time are shown in Table 6.15.

Table 6.14 LCD FPGA Connections (DE2)

LCD Function	Cyclone II Pin
RS (Register Select)	K1
R/W (Read / Write)	K4
E (Enable Signal)	K3
DB0	J1
DB1	J2
DB2	H1
DB3	H2
DB4	J4
DB5	J3
DB6	H4
DB7	H3

Table 6.15 LCD Instructions (DE2)

D[7:0]	Execute Time	Description
0000_0100	40 μs	Turn Off Display
0000_0001	1.64 ms	Clear Display
0000_001-	1.64 ms	Curser Home
0000_1111	40 μs	Display on, Curser on, Blink
0011_00--	40 μs	8 bit data, 1 line, 5×7 matrix
0011_10--	40 μs	8 bit data, 2 line, 5×7 matrix

After initializing the LCD (executing the first four instructions of Table 6.15), 8-bit ASCII data are written into it by setting RS, R/W to **1** and **0** respectively, placing ASCII data on D[7:0], and issuing the *E* signal and allowing 40 us. Each time a data byte is written into the LCD, the cursor moves one place to the right. Moving the cursor back to the home position requires execution of the third instruction shown in this table.

6.4.2.4 PS/2 Keyboard and Mouse Connector. DE2 has a 6-pin mini-din type PS/2 connector for connecting a mouse or a keyboard. Figure 6.50 shows a PS/2 connector and its DE2 pin connections. Connections of PS/2 bi-direction clock and data pins to Cyclone II pins are shown in Table 6.16.

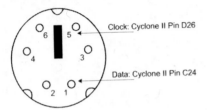

Clock: Cyclone II Pin D26

Data: Cyclone II Pin C24

Figure 6.50 Mini-din PS/2 Connector (DE2)

Table 6.16 PS/2 Pin Connections (DE2)

PS/2 Signal	PS/2 Pin	Cyclone II Pin
Data	1	C24
Clock	5	D26

6.4.2.5 VGA Standard Connector. The standard VGA connector of DE2 (Figure 6.51) is of the standard 15-pin D-type connector. This connector has three pins for color and two pins for horizontal and vertical synchronizations. Horizontal and vertical synchronization inputs are directly connected to Cyclone II pins of DE2. The color inputs come from a triple DAC (Digital to Analog Converter) device that takes 10 bit digital data and generates analog signals for intensity of each of the three colors.

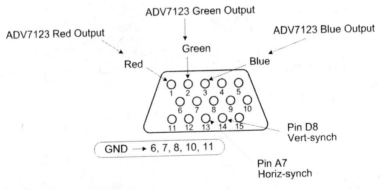

Figure 6.51 Standard VGA Connector (DE2)

As shown in Figure 6.51, Pins 1, 2 and 3 of the VGA connector are connected to RGB color outputs of the on-board ADV7123. This device has thirty color inputs (10 for each color) that are directly connected to Cyclone II pins. In addition, it has Blank, Synch and Clock inputs that are also driven by DE2 Cyclone II. Table 6.17 shows connection of VGA related pins to DE2 Cyclone II.

Table 6.17 VGA Pin Connections (DE2)

VGA Signals	Cyclone II Pins									
VGA_R [9:0]	E10	F11	H12	H11	A8	C9	D9	G10	F10	C8
VGA_G [9:0]	D12	E12	D11	G11	A10	B10	D10	C10	A9	B9
VGA_B [9:0]	B12	C12	B11	C11	J11	J10	G12	F12	J14	J13
VGA_CLK	B8									
VGA_BLANK	D6									
VGA_HS	A7									
VGA_VS	D8									
VGA_SYNC	B7									

6.4.2.6 FLASH Memory Device. U20 that is located below the Cyclone II FPGA on the DE2 board (see Figure 6.49) is a 4-Mega byte FLASH. Cyclone II pin connections to this device are shown in Table 6.18.

Table 6.18 FLASH Cyclone II Pin Connections (DE2)

FLASH Signal	Cyclone Pin			
A [21:20]	Y14	Y15		
A [19:16]	AA15	AB15	AC15	AE16
A [15:12]	AD16	AC16	W15	W16
A [11:08]	AF17	AE17	AC17	AD17
A [07:04]	AA16	Y16	AF18	AE18
A [03:00]	AF19	AE19	AB18	AC18
DQ [07:04]	AE21	AF21	AC20	AB20
DQ [03:00]	AE20	AF20	AC19	AD19
Write Enable (WE_n)	AA17			
Reset (RST_n)	AA18			
Chip Enable (CE_n)	V17			
Output Enable (OE_n)	W17			

To perform read and write operations, active low CE_n and OE_n inputs of FLASH (Cyclone II pins V17 and W17) must be **0**. The WE_n must be **1** for reading and **0** for writing.

6.4.2.7 SRAM Device. The U18 chip on the lower left side of the FPGA of DE2 is the on board SRAM of this board. This is a 512-K byte asynchronous SRAM. Table 6.19 shows Cyclone II pin connections to this device.

Writing is done by applying appropriate address and data, holding CE_n and OE_n low and applying a **0** to WE_n. For reading from the memory WE_n must be inactive (**1**). LB_n and UB_n enable lower and higher bytes, respectively.

Table 6.19 SRAM-Cyclone II Pin Connections (DE2)

SRAM Signal	Cyclone II Pin			
Addr [17:16]	AC8	AB8		
Addr [15:12]	Y10	W10	W8	AC7
Addr [11:08]	V9	V10	AD7	AD6
Addr [07:04]	AF5	AE5	AD5	AD4
Addr [03:00]	AC6	AC5	AF4	AE4
DQ [15:12]	AC10	AC9	W12	W11
DQ [11:08]	AF8	AE8	AF8	AE7
DQ [07:04]	Y11	A11	AB10	AA10
DQ [03:00]	AA9	AF6	AE6	AD8
Write Enable (WE_n)	AE10			
Low Byte (LB_n)	AE9			
Upper Byte (UB_n)	AF6			
Chip Enable (CE_n)	AC11			
Output Enable (OE_n)	AD10			

6.4.2.8 SDRAM Memory Device. The U17 chip of DE2 is an 8 Meg Byte Synchronous Dynamic RAM. This memory is located on the DE2 board on the left side of Cyclone II. The memory is organized as a 1Meg × 16-bit × 4-bank memory. Memory read and write operations are clocked with the rising edge the clock (*clk*) input. Pin connections of this device are shown in Table 6.20.

Table 6.20 SDRAM-Cyclone II Pin Connections (DE2)

SDRAM Signal	Cyclone Pin			
AD [11:08]	V5	Y1	W3	W4
AD [07:04]	U5	U7	U6	W1
AD [03:00]	W2	V3	V4	T6
Bank [1:0]	AE3	AE2		
CAS_n	AB3			
RAS_n	AB4			
DQ [15:12]	AA5	AC1	AC2	AA3
DQ [11:08]	AA4	AB1	AB2	W6
DQ [07:04]	V7	T8	R8	Y4
DQ [03:00]	Y3	AA1	AA2	V6
Write Enable (WE_n)	AD3			
Lower Byte (LDQM)	AD2			
Upper Byte (UDQM)	Y5			
Chip Select (CS_n)	AC3			
Clock Enable (CKE)	AA6			
Clock (CLK)	AA7			

6.4.2.9 DE2 Clock Circuitry.

The DE2 board provides two clocks that have permanent connections to Cyclone II pins. In addition, an external pin on the lower left corner of the board has a permanent connection to the FPGA for any external clock that may be needed for a design. Table 6.21 shows Cyclone II clock pin connections.

Table 6.21 DE2 Clocks for Cyclone II

Clock	Frequency (MHz)	Cyclone Pin
CLOCK_27	27	D13
CLOCK_50	50	N2
EXT_CLOCK	External	P26

6.4.2.10 Other DE2 Devices.

As previously mentioned, there are several other more advanced devices on the DE2 board. We limited our discussion here only to the very basic interface devices. The DE2 user's manual and datasheets for the individual devices are provided on the CD that accompanies this book and Altera's DE2 System CD, and provide complete information about using various DE2 resources.

6.4.3 Programming DE2 Cyclone II

After simulation and synthesis, the next step in design is to program the target FPGA in a development board. This is done by a computer running Quartus II, connected by USB Blaster to UP3 or DE2. Programming the FPGA of our development board and using the existing resources of this board for testing our design is the final phase of design prototyping. This phase requires pin assignment, programmer setup, and device programming. In what follows, we will show how our serial adder example is programmed into the Cyclone II FPGA of a DE2 board. Except for the pin connections, the procedure is the same for a UP3 board.

6.4.3.1 Pin Assignment.

The last sections discussed available IO devices and mapping of their pins to Cyclone II pins. In order to take advantage of these IO devices, we have to assign ports of our design to Cyclone II pins connected to the corresponding devices.

Our serial adder of Figure 6.36 has inputs *ain*, *bin*, *reset*, *start* and *clock*, and output *ready* and *result [7:0]*. Toggle switches SW17 and SW16 (Table 6.10), and push buttons KEY3 and KEY2 (Table 6.11) are used for *ain*, *bin*, *reset*, and *start*, respectively. The 50 MHZ clock of DE2 is used for the *clock* input of the serial adder (see Table 6.21). The *ready* output goes on green LED LEDG0 (Table 6.12). Bits

7 through 0 of *result* are displayed on red LEDs, LEDR17 through LEDR10 (Table 6.12). Table 6.22 shows port connections of *serial_adder* to DE2 devices and Cyclone II pins.

Table 6.22 *serial_adder* Port Connection (DE2)

Ports	DE2 Device				Cyclone Pin			
ain	SW17				V2			
bin	SW16				V1			
reset	KEY3				W26			
start	KEY2				P23			
clock	50	MHz	Clock		N2			
ready	LEDG0				AE22			
result [7:4]	LEDR17	LEDR16	LEDR15	LEDR14	AD12	AE12	AE13	AF13
result [4:0]	LEDR13	LEDR12	LEDR11	LEDR10	AE15	AD15	AC14	AA13

In order to program our design for the pin specifications shown in this table, Quartus II Assignment Editor window or Pin Planner can be used. The Assignment Editor window can also be opened by selecting it from standard Quartus II toolbar (Figure 6.24). In the Edit part of the Assignment Editor window select ports of *serial_adder* under the To column and select Cyclone II pin number under the Value column. The Assignment Editor window, after assigning all pins as specified in Table 6.22 is shown in Figure 6.52.

To	Assignment Name	Value
ain	Location	PIN_V2
bin	Location	PIN_V1
reset	Location	PIN_W26
clock	Location	PIN_N2
ready	Location	PIN_AE22
result[7]	Location	PIN_AD12
result[6]	Location	PIN_AE12
result[5]	Location	PIN_AE13
result[4]	Location	PIN_AF13
result[3]	Location	PIN_AE15
result[2]	Location	PIN_AD15
result[1]	Location	PIN_AC14
result[0]	Location	PIN_AA13
start	Location,	PIN_P23

Figure 6.52 Assigning *serial_adder* Pins Using Assignment Editor

6.4.3.2 Programmer Setup.

The first time Quartus II is to program a device, it has to be setup for use of Byte Blaster, or USB Blaster, or any other programming method. We show how this is done for the USB Blaster. This procedure is no different than that for the Byte Blaster. DE2 only uses its onboard USB Blaster.

Programming Cyclone II begins with activating the **Programmer** tool from the **Tools** menu as shown below:

{Main}: Tools ⇨ Programmer

This window allows you to setup the programming hardware, select programming mode, and start programming Cyclone II.

In this window, press the **Hardware Setup...** button to open the **Hardware Setup** window as shown in Figure 6.53. From the list of available hardware items in this window select the hardware you are using for programming (in our case, USB Blaster), and close this window.

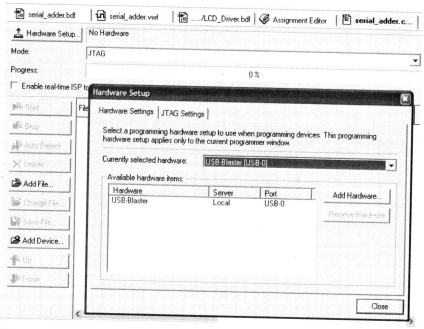

Figure 6.53 Programming Hardware Setup

This procedure is done only once and need not be repeated unless you reinstall Quartus II, or want to change your programming hardware.

6.4.3.3 Programming Cyclone II.

The final setup in implementing your hardware on a Cyclone is selecting a program file and starting the programmer. This is done in the programming window after hardware setup is complete.

For this purpose, we have to decide whether we are programming Cyclone II in AS or JTAG mode. The AS (Active Serial) programs the on-board FLASH and is non-volatile, while the JTAG mode programs the Cyclone II itself and is volatile. The AS mode uses *.pof* programming files, while to JTAG mode uses *.sof* files. We use the JTAG mode and need *serial_adder.sof* file for this purpose.

The JTAG programming mode is selected in the programming window by clicking **Mode:** and selecting JTAG. Following this, the program file must be selected. For this purpose click on **Add File...** to open the window for file selection. This is shown in Figure 6.54.

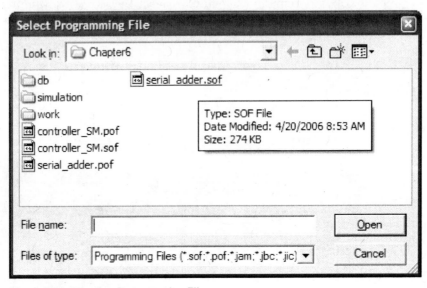

Figure 6.54 Selecting Programming File

We select *serial_adder.sof* file that is the right file for JTAG programming mode. After the file is selected it appears in the white space in the programming window. Check the **Program/Configure** column next to the file and click on **start** to start programming the *serial_adder* into Cyclone. Figure 6.55 shows the Quartus II window after completion of device programming.

Figure 6.55 Device Programming

6.4.3.4 Testing *serial_adder*. The serial adder that is programmed into Cyclone II can now be tested. The setup discussed above makes it difficult to test our serial adder because of its fast clock. The 50 MHz clock runs the complete addition process before we even have a chance to set its input data. To correct this problem, we should use one of the push buttons for the circuit clock, and manually operate this clock while setting appropriate input data. Readers are encouraged to go through this process and verify the operation of the serial adder.

In the next chapter we develop several library components for better utilization of UP3 and DE2 resources. For example, the LCD display with the help of a hardware driver can be used for displaying numbers and characters.

6.5 Summary

This chapter showed how various parts of a design could be implemented and tested using Altera simulation tools, synthesis programs, and two of Altera's development boards. We discussed a design that, in spite of its small size, had components that were implemented with various mechanisms available in Altera's tool set. This chapter can be regarded as a complete tutorial for using ModelSim, Quartus II, UP3 and DE2 boards. In addition, the presentation of two Altera development boards shows readers how an FPGA is used in a board level design and how various components such as memories and interface devices operate.

7 Design of Utility Hardware Cores

The previous chapter showed how various design tools could be utilized for design, implementation, and prototyping a complete design. We showed how existing library components could be used in a new design. This chapter discusses creation and managements of general purpose hardware components that can be useful in more complex designs. We refer to such components as utility cores that are placed in a library that can be accessed by various designers.

Utility cores cover a wide range of complexity, and are different from one design to another. Our main purpose of covering this material here is to introduce the concept of hardware components that are packaged into a library and are used by other designs. At the same time we introduce several common interfaces and familiarize readers with operation of devices such as keyboards and displays.

We begin this chapter with a discussion of libraries, components of a library, and accessing these components. The sections that follow this discussion introduce utility hardware components ranging from a simple switch debouncer to a VGA display adaptor. Every component designed will be placed in our library of utility components and a tester will be developed for it.

7.1 Library Management

The utility cores that we develop in this chapter are primarily IO interface drivers that will be used in the next two chapters and in various design projects. We create a directory for our library components and place each of the library components as projects in this directory. For designs implemented on the UP3 board, the directory is

UP3Library that must be added as a "User Library" to new projects that use components of this library. We use *DE2Library* for DE2 designs.

In addition to *UP3Library*, another library (*UP3LibraryTesters*) is created that contains testers for the components or *UP3Library*. Projects created in *UP3LibraryTesters* are examples of utilization of library elements of *UP3Library*. Tester projects verify basic operations of our library components and in many cases combine several of *UP3Library* components into a tester. Components of our *UP3Library* range from a simple switch debouncer to a VGA controller.

Similarly, we create a test library for our designs of the DE2 board. This library is named *DE2LibraryTesters*. most designs in UP3 and DE3 libraries are the same, except that the former uses Cyclone and the latter uses the Cyclone II FPGA.

7.2 Basic IO Device Handling

This section develops interfaces for utilization of push buttons for providing control and data signals of a design and for single stepping through a design. In an example at the end of this section we show a simple design that uses LEDs, DIP switches, and Pushbuttons.

7.2.1 Debouncer

Pushbuttons are mechanical switches and are not debounced. This means that when you press a pushbutton, it makes several contacts before it stabilizes. The result is that when you press a pushbutton that is **1** in the normal position, its output changes several times between logic **0** and **1** before it becomes **0**, and when you release it, it again switches several times between these logic values before it becomes **1**. Figure 7.1 shows a pushbutton contact bounce.

In some cases, pushbuttons are debounced using a Schmitt Trigger circuit. If not, a logical circuit can be designed for debouncing a pushbutton. UP3 pushbuttons are not debounced, but those of the DE2 board are.

Pushbutton bouncing causes no problem if it is used as an input of a combinational circuit for test purposes. In this case, you should give enough time for all changes to propagate before reading the switch's output. However, in sequential circuits with a fast clock, each of the bounces between **0** and **1** logic values may be regarded as an actual logic value. For example, for a counter with a fast clock for which a mechanical pushbutton is used as a count input, pressing the pushbutton may cause several counts.

Figure 7.1 Contact Bounce in a Pushbutton

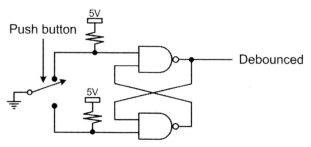

Figure 7.2 Debouncing a Single-Pole Double-Throw (SPDT) Switch

A Single-Pole Double-Throw (SPDT) mechanical switch such as that shown in Figure 7.2 can easily be debounced by an SR-latch also shown in this figure. However, a UP3 pushbutton is Single-Pole Single-Throw (SPST) and its only available terminal is the output that connects to Gnd for logic **0** or to Vdd through a pullup resistor for logic **1**.

Debouncing UP3 pushbuttons requires a slow clock to sample the switch output before and after it is pressed or released. The clock should be slow enough to bypass all the transitional changes that occur on its output terminal. In what follows we show the generation of a switch debouncer and its necessary clock. For the design of the former part we use schematic entry at the gate level, and for the latter part we use schematic entry using Quartus II megafunctions.

Although debouncing DE2 pushbuttons is not necessary, this first project is a good start and provides general information for setting up a design and testing it. You can still use the debouncer circuit that we develop here with DE2 pushbuttons, even though the Schmitt Trigger circuit exits.

7.2.1.1 Debouncer-Gate Level Entry.

The *debouncer* project is created in *UP3Library*. We use schematic entry at the gate level for this design. The Quartus II schematic of *debouncer* is shown in Figure 7.3. The inputs of this circuit are *PushButton* and *SlowClock*.

One of the flip-flops used here is triggered on the rising edge of *SlowClock* and the other is triggered on the falling edge of this clock. Since the output of this circuit is generated by ANDing the two flip-flop outputs, both flip-flops must see logic **1** on their inputs before the output of the circuit becomes **1**. This means that the pushbutton connected to the *PushButton* input of this circuit must stay high for the entire duration of the slow clock for the circuit output to become **1**.

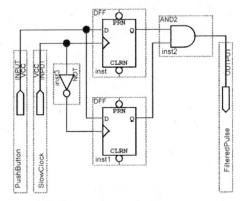

Figure 7.3 Debouncer Schematic

This design is entered in the Quartus II environment using its block diagram editor. Flip-flops used here are part of the Quartus II library of primitives. To use this circuit in other designs, a symbol must be created for it. Figure 7.4 shows the symbol after it was created by Quartus II and modified using the symbol editor utility of Quartus II.

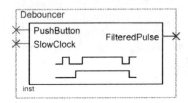

Figure 7.4 Debouncer Symbol

This part of our design can be simulated, but the real test of this circuit is using it with UP3 pushbuttons. This requires the use of a slow clock that will be created next. UP3 has a slow 14.318 MHZ clock that becomes available on pin 152 of Cyclone when pins 6 and 8 of JP3 are connected. The slowest clock available on a DE2 board is its 27 MHz clock on pin D13.

7.2.1.2 Slow Clock Using Megafunctions.

Obviously a 14 MHz clock is too fast for filtering transitions in pushbuttons. Dividing this clock by 2^{18} produces a 54 Hz clock that will be more adequate for filtering slow mechanical transitions. We use an 18-bit counter for dividing the 14 MHz clock of UP3. The divider circuit is implemented by use of a counter megafunction.

The divider output is the most significant bit of the 18-bit count output. Figure 7.5 shows the circuit of the divider circuit and its corresponding symbol. The waveform shown in the symbol of this circuit has been manually added to the default symbol that is automatically generated by Quartus II.

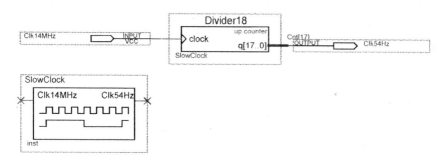

Figure 7.5 Slow Clock Generation

7.2.1.3 A Debounced Switch Using Completed Parts. By putting together the hardware of Figure 7.4 and Figure 7.5, hardware for debouncing a switch is generated. For this hardware we generate the *CleanPulse* project in the *UP3Library* directory, and in its schematic we use symbols for *Debouncer* (Figure 7.4) and *SlowClock* (Figure 7.5).

The schematic of *CleanPulse* is shown in Figure 7.6. For using this circuit, its *FastClock* input must be driven by clock at Pin 152 of Cyclone, and the output of a pushbutton being debounced must be connected to its *PushButton* input.

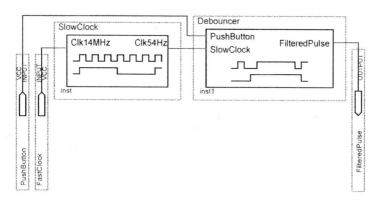

Figure 7.6 *CleanPulse* Schematic

The *SlowClock* part of this circuit can be shared for debouncing multiple pushbuttons. For single pushbutton debouncing a symbol is created that is shown in Figure 7.7.

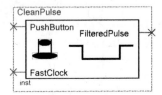

Figure 7.7 *CleanPulse* **Symbol**

7.2.2 Single Stepper

Debugging a sequential circuit requires stepping through its states and examining its signals in every state. Single stepping cannot be done in the normal mode of operation where clock frequencies are 14 MHz or higher. This section shows the design of a circuit that generates a single clock pulse on its output for every time its pushbutton input is pressed. We refer to the circuit we are designing as *Clean1Pulser*. This circuit first debounces a pushbutton and then generates one pulse for every time the pushbutton is pressed.

Since DE2 pushbuttons do not require debouncing, only the part of *Clean1Pulser* circuit that puts out one pulse is required for designs on this board.

7.2.2.1 Design of a One-Pulser. Generally, when a pushbutton is manually operated, the duration of the time that the switch is kept pressed is in the order of seconds. Compared with clock rates of 50 MHz or even 14 MHz, the time a switch is kept pressed is way too long for observing steps of operation of a design.

A circuit that we refer to as *OnePulser* has a slow, but clean and filtered, input (*LongPulse*), and a fast *clk* input. When *LongPulse* becomes **0**, the output of *OnePulser* becomes **0** for an entire period of *clk*.

Figure 7.8 shows the state diagram of the *OnePulser* circuit. While in *S0*, the *SinglePulse* output of this circuit is **1**. When *LongPulse* input becomes **0**, on the active edge of the clock the output becomes **0** in state *S1* and immediately with the next clock it becomes **1** again. This puts a **0** of one clock period on *SinglePulse*. Staying in *S2* while the *LongPulse* input is **0**, prevents more pulses to appear on the output until this input is released and pressed again.

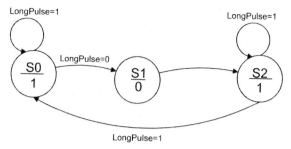

Figure 7.8 *OnePulser* State Machine

The Verilog code of *OnePulser* that is synthesized into the *OnePulser* library component is shown in Figure 7.9. The symbol for this component is shown in Figure 7.10.

```
module OnePulser (input clk, LongPulse,
                  output reg SinglePulse);
  reg [1:0] p_state, n_state;

  always @(posedge clk) p_state <= n_state;

  always @(p_state, LongPulse) begin
    n_state = 2'b0;
    SinglePulse = 1'b1;
    case (p_state)
      2'd00: n_state = (LongPulse) ? 2'd00 : 2'd01;
      2'd01: begin
               n_state = 2'd02;
               SinglePulse = 1'b0;
             end
      2'd02: n_state = (~LongPulse) ? 2'd02 : 2'd00;
      default: n_state = 2'd00;
    endcase
  end
endmodule
```

Figure 7.9 *OnePulser* Verilog Code

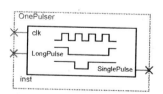

Figure 7.10 *OnePulser* Block Symbol

7.2.2.2   Clean One Pulser. Using *CleanPulse* (debouncer) and *OnePulser* library components, the design of the *Clean1Pulser* is shown in Figure 7.11. This circuit is used for single stepping a sequential circuit. Connecting a UP3 pushbutton to the *PushButton* input of this circuit causes a positive or negative pulse of *FastClock* duration on the circuits output. Using this circuit with a DE2 pushbutton still works the same way. However, since DE2 pushbuttons (KEY3, KEY2, KEY1, and KEY0) are already debounced, using the *CleanPulse* circuit is not needed.

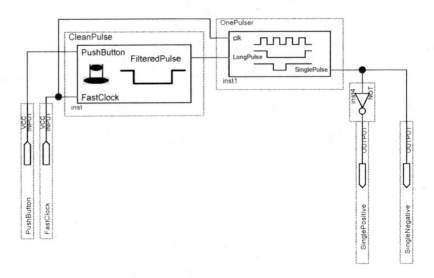

Figure 7.11 *Clean1Pulser* **Circuit Diagram**

The block symbol for *Clean1Pulser* is shown in Figure 7.12. When used with a pushbutton, each time the pushbutton is pressed a filtered pulse that can be used for clocking a sequential circuit is generated.

7.2.3 Utilizing UP3 Basic IO

This section presents a counter design that uses utility library components discussed above. This simple example demonstrates utilization of the library components we have developed so far.

The counter we are designing is an up/down modulo-16 4-bit binary counter, with parallel data input (*data[3:0]*), an asynchronous load input (*aload*), an *updown* control, and a rising edge sensitive clock. We used Quartus II configurable *lpm_counter* for this design. The Quartus II schematic for the tester of this design is shown in Figure 7.12.

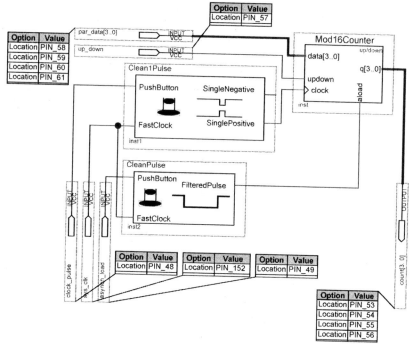

Figure 7.12 Testing a Counter with Debouncers

The clock input of *Mod16Counter* is driven by an instance of *Clean1Pulse*. The *FastClock* input of *Clean1Pulse* is driven by *sys_clk* that is tied to pin 152 of Cyclone. UP3 jumper JP3.6 and JP3.8 are connected to provide a clock frequency of 14.318 on pin 152. The *PushButton* input of *Clean1Pulse* connects to SW4 pushbutton. Every time SW4 is pressed a positive pulse will be applied to *clock* input of *Mod16Counter*.

The *aload* input of *Mod16Counter* that asynchronously loads *par_data* input into the counter is driven by a *CleanPulse* library module. This component debounces its *PushButton* input. This input is tied to pin 49 of Cyclone that is driven by SW5 pushbutton. The *updown* input of the counter that selects the count direction, is directly tied to SW6 pushbutton. Because this select input is synchronous, debouncing the UP3 pushbutton is not necessary.

The four DIP switches of UP3 connect to the parallel input (*par_data[3..0]*) of the counter. The outputs of the counter are displayed on D3 to D6 LEDs of UP3.

The count sequence of *Mod16Counter* is verified by pressing SW4. When this pushbutton is pressed, the count increases or decreases depending on *updown*. The binary count of the circuit is displayed on the UP3 LEDs.

7.2.4 Utilizing DE2 Basic IO

In this section we use DE2 push buttons to operate a counter in the single-step mode. The output of the counter will be displayed in binary using DE2 LEDs. This simple example demonstrates utilization of the *OnePulser* library components that we have developed for our library.

The counter we are designing is an up/down modulo-16 4-bit binary counter, with parallel data input (*data[3:0]*), an asynchronous load input (*aload*), an *updown* control, and a rising edge sensitive clock. We used Quartus II configurable *lpm_counter* for this design. The Quartus II schematic for the tester of this design is shown in Figure 7.13.

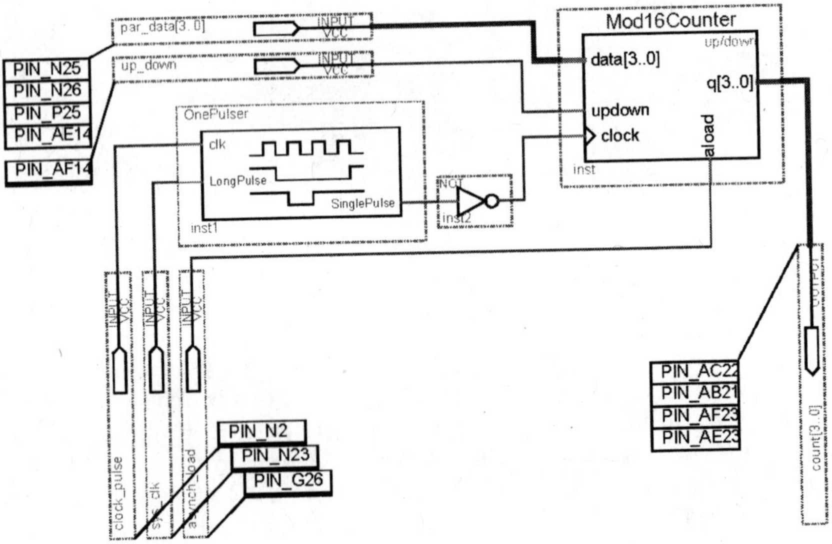

Figure 7.13 Single-Stepping a Counter on DE2

The clock input of *Mod16Counter* is driven by an instance of *OnePulser*. The *clk* input of *OnePulser* is driven by *clock_pulse* that is tied to pin N2 of Cyclone II (the 50 MHz clock). The *LongPulse* input of *OnePulser* connects to pushbutton KEY1. Every time KEY1 is pressed a positive pulse will be applied to *clock* input of *Mod16Counter*. The *aload* input of *Mod16Counter* asynchronously loads *par_data* input into the counter. This input is connected to DE2 pushbutton KEY0. The *updown* input of the counter that selects the count direction is directly tied to DE2 toggle switch SW4. Parallel inputs of the counter (*par_data[3..0]*) are tied to four toggle switches SW3 to SW0. The outputs of the counter are displayed on four right-most red LEDs.

The count sequence of *Mod16Counter* is verified by pressing and holding down KEY0 and pressing and releasing KEY1. With each press and release of KEY1, the counter counts up or down depending of SW4 position.

7.3 Frequency Dividers

In many of the designs that we will discuss in the sections that follow, timers of varying durations, or slow clocks are needed. For this purpose, we develop frequency dividers. A frequency divider is a counter with a carry out output. An n-bit counter divides its input clock by 2^n. This means that for every 2^n pulses on the circuit clock input one pulse appears on counter's carry output.

Frequency dividers will be used in LCD display, keyboard interface, and VGA driver that are discussed in the following sections. The debouncer circuit of the previous section also used a frequency divider. A toggle flip-flop is a divide-by-2 circuit.

7.4 Seven Segment Displays

Many development boards including Altera's DE2 include seven-segment displays (SSD). An SSD consists of seven LEDs organized in such a way to display digits 0 through 9 and letters A though F. An SSD is also referred to as a HEX display. Figure 7.14 shows an SSD display and numbering of its LEDs.

Figure 7.14 SSD Segment Numbers

7.4.1 SSD Driver

Figure 7.15 shows Verilog code of an SSD driver. This code takes a 4-bit input and generates SSD code for driving LEDs of a display according to Figure 7.14. The Quartus II symbol for this unit has four inputs and seven outputs and will be shown in the circuit that we will discuss next.

```verilog
module SevenSegmentDisplay (input [3:0] HEXin,
                            output [6:0] SSDout);

    assign SSDout =
        HEXin == 4'b0000 ? 7'b1000000 :
        HEXin == 4'b0001 ? 7'b1111001 :
        HEXin == 4'b0010 ? 7'b0100100 :
        HEXin == 4'b0011 ? 7'b0110000 :
        HEXin == 4'b0100 ? 7'b0011001 :
        HEXin == 4'b0101 ? 7'b0010010 :
        HEXin == 4'b0110 ? 7'b0000010 :
        HEXin == 4'b0111 ? 7'b1111000 :
        HEXin == 4'b1000 ? 7'b0000000 :
        HEXin == 4'b1001 ? 7'b0010000 :
        HEXin == 4'b1010 ? 7'b0001000 :
        HEXin == 4'b1011 ? 7'b0000011 :
        HEXin == 4'b1100 ? 7'b1000110 :
        HEXin == 4'b1101 ? 7'b0100001 :
        HEXin == 4'b1110 ? 7'b0000110 :
        HEXin == 4'b1111 ? 7'b0001110 :
                           7'b1111111 ;

endmodule
```

Figure 7.15 SSD Driver Verilog Code

7.4.2 Testing DE2 SSD Driver

Testing the SSD driver is achieved by using it at the output of the modulo-16 counter of Figure 7.13. This test circuit is shown in Figure 7.16. With the pin assignment shown, the counter output is displayed on the right-most HEX display of the DE2 development board.

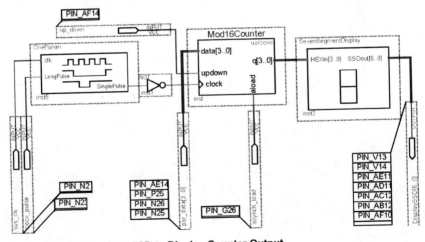

Figure 7.16 Using a DE2 SSD to Display Counter Output

7.5 LCD Display Adaptor

The UP3 and DE2 development boards come with a 16 character LCD display. The display has internal memory for character matrix definition, and a standard character set. To use this display, it has to be initialized, and then write into its ports that are connected to Cyclone pins. A simple interface is described here. Although, this interface does not utilize all capabilities of this particular. LCD display, it does show key issues in operating such a device.

The datasheet of the LCD display has details of its operation. Commands and data can be written into this device, and its status can be read when it is in the command mode. Our interface for this device initializes it and readies it for being written into for display.

7.5.1 Writing into LCD

Chapter 6 showed pin connections and operation of an LCD display unit. To write data or command into the LCD, the data or command byte must be placed on its *DB input*, *RS* must be set to 1 or 0 for data or command, *RW* must be set to 1 or 0 for read or write, and its *E* input (enable) must become 0 for at least 500 μs.

The Verilog code of LCD interface circuit is shown in Figure 7.17. This circuit has an *slc* input that is driven by a slow clock. This input decides on the duration of outputs generated by this Verilog code. Every time *wrt* becomes 1 for any length of time, a pulse whose duration is equal to that of *slc* appears on the *en_n* output of this circuit.

```
module write_synch (input clk, wrt, slc,
                    output en_n, busy);
    // slc: slow clock; en_n: active low enable
    reg [1:0] p_state, n_state;
    always @(posedge clk)
        case (p_state)
            2'b00: if (wrt) n_state <= 2'b01;
                   else n_state <= 2'b00;
            2'b01: if (slc) n_state <= 2'b10;
                   else n_state <= 2'b01;
            2'b10: if (slc) n_state <= 2'b10;
                   else n_state <= 2'b11;
            2'b11: if (wrt) n_state <= 2'b11;
                   else n_state <= 2'b00;
        default : n_state <= 2'b00;
        endcase
    assign busy = (p_state != 2'b00) ? 0 : 1;
    assign en_n = (p_state == 2'b10) ? 0 : 1;
endmodule
```

Figure 7.17 Producing Enable for LCD

The *clk* input of this module should be connected to any available clock on the development board. The *wrt* input is the write pulse that is at least as long as the duration of *clk*. The *slc* (slow clock) input is a slow clock with duration of 500 ns or more. The LCD display requires an enable cycle of at least 500 ns. A clock frequency of 1 MHz on *slc* satisfies this requirement. The outputs of the *write_synch* module are *en_n* and *busy*. The *busy* output becomes **1** when *wrt* of **1** is detected and remains at this level until *en_n* is inactive.

Figure 7.18 shows the state machine that corresponds to the Verilog code of Figure 7.17. State S0 waits for *wrt* to become **1**. Following this event, when *slc* becomes **1**, the *en_n* output becomes **0** and remains **0** for as long as *slc* is **1**. At this point, the machine goes into *S3* waiting for *wrt* to become **0** before returning to *S0*.

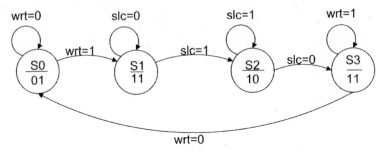

Figure 7.18 LCD Enabling State Machine

The Verilog module described above is incorporated into the *LCD_Driver* circuit shown in Figure 7.19. This circuit is implemented as a Quartus II schematic. The *RW* output is tied to ground since we will only be using this circuit for writing into the LCD and not reading from it.

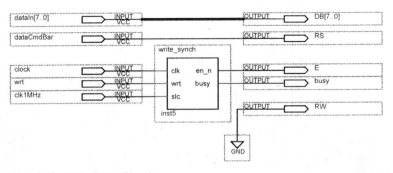

Figure 7.19 *LCD_Driver* Circuit

7.5.2 LCD Initialization

Before display data can be written into the LCD display, it has to be reset and initialized. The minimum set of instructions that are needed for this purpose are 2'h04, 2'h01, 2'h02, and 2'h0F. The first instruction initializes the display. The next two instructions turn clear LCD display and set curser at the home position. The last instruction turns on the display, turns on the curser, and causes the curser to blink.

The procedure for writing commands into the LCD is similar to that of writing data for display, except that for commands, the RS input of LCD must be **0**.

The *LCD_Driver* of Figure 7.19 writes 8-bit commands into the LCD when its *dataCmdBar* input is **0**. On the other hand, a separate hardware must be used for generating the above mentioned instructions and applying them to the *LCD_Driver* when an initialization pulse is received.

We have developed *initializer* and *Init_ROM* modules for initializing the LCD. The *initializer* module waits for a **1** on its *init* input, and when received it generates the sequence of 00, 01, and 10 on its *addr* output. With each address, it issues *wrt* and waits for busy to become **0** before issuing the next address. The *Init_ROM* module is an unclocked ROM that contains the 8'h04, 8'h01, 8'h02, and 8'h0F commands in its 0, 1, 2, and 3 locations.

Figure 7.20 shows the Verilog code of the *initializer* module and Figure 7.21 shows the *Init_ROM* module.

```
module initializer(reset, clk, init, busy, wrt, sel, addr);

    input reset, clk, init, busy;
    output wrt, sel;
    output [1: 0] addr;

    reg[1: 0] ps, ns;
    reg[1: 0] count;

    wire countRst, countEn;

    always @(posedge clk or posedge reset) begin
        if(reset) begin
            ps <= 2'b00;
            count <= 2'b00;
        end
        else begin
            ps <= ns;
            if(countRst) count <= 2'b00;
                else if(countEn) count <= count + 1;
        end
    end
```

```
always @(ps or init or busy or count) begin
   case(ps)
      2'b00: if(init) ns = 2'b01;
                else ns = 2'b00;
      2'b01: if(busy) ns = 2'b01;
                else ns = 2'b11;
      2'b11: ns = 2'b10;
      2'b10: if(count == 2'b10) ns = 2'b00;
                else ns = 2'b01;
      default: ns = 2'b00;
   endcase
end
assign countRst = (ps == 2'b00);
assign countEn = (ps == 2'b10);
assign sel = (ps != 2'b00);
assign wrt = (ps == 2'b11);
assign addr = count;

endmodule
```

Figure 7.20 *initializer* **Module**

```
module Init_ROM (input [1:0] Addr, output [7:0] Q);
   assign Q = (Addr==2'b00) ? 8'h04 :
              (Addr==2'b01) ? 8'h01 :
              (Addr==2'b10) ? 8'h02 :
                              8'h0F ;
endmodule
```

Figure 7.21 Command ROM

Together, modules of Figure 7.20 and Figure 7.21, execute a sequence of four instructions. At the same time that an instruction appears on the Q output of *Init_ROM*, the *wrt* output of *initializer* becomes **1**. The next instruction is applied when the *busy* output of *LCD_Driver* become **0**.

7.5.3 Display Driver with Initialization

Putting together our driver (Section 7.5.1) and initializer (Section 7.5.2) gives us a display driver with initialization. This driver that is shown in Figure 7.22 uses *LCD_Driver* to write data or command to the LCD. The *init* input causes the *initializer* to take control of the multiplexers and send instructions from *Init_ROM* to *LCD_Driver*. When all instructions have been sent, *sel* becomes **0**, and data from *datain* port of the *LCD_DriverInit* will be routed to the *LCD_Driver*.

As shown in Figure 7.22, the LCD driver requires a slow (*clk1MHz*) input for the timing of the LCD display.

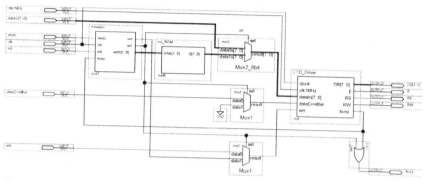

Figure 7.22 LCD Driver with Initialization

7.5.4 Testing the LCD Driver (UP3)

Figure 7.23 shows a test circuit for our LCD driver. A simple test of this circuit is done by writing 8-bit constants into it. For this purpose we have connected the four most significant bits of the data input of *LCD_DriverInit* to **1000** constant and its least significant bits to the UP3 DIP switches. As shown in Figure 7.23 a divide-by-64 circuit generates LCD slow clock from the input 48 MHz clock. The *Clean-Pulse* unit shown in this figure filters push-button that drives the write input of the LCD driver. This test hardware verifies the operation of our LCD interfaces.

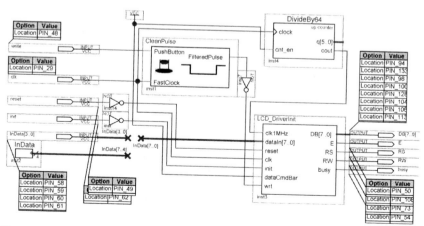

Figure 7.23 LCD Driver Tester (UP3)

7.5.5 Testing the LCD Driver (DE2)

The test circuit of Figure 7.24 tests the LCD display of the DE2 board. Testing is done by writing 8-bit constants into the LCD. For this purpose we have connected the data input of *LCD_DriverInit* to DE2 toggle switches, SW7 through SW0. As shown in Figure 7.23 a divide-by-64 circuit generates LCD slow clock from the input 50 MHz clock. The write input of the LCD driver is driven by a DE2 pushbutton. This test hardware verifies the operation of our LCD interfaces.

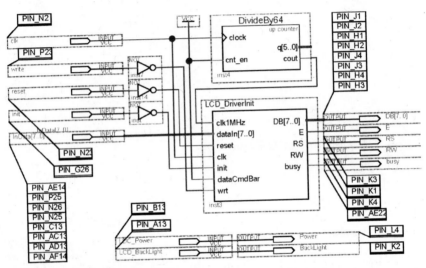

Figure 7.24 LCD Driver Tester (DE2)

7.6 Keyboard Interface Logic

As another library component, this section shows the design of keyboard interface logic. This interface is implemented on a Cyclone and Cyclone II and can be used for inputting data through a PS2 port. We will show how keyboards work and how they transmit serial data representing keys pressed. We will design an interface logic that communicates with a keyboard and collects the serially transmitted data from a keyboard.

7.6.1 Serial Data Communication

Figure 7.25 shows the interface that we are designing in this section. This device communicates with they keyboard through two bidirectional signals. Keys pressed are turned into 8-bit ASCII and are made available for the processing device.

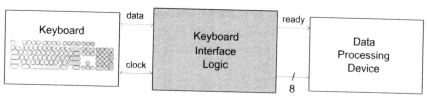

Figure 7.25 Keyboard Interface Logic

Data communication between the keyboard and the host system is synchronous serial over bi-directional *clock* and *data* lines. Keyboard sends commands and key codes, and the system sends commands.

Either the system or the keyboard drive the *data* and *clock* lines, while clocking data in either direction is provided by the keyboard clock. When no communication is occurring, both lines are high. Figure 7.26 shows the timing of keyboard serial data transmission.

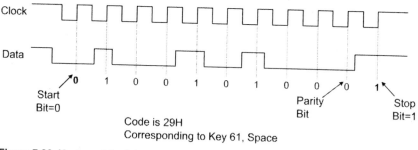

Figure 7.26 Keyboard Serial Data Transmission

7.6.1.1 Serial Data Format.

Data transmission on the data line is synchronized with the clock; data will be valid before the falling edge and after the rising edge of the clock pulse. Serial data transmission begins with the data line dropping to **0**. This bit value is taken on the rising edge of the clock and considered as the start-bit. On the next eight clock edges, data is transmitted in low to high order bit. The next data bit is the odd-parity bit, such that data bits and the parity bit always have odd number of ones. The last bit on the data line is the stop-bit that is always **1**. After the stop-bit, the data line remains high until another keyboard key is pressed and a new transmission begins. For every key pressed, eleven bits are transmitted.

When the keyboard sends data to or receives data from the system it generates the clock signal to time the data. The system can prevent the keyboard from sending data by forcing the clock line to **0**, during this time the data line may be high or low. When the system sends data to the keyboard, it forces the data line to **0** until the keyboard starts to clock the data stream.

7.6.1.2 Keyboard Transmission. When the keyboard is ready to send data, it first checks the status of the clock to see if it is allowed to transmit data. If the clock line is forced to low by the system, data transmission to the system is inhibited and keyboard data is stored in the keyboard buffer. If the clock line is high and the data line is low, the keyboard is to receive data from the system. In this case, keyboard data is stored in the keyboard buffer, and the keyboard receives system data. If the clock and data lines are both high the keyboard sends the start-bit, 8 data bits the parity bit and the stop-bit. Figure 7.27 summarizes these communication modes.

During transmission, the keyboard checks the clock line for low level at least every 60 μseconds. If the system forces the clock line to **0** after the keyboard starts sending data, a condition known as line contention occurs, and the keyboard stops sending data. If line contention occurs before the rising edge of the 10th clock pulse, the keyboard buffer returns the clock and data lines to high level.

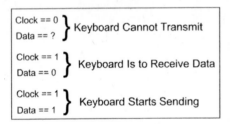

Figure 7.27 Keyboard Communication Modes

7.6.1.3 System Transmission. The system sends 8-bit commands to the keyboard. When the system is ready to send a command to the keyboard, it first checks to see if the keyboard is sending data. If the keyboard is sending, but has not reached the 10th clock signal, the system can override the keyboard output by forcing the keyboard clock line to **0**. If the keyboard transmission is beyond the 10th clock signal, the system receives the transmission.

If the keyboard is not sending or if the system decides to override the output of the keyboard, the system forces the keyboard clock line to **0** for more than 60 μseconds while preparing to send data. When the system is ready to send the start bit, it allows the keyboard to drive the clock line to **1** and drives the data line to low. This signals the keyboard that data is being transmitted from the system. The keyboard generates the clock signals and receives the data bits, parity and the stop-bit. After the stop-bit, the system releases the data line. If the keyboard receives the stop-bit it forces the data line low to signal the system that the keyboard has received its data.

Upon receipt of this signal, the system returns to a ready state, in which it can accept keyboard output or goes to the inhibited state until it is ready. If the keyboard does not receive the stop-bit, a framing error has occurred, and the keyboard continues to generate clock signals until the data line becomes high. The keyboard then makes the data line low and requests a resending of the data. A parity error will also generate a re-send request by the keyboard.

In our implementation of keyboard interface logic, system transmission is not considered; i.e., flow of data is always from the keyboard to the data processing drive (Figure 7.25).

7.6.2 Power-On Routine

The keyboard logic generates a power-on-reset signal when power is first applied to it. The timing of this signaling is between 150 milliseconds and 2.0 seconds from the time power is first applied to the keyboard.

Following this signaling, basic assurance test is performed by the keyboard. This test consists of a keyboard processor test, a checksum of its ROM, and a RAM test. During this test, activities on the clock and data lines are ignored. The keyboard LEDs are turned on at the beginning and off at the end of the test. This test takes a minimum of 300 milliseconds and a maximum of 500 milliseconds. Upon satisfactory completion of the basic assurance test, a completion code (hex AA) is sent to the system, and keyboard scanning begins.

7.6.3 Codes and Commands

A host system may send 8-bit commands to the keyboard, while a keyboard may send commands and key codes to the system.

7.6.3.1 System Commands. System commands may be sent to the keyboard at any time. The keyboard will respond within 20 milliseconds, except when performing the basic assurance test (BAT), or executing a Reset command. System commands and their hexadecimal values are shown in Table 7.1.

7.6.3.2 Keyboard Commands. Table 7.2 shows the commands that the keyboard may send to the system and their hexadecimal values.

Table 7.1 System Commands to Keyboard

Command	Hex
Set/Reset Status Indicators	ED
Echo	EE
Invalid Command	EF
Select Alternate Scan Codes	F0
Invalid Command	F1
Read ID	F2
Set Typematic Rate/Delay	F3
Enable	F4
Default Disable	F5
Set Default	F6
Set All Keys – Typematic	F7
- Make/Break	F8
- Make	F9
- Typematic/Make/Break	FA
Set Key Type – Typematic	FB
- Make/Break	FC
- Make	FD
Resend	FE
Reset	FF

Table 7.2 Keyboard Commands to System

Command	Hex
Key Detection Error/Overrun	00
Keyboard ID	83AB
BAT Completion Code	AA
BAT Failure Code	FC
Echo	EE
Acknowledge (ACK)	FA
Resent	FE

7.6.3.3 Keyboard Codes. Keyboards are available for several languages and settings. The keyboard that is most common for the English language is one with 104 keys shown in Figure 7.28. Keys of this keyboard are identified by numbers, and for every key there is a scan code. Several scan codes are available, and the default scan code is Scan Code 2 that we will discuss here.

Keyboard scan codes consist of a Make and a Break code. The Make code identifies the key pressed and the Break code indicates the release of a key. For most keys the Break code is F0 followed by the Make code. For example when the Space bar (key 61) is pressed and released, hexadecimal codes 29, F0 and 29 are transmitted from the keyboard to the system via the data serial line. If this key remains pressed, the Make code (29) is continuously transmitted until it is released. Make codes for Scan Code 2 are shown in Table 7.3.

Figure 7.28 Standard 104-key Keyboard and Key Number

Table 7.3 Keyboard Scan Codes and Corresponding ASCII Characters

Key Numb	Make Code	ASCII No Shift	ASCII Shift	Character No Shift	Character Shift	Key Numb	Make Code	ASCII No Shift	ASCII Shift	Character No Shift	Character Shift
1	0E	96	126	`	~	34	2B	70	102	F	f
2	16	49	33	1	!	35	34	71	103	G	g
3	1E	50	64	2	@	36	33	72	104	H	h
4	26	51	35	3	#	37	3B	74	106	J	j
5	25	52	36	4	$	38	42	75	107	K	k
6	2E	53	37	5	%	39	4B	76	108	L	l
7	36	54	94	6	^	40	4C	59	58	;	:
8	3D	55	38	7	&	41	52	39	34	'	"
9	3E	56	42	8	*	42	5D				
10	46	57	40	9	(	43	5A	13	Enter	Enter	Enter
11	45	48	41	0	)	44	12		Shift	Shift	Shift
12	4E	45	95	-	_	45	61				
13	55	61	43	=	+	46	1A	90	122	Z	z
15	66	08	08	BS	BS	47	22	88	120	X	x
16	0D	09	09	Tab	Tab	48	21	67	99	C	c
17	15	81	113	Q	q	49	2A	86	118	V	v
18	1D	87	119	W	w	50	32	66	98	B	b
19	24	69	101	E	e	51	31	78	110	N	n
20	2D	82	114	R	r	52	3A	77	109	M	m
21	2C	84	116	T	t	53	41	44	60	,	<
22	35	89	121	Y	y	54	49	46	62	.	>
23	3C	85	117	U	u	55	4A	47	63	/	?
24	43	73	105	I	i	57	59			Shift	Shift
25	44	79	111	O	o	58	14			Ctrl	Ctrl
26	4D	80	112	P	p	59	E0 1F			Win	Win
27	54	91	123	[	{	60	11			Alt	Alt
28	5B	93	125	]	}	61	29	32	32	Space	Space
29	5D	92	124	\			62	E0/11			Alt	Alt
30	58			Caps	Caps	63	E0 27			Win	Win
31	1C	65	97	A	a	64	E0/14			Ctrl	Ctrl
32	1B	83	115	S	s	65	E0 2F			Menu	Menu
33	23	68	100	D	d						

The Make and Break arrangement, makes it possible for the system to identify multiple keys pressed and the order in which they have been pressed. For example, if one presses and holds down the Left-Shift key (key number 44), 12 Hex is continuously sent to the system. While this is happening, if key number 9 (the 8/* key) is pressed and released, 3E, F0 and 3E codes are transmitted. The receiving system identifies this sequence of events as the intention to enter an asterisk (*).

7.6.4 Keyboard Interface Design

This section discusses a keyboard interface for reading scan data from the keyboard and producing ASCII codes of the keys pressed. Code Set 2 is assumed, and the interface only handles data transmission from the keyboard. The interface reads serial data from the keyboard, detects the Make code when a key is pressed and looks up the Make code in an ASCII conversion table. For simplicity, the look-up table only handles upper-case characters.

7.6.4.1 Collecting the Make Code. The first part of our interface connects to the keyboard data and clock lines and when a key is pressed, it outputs an 8-bit scan code. The *KBdata*, *KBclock* inputs are for the keyboard data and clock inputs, and the 8-bit *ScanCode* is the main output of this part.

This part also uses a fast synchronizing clock, *SYNclk*, and a keyboard reset input, *KBreset*. In addition to the *ScanCode* output, this part outputs a signal to indicate that a scan code is ready (*ScanRdy*) and another output to indicate that a key has been released (*KeyReleased*). These outputs make distinction between Make and Break states. When a key is pressed three positive pulses appears on *ScanRdy*, one for the Make code and two for the Break codes. The *KeyRealesed* output only becomes **1** once when all three codes have been transmitted.

Figure 7.29 shows the outline of the Verilog code for collection of bits of *KBdata* and generation of scan code and the corresponding handshaking outputs. As shown in this figure, *KBClock* that is synchronized with *SYNclk* is generated from *KBclk*. This synchronized clock is used in state machines that follow instead of *KBclk*. The code consists of three state machines.

The first state machine generates a signal that remains **1** from the detection of start-bit to the end of stop-bit. Another state machine waits for start-bit and collects data bits on the rising edge of *KBClock*. The last state machine waits for the completion of the Break Codes and issues the *KeyReleased* output.

```
module KB_ScanCode(input KBclk, KBdata, ResetKB, SYNclk,
    output ScanRdy, output KeyReleased,
    output reg [7:0] ScanCode);

    reg KBClock; // KB Synchronized Clock

    always @(posedge SYNclk) KBClock <= KBclk;

    //Detect Start and Stop

    //Collect data bite

    //Wait for key release

endmodule
```

Figure 7.29 Keyboard Scan Code Generation

Figure 7.30 shows the part of code of *KB_ScanCode* module for issuing *StartBitDetected*. The state machine shown uses the fast system *SYNclk*. The machine waits for *KBdata* to become **0** to issue *StartBitDetected*. This signal remains **1** until stop bit is detected and data bit collection is complete. The *completed* signal issued by the data collection state machine resets *StartBitDetected*.

```
wire Completed;

reg StartBitDetected;

always @(posedge SYNclk)
    if (ResetKB) StartBitDetected <= 1'b0;
    else
        case (StartBitDetected)
            0: if (KBdata) StartBitDetected <= 1'b0;
               else StartBitDetected <= 1'b1;
            1: if (~Completed)StartBitDetected <= 1'b1;
               else StartBitDetected <= 1'b0;
        endcase
```

Figure 7.30 Detect Start and Stop

The part of the Verilog code of Figure 7.29 that is responsible for collection of data bits transmitted by the keyboard is shown in Figure 7.31. The clock used for this state machine is *KBClock*. The machine waits for *StartBitDetected* in its *Idle* state. When detected, it cycles through *Bit1* to *Bit8* states as data bits appear on *KBdata* on the rising edge of *KBClock*. When all bits have been detected, it goes to the *stop* state.

```
reg [3:0] p_state, n_state;
parameter [3:0] Idle=0,
                Bit1=1, Bit2=2, Bit3=3, Bit4=4,
                Bit5=5, Bit6=6, Bit7=7, Bit8=8,
                OPar=9, Stop=10;
always @(posedge KBClock)
   if (ResetKB) p_state <= Idle;
   else p_state <= n_state;
always @(p_state, StartBitDetected, KBdata) begin
   case (p_state)
      Idle: if (~StartBitDetected) n_state <= Idle;
            else n_state <= Bit1;
      Bit1,Bit2,Bit3,Bit4,Bit5,Bit6,Bit7,Bit8, OPar:
            n_state <= p_state + 1;
      Stop: n_state <= Idle;
      default: n_state <= Idle;
   endcase
end
assign Completed = (p_state == Stop) ? 1 : 0;
assign ScanRdy = (p_state == Stop);
always @(posedge KBClock)
   if (p_state >= Bit1 & p_state <= Bit8)
      ScanCode <= {KBdata, ScanCode[7:1]};
```

Figure 7.31 Data Bit Collection

The *completed* output of this part of the circuit is issued when the machine reaches the *stop* state. Also note in Figure 7.31 that *ScanRdy* is issued the same way. The last part of Figure 7.31 shows a shift register that clocks data bits as they appear on *KBdata* on the rising edge of *KBClock*. When all bits are shifted, the *ScanCode* register will have the keyboard scan code. For every key pressed this happens three times for Make and Break codes.

```
reg [1:0] p_release, n_release;
parameter [1:0] WaitF0 = 0, WaitRPT = 1, Released = 2;
always @(posedge KBClock) p_release <= n_release;
always @(p_release, ScanCode, ScanRdy)
   case (p_release)
      WaitF0  : if ((ScanCode == 8'hF0) && (ScanRdy))
                   n_release = WaitRPT;
                else n_release = WaitF0;
      WaitRPT : if (ScanRdy) n_release = Released;
                else n_release = WaitRPT;
      Released: n_release = WaitF0;
      default:  n_release = WaitF0;
   endcase
assign KeyReleased = (p_release == Released) ? 1 : 0;
```

Figure 7.32 Waiting for Key Release

The last part of the Verilog code of Figure 7.29 is a state machine that issues *KeyReleased* when a complete set of Make and Break codes are transmitted by the keyboard. This part that is a state machine triggered by *KBClock* and is shown in Figure 7.32. This machine cycles through *WaitF0*, *WaitRPT*, and *Released* states as Break codes are transmitted. The F0 code that is transmitted appears on *ScanCode* when *ScanRdy* becomes **1**. When this is detected the machine goes from *Wait_F0* to *WaitRPT*. In the ÷*WaitRPT* state, the machine waits for the repeat of the code of the key pressed. This code is accompanied by another pulse on *ScanRdy*. When detected, the machine moves into the *released* state in which the *KeyReleased* output is issued.

7.6.4.2 ASCII Look-Up. The ASCII lookup part of our keyboard interface is a ROM of Quartus II megafunctions with 8 address lines and word length of 8 bits. Hexadecimal locations 0D through 66 of this ROM are defined to contain ASCII codes for scan codes that correspond to ROM addresses. This megafunction is defined to use the *KbASCII.mif* memory initialization file, a portion of which is shown in Figure 7.33. The *ScanCode* output of Figure 7.29 connects to the address input of this ROM, and ASCII codes corresponding to input addresses appear on its output. This lookup ROM implemented as a megablock is defined as *Scan2ASCII* block symbol.

```
DEPTH = 128;
WIDTH = 8;
ADDRESS_RADIX = HEX;
DATA_RADIX = DEC;
% Keyboard Scan Code to ASCII %
CONTENT
BEGIN
% Set 2:    ASCII    ;         Key   Char    %
%------+---------------+--------------------%
0D    :    09    ;    %    16    Tab     %
0E    :    96    ;    %    1             %
11    :    0     ;    %    60    Alt     %
12    :    0     ;    %    44    Shift   %
14    :    0     ;    %    58    Ctrl    %
15    :    81    ;    %    17    Q       %
16    :    49    ;    %    2     1       %
1A    :    90    ;    %    46    Z       %
1B    :    83    ;    %    32    S       %
1C    :    65    ;    %    31    A       % .
. . . .
66    :    08    ;    %    15    BS      %
END;
```

Figure 7.33 ASCII Conversion Memory Initialization File

7.6.4.3 Keyboard Driver.

The complete keyboard driver that takes a key press and generates the corresponding ASCII code consists of the *KB_ScanCode* Verilog code of Figure 7.29 and the *Scan2ASCII* lookup ROM. Figure 7.34 shows the block diagram of our keyboard driver.

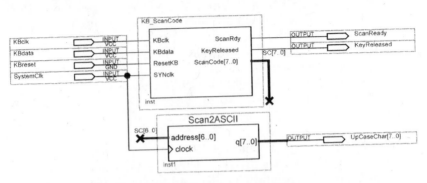

Figure 7.34 Schematic Diagram of *KB_Driver* Keyboard Driver

7.6.4.4 Testing the Keyboard Driver.

We use the LCD display of Section 7.5 to test our keyboard driver of Figure 7.34 shows this test-bench. The character output of *KB_Driver* connects to *dataIn* of the LCD display. Since we need to see only the character that has been pressed (and not the Break Codes), we use the *KeyReleased* output of the *KB_Driver*. This output connects to *wrt* input of *LCD_DriverInit*. As before, the *DivideBy64* circuit generates the slow clock needed by the LCD display.

The diagram of Figure 7.35 shows pin assignments for the keyboard circuit implemented on the UP3 development board. Figure 7.36 shows this same circuit implemented with a DE2 Cyclone II.

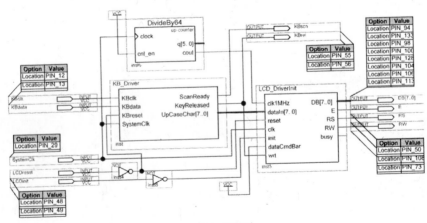

Figure 7.35 Testing the Keyboard Interface (UP3)

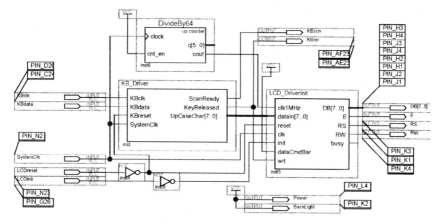

Figure 7.36 Testing the Keyboard Interface (DE2)

7.7 VGA Interface Logic

In this section we discuss the design of VGA interface logic. The VGA driver discussed is capable of displaying characters from a display RAM on a standard VGA monitor. After discussing how a VGA monitor works, we show hardware for driving it. The design methodology presented here uses Verilog blocks, megafunctions, memories and schematic capture. A simple testbench at needed will show the basic operation of this unit.

7.7.1 VGA Driver Operation

A standard VGA monitor consists of a grid of pixels that can be divided into rows and columns. A VGA monitor contains at least 480 rows, with 640 pixels per row, as shown in Figure 7.37. Each pixel can display various colors, depending on the state of the red, green, and blue input signals. These signals are analog signals and determine the intensity of their corresponding colors.

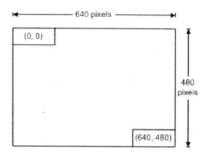

Figure 7.37 VGA Monitor

7.7.1.1 Pixel Sweeping.
In a VGA monitor pixels are refreshed with proper color one-by-one from location (0, 0) to (640, 480). This is done from upper left of the monitor to its lower right. To eliminate any screen flickering, the refreshing must be done such that the entire screen is refreshed, i.e., all screen is swept in less than 0.02 second. For a better than maximum refresh time, of we choose a refresh cycle of 1/60 of a second. With this rate, and considering the overhead time of moving scan from one line to another, and one screen to next, a frequency of 25.175 MHz is required for refreshing each pixel. In most VGA monitors the refresh time can range anywhere from 0.02 to 1/160 of a second. This translates to a pixel frequency of 21 MHz to 67 MHz. Higher pixel resolutions require higher frequencies. Discussion of VGA timing that follows justifies pixel frequencies and refresh rates.

7.7.1.2 VGA Timing.
The discussion of VGA timing that we are presenting here is based on the 60 Hz screen refresh rate with a clock frequency of 25.175 MHz. After the completion of this discussion we will show how the timings described translate to other refresh rates.

For the VGA monitor to work properly, it must receive data at specific times with specific pulses. Horizontal and vertical synchronization pulses must occur at specified times to synchronize the monitor while it is receiving color data.

Figure 7.38 shows the timing waveform for the color information with respect to the horizontal synchronization signal. Times shown are for the standard pixel frequency of 25.175 MHz for 640×480 pixel resolution.

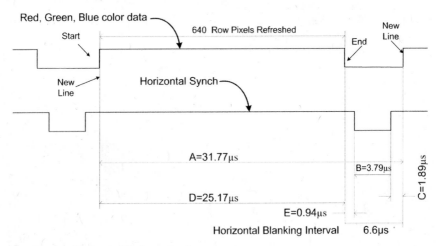

Figure 7.38 Horizontal Refresh Cycle

As shown, the row pixel time interval (D) is 25.17 μs in which 640 pixels are refreshed (one pixel per every clock cycle). Considering 640 cycles in 25.17 μs, the 6.6 μs horizontal blanking interval (B+C+E) becomes 160 cycles. This makes complete sweeping of a row of a VGA monitor to take place in 800 cycles of the 25.175 MHz clock, or equivalent to 800 pixels. Figure 7.39 shows horizontal refresh parameters in time and equivalent pixel values.

Parameters	A	B	C	D	E
Time for 25.175 MHz	31.77 μs	3.77 μs	1.89 μs	25.17 μs	0.96 μs
Cycle	800	95	48	640	24

Figure 7.39 Horizontal Parameters

Figure 7.40 shows the timing waveform for the color information with respect to the vertical synchronization signal. As discussed above, a complete horizontal sweep (a line of pixels) takes 800 clock cycles that is equivalent to 31.77 μs for 25.175 MHz clock.

A complete screen of 480 pixel lines takes 480×800 clock cycles (31.77 μs × 480 = 15.25 ms) to sweep, plus an additional band guard (P+Q+S) of 1.434 ms. Considering that the time of sweeping a pixel line is 31.77 μs, the duration of the band guard becomes equivalent to 45 lines.

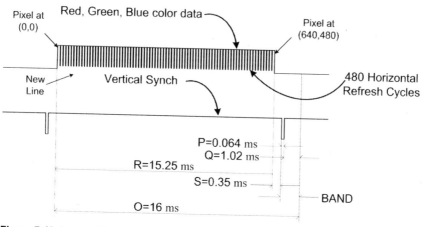

Figure 7.40 Vertical Refresh Cycle

This translates to 480+45=525 lines of 800 cycles each. Figure 7.41 shows vertical refresh parameters in time and equivalent line values. The values given are for a clock frequency of 25.175 MHz.

Parameters	O	P	Q	R	S
Time For 25.175 MHz	16 ms	0.064 ms	1.02 ms	15.25 ms	0.35 ms
Horizontal Cycle	525	2	32	480	11

Figure 7.41 Vertical Parameters

As previously discussed, VGA monitors allow refresh clock frequencies of 21 to 67 MHz. For a 25 MHz clock, the following parameters are calculated:

$T_{Pixel} = 1/f_{clk} = (1/25)$ μs

$T_{PixelRow} = 1/f_{clk} \times 640 = 25.6$ μs

$T_{Row} = A = B+C+D+E$
$= (1/f_{clk}) \times 800 = 32.0$ μs
$T_{PixelLine} = (T_{Row} \times 480 \text{ rows}) = 15.36$ ms

$T_{Screen} = O = P+Q+R+S$
$= (T_{Row} \times 525 \text{ rows}) = 16.8$ ms

Where:

T_{Pixel} = Time required to update a pixel

f_{clk} = Clock frequency = 25 MHz

$T_{PixelRow}$ = Time required to update row pixels

T_{Row} = Time required to sweep a row including its guard band

$T_{PixelLine}$ = Time required to update screen lines

T_{Screen} = Time required to update screen including its guard band

The monitor writes to the screen by positioning a pixel using vertical and horizontal synchronization signals. The red, green, blue color inputs of the screen are driven when the screen is at the expected location. These signals send the correct color data to the pixel.

7.7.2 Monitor Synchronization Hardware

The hardware required for VGA signal generation must keep track of the number of clock cycles (equivalent pixels), and issue signals according to the timing waveforms of Figure 7.38 and Figure 7.40. The Verilog code of Figure 7.42 uses the *SynchClock* clock signal to generate *Hsynch* (*Horizontal Synch* of Figure 7.38), *Vsynch* (*Vertical Synch* of Figure 7.40), and *Red*, *Green*, and *Blue* color data.

```
module MonitorSynch (
    input PixelOn,
    input RedIn, GreenIn, BlueIn, Reset, SynchClock,
    output Red, Green, Blue, Hsynch, Vsynch,
    output [9:0] PixelRow, PixelCol);

    reg [9:0] Hcount, Vcount;

    always @(posedge SynchClock) begin
        if (~Reset) Hcount = 0;
        else
            if (Hcount == 799) Hcount =0;
            else Hcount <= Hcount + 1;
    end
    always @(posedge SynchClock) begin
        if (~Reset) Vcount = 0;
        else
            if (Vcount >= 525 && Hcount >= 756) Vcount = 0;
            else if (Hcount == 756) Vcount <= Vcount + 1;
    end

    assign Hsynch = (Hcount>= 661 && Hcount <= 756) ? 0 : 1;
    assign Vsynch = (Vcount>= 491 && Vcount <= 493) ? 0 : 1;
    assign {Red, Green, Blue}=(Hcount<=640 && Vcount<480) ?
        {PixelOn&RedIn, PixelOn&GreenIn, PixelOn&BlueIn} : 0;
    assign PixelCol = (Hcount <= 640) ? Hcount : 0;
    assign PixelRow = (Vcount <= 480) ? Vcount : 0;
endmodule
```

Figure 7.42   Monitor Synchronization Hardware

The code shown uses color specifications from *RedIn*, *GreenIn* and *BlueIn* input signals and during the time periods specified by parameter *D* in Figure 7.38 and parameter *R* in Figure 7.40, puts them on the *Red*, *Green* and *Blue* output signals. At any pint in time, the Verilog code of Figure 7.42 outputs the position of the pixel being updated in its 10-bit *PixelRow* and *PixelCol* output vectors.

Two always blocks that are synchronized with *SynchClock* keep track of horizontal and vertical counts (*Hcount* and *Vcount*). *Hcount* is associated with parameter *A* of Figure 7.38 or Figure 7.39 and *Vcount* with parameter *O* of Figure 7.40 or Figure 7.41.

Considering the very first pixel at (0, 0) position, the counting of the horizontal pixels begins at the beginning of the *D* region of the waveform of Figure 7.38. Therefore as the Verilog code shows, *Hsynch* becomes 0 when *Hcount* is between 661 and 756 (this is the *B* region). Likewise, considering the beginning of region *R* as the 0 point, the *P* region in Figure 7.40 begins at *Vcount* of 491 and ends at 493. Therefore, as the code shows, *Vsynch* is **0** during such *Vcount* values. With the (0, 0) point defined as such, pixels are active while *Hcount* is between 0 and 640 and *Vcount* is between 0 and 480. During these count periods output colors are active and *PixelRow* and *PixelCol* outputs reflect *Hcount* and *Vcount* respectively.

The Verilog code of Figure 7.42 is defined as a block that will be used in our implementation of a character display design.

7.7.3 Character Display

The design we are considering in this section is a character display hardware that outputs an address to a character display memory and inputs an ASCII code representing the character to display. We assume the display memory has 4800 ASCII characters that will be displayed in 60 rows of 80 characters. Considering the 480 by 640 resolution, this makes each character occupy a matrix of pixels.

In addition to the synchronization module of the previous section (*MonitorSynch* module), the character display hardware has a character pointer and a pixel generation hardware. The character pointer finds the character address at a specified screen pixel location, and the pixel generation hardware finds the pixel value (1 or 0) for the character that is being displayed.

Figure 7.43 shows the complete *CharacterDisplay* hardware. The elements of this diagram are discussed next.

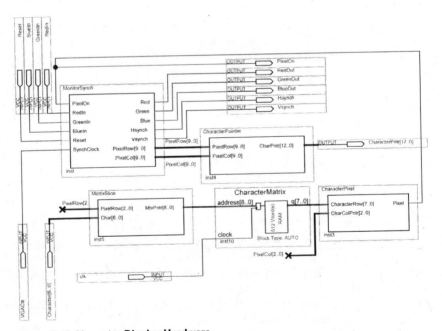

Figure 7.43 Character Display Hardware

7.7.3.1 Character Pointer. The character pointer hardware is the block shown in upper right part of Figure 7.43. *MonitorSynch* provides X and Y pixel coordinates for this hardware. *CharacterPointer* takes these coordinates and generates a 13 bit address pointing to one of the 4800 characters at the screen location. Because each character consists of 8 row and 8 column pixels, 64 different X and Y coordinates map to the same character address. The Verilog code for generating this mapping is the *CharacterPointer* module of Figure 7.44.

```verilog
module CharacterPointer (
    input [9:0] PixelRow, PixelCol, output [12:0] CharPntr);
    wire [6:0] ScreenLine, ScreenPos;

    assign ScreenLine = PixelRow [9:3];
    // 60 Lines=480/8 (8 is Character Pixel Hight)

    assign ScreenPos = PixelCol [9:3];
    // 80 Positions=640/8 (8 is Character Pixel Width)

    assign CharPntr =   ScreenLine*80 + ScreenPos;
endmodule
```

Figure 7.44 Finding Character Position from a Pixel Position

As shown, *ScreenLine* and *ScreenPos* ignore three low order bits of pixel coordinates, which causes the corresponding screen location to map to the character at that location. The *CharPntr* output of this module provides a one-dimensional pointer for the display memory.

7.7.3.2 Pixel Generation Hardware. The lower part of Figure 7.43 is responsible for generation of a specific pixel value (1 or 0) for the specific X-Y position of screen and the character being displayed. Inputs to this part are ASCII code of the character being displayed (*Character[6..0]*) and coordinates within the 8×8 pixel area of the character (*PixelRow[2..0]* and *PixelCol[2..0]*). The three parts of pixel generation are *MatrixSlice*, *CharacterMatrix* and *CharacterPixel*.

The *MatrixSlice* module (Figure 7.45) takes the ASCII code of the input character and subtracts 32 from it to make printable character codes begin from 0. It then appends three bits of *PixelRow* to its right to form an address for the pixel row of the corresponding character.

```verilog
module MatrixSlice (input [2:0] PixelRow, input [6:0] Char,
                    output [8:0] MtxPntr);
    assign MtxPntr = {Char - 32, PixelRow[2:0]};
endmodule
```

Figure 7.45 Matrix Slice Verilog Code

The output of this module looks up a row of the character being displayed from the *CharacterMatrix* component. In our simple design we use 8×8 character resolution and only support ASCII characters from 32 to 95. With these 64 supported characters, our character matrix becomes an 8-bit memory of 512 words, in which every 8 consecutive words define a character. For example, as shown in Figure 7.46, pixels for character "5" with ASCII code of 53 decimal, begin at address 0A8 Hex that is (53-32)×8.

```
0A8  :  01111110 ;  %   * * * * * *   %
0A9  :  01100000 ;  %   * *           %
0AA  :  01111100 ;  %   * * * * *     %
0AB  :  00000110 ;  %             * * %
0AC  :  00000110 ;  %             * * %
0AD  :  01100110 ;  %   * *     * *   %
0AE  :  00111100 ;  %     * * * *     %
0AF  :  00000000 ;  %                 %
```

Figure 7.46 Character Matrix for Character "5"

The *CharacterMatrix* component is a RAM that is implemented with an Altera LPM and is mapped into the on-chip memory of our FPGA. This component is called LPM_RAM and is available under the *storage* category of Altera megafunctions. Using the megafunction wizard we configure this component as an 8-bit memory with nine address lines. During the configuration process we are asked to enter the Memory Initialization File name (*.mif*), for which we use *CharMtx.mif*. Using the *mif* format, pixel values (similar to those shown in Figure 7.46 for character "5") are defined for ASCII characters from 32 to 95. Figure 7.47 shows the beginning and end of this file, from which its complete format can be seen.

The output of *CharacterMatrix* is *q[7..0]*. This output has a slice of the character that is being displayed. For example for row 2 of character "5", *q[7..0]* is **01111100**.

The last component shown in Figure 7.37 that is responsible for pixel generation is *CharacterPixel*. This component takes a character row and its column pointer (*PixelCol [2:0]*) and looks up the pixel to be displayed. The Verilog code of *CharacterPixel* is shown in Figure 7.48. This code uses *CharColPntr* as index to look into *CharacterRow* in reverse bit order.

The complete schematic of our character display hardware is shown in Figure 7.43. The *MonitorSynch* module continuously sweeps across the 640×480 pixel screen and refreshes pixel with colors specified by its three color inputs. At the same time it reports the position of the pixel being refreshed to *CharacterPointer*. Based on these coordinate, this module calculates the address of the character that is being displayed. After the character is looked up refreshed is found. This pixel value allows color inputs to be used by the *MonitorSynch* module for painting the pixel. The complete design shown in Figure 7.43 is referred to as *CharacterDisplay*.

```
DEPTH = 512;
WIDTH = 8;
ADDRESS_RADIX = HEX;
DATA_RADIX = BIN;
% Character Matrix ROM,                %
% addressed by PixelGeneration module %
CONTENT
BEGIN
   % ASCII 0010_0000 to 0010_1111  %
   000  : 00000000 ; %                %
   001  : 00000000 ; %                %
   002  : 00000000 ; %                %
   003  : 00000000 ; %                %
   004  : 00000000 ; %                %
   005  : 00000000 ; %                %
   006  : 00000000 ; %                %
   007  : 00000000 ; %                %

   . . .

   1F8  : 00000000 ; %                %
   1F9  : 00010000 ; %        *       %
   1FA  : 00110000 ; %       **       %
   1FB  : 01111111 ; %    *******     %
   1FC  : 01111111 ; %    *******     %
   1FD  : 00110000 ; %       **       %
   1FE  : 00010000 ; %        *       %
   1FF  : 00000000 ; %                %
END;
```

Figure 7.47 Character Matrix *mif* File

```verilog
module CharacterPixel (
    input [7:0] CharacterRow,
    input [2:0] CharColPntr, output Pixel);
    wire [2:0] indx =
            {CharColPntr[2], CharColPntr[1], CharColPntr[0]};
    wire [7:0] Vector =
            {CharacterRow[0], CharacterRow[1],
             CharacterRow[2], CharacterRow[3],
             CharacterRow[4], CharacterRow[5],
             CharacterRow[6], CharacterRow[7] };
    assign Pixel = Vector [indx];
endmodule
```

Figure 7.48 Looking up Character Pixel

7.7.4 VGA Driver for Text Data

The previous section discussed the complete design of *CharacterDisplay* hardware. This hardware outputs the address of one of the 4800 characters that is to appear on the screen, looks up its ASCII code, and generates pixel colors and horizontal and vertical synchronization signals. This section shows a simple VGA driver that provides data to be displayed to our *CharacterDisplay* hardware.

The complete schematic of our *VGA_Driver* is shown in Figure 7.49. On the right hand side is *CharacterDisplay* that generates character address, monitor synch, and pixel information. On the left hand side is *DisplayMemory* that is a dual-port read/write memory. Character address from *CharacterDisplay* goes to its read address. The write address, data to be written, and the write enable of this memory are provided externally. While *CharacterDisplay* is displaying the current contents of the memory, new data for display can be written into this memory.

Our VGA-Driver has two clock inputs, *VGAclk* and *Memoryclk*. The latter must be a faster clock so that address output from *CharacterDisplay* can be used to lookup the display character from the Display Memory.

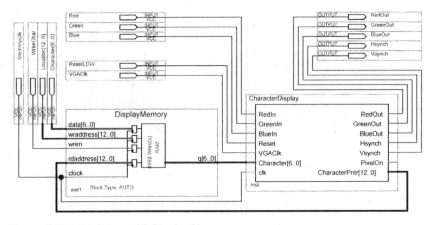

Figure 7.49 VGA Driver with Display Memory

7.7.5 VGA Driver Prototyping (UP3)

Figure 7.50 shows a simple tester for our VGA driver. This circuit causes the initial data in *DisplayMemory* to be displayed on the VGA monitor. Changing display data can only happen by changing contents of *DisplayRAM.mif* file which is loaded into this memory.

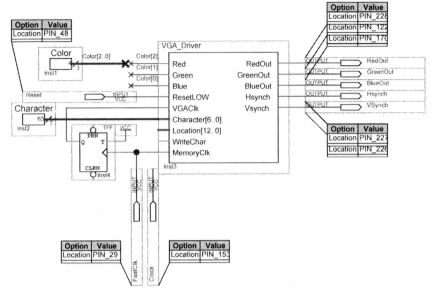

Figure 7.50 VGA Driver Tester (UP3)

Note the use of the T-type flip-flop for generating a slower clock for the VGA driver than that of the memory. This circuit is implemented on a UP3 development board and verifies the operation of our VGA adapter. A more elaborate testbench would have a counter to set memory display memory locations to desired characters. We leave this as an exercise.

7.7.6 VGA Driver Prototyping (DE2)

Figure 7.51 shows our VGA driver that is programmed into the Cyclone II FPGA of a DE2 development board. Initially, the *Display-Memory* is initialized with the contents of the *DisplayRAM.mif* file. The data in this file will be displayed on the monitor.

As discussed in Chapter 6, the DE2 board has a triple DAC that takes 10-bit color data for each of the red, green and blue colors and generates analog inputs for the color inputs of a display. As evident from Figure 7.51, we have not taken advantage of all the 30 bits that are available for color specification. Instead, we have used the most significant bit of the 10-bit color input of each color and have driven it with 1 or 0. This means that we can only display eight colors at fixed intensities.

This design uses a 50 MHz clock input that is divided by two by using a toggle flip-flop. The 25 MHz output of the flip-flop is used for the VGA clock.

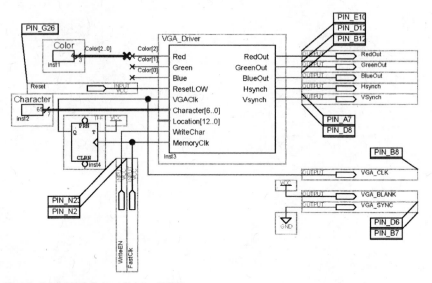

Figure 7.51 VGA Driver Tester (DE2)

The design discussed here verifies the operation of the VGA driver that we developed. This driver has a memory, and we have provided address lines and character inputs for addressing screen locations and writing into them. The reader is encouraged to take advantage of these utilities and implement a more elaborate design using this VGA driver.

7.8 Summary

This chapter presented several designs that mainly consisted of drivers for various peripherals. By use of these examples we achieved several goals. First, we were able to show designs that used the Cyclone or Cyclone II FPGA and their corresponding development boards. These designs elaborated some of the Verilog and logic design concepts of the earlier parts of this book. Secondly, we showed how designs could be packaged into tested user defined cores and blocks. The other goal achieved in this chapter was showing how peripherals like keyboards and VGA monitors operate, and how interfaces are designed.

8

Design with Embedded Processors

So far in this book, we have learned the basics of hardware design, how hardware is implemented, what tools are available, and how a hardware component can be utilized as part of a larger system. An alternative way of implementing a hardware function is by programming an existing processor to perform the function. The existing processor used as such is the embedded processor.

Design with an embedded processor requires 1) selection, configuration, or the design of the embedded processor, 2) design and configuration of processor memory and interfaces, and 3) development of the software that the processor is to run to perform the hardware function being implemented. This chapter focuses on these topics. We start by discussion of these topics and further elaboration of each. We will then show a complete filter design done with an embedded processor. The design of filter will be done manually to show the details of an embedded design process. After completion of this presentation, we show a microcontroller implemented by an embedded core.

8.1 Embedded Design Steps

As mentioned above, implementation of a hardware function with an embedded processor requires selection of the processor, design of its interfaces, and writing the program that it runs. We will discuss the details of these steps here.

In large designs use of proper design automation tools for performing these steps is essential. However, our discussion here is generic and independent of any tool or environment. The next chapter shows this design process using Altera's FPGA based tools.

Steps involved in design of an embedded processor for implementing a hardware function are similar to those of a microcontroller based system, except that embedded systems offer more flexibility and customization in each step of the work. The steps involved for implementing a hardware function in an embedded system, or a microcontroller system, are the selection of the processor, design of its interfaces and external bussing structure, and writing the program implementing the hardware function. Figure 8.1 shows where in a design process these steps apply.

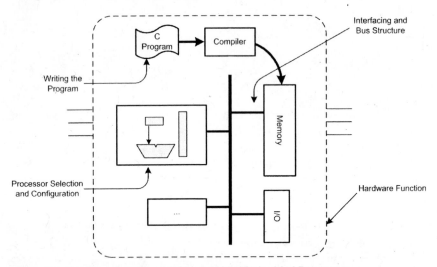

Figure 8.1 Hardware Function Implemented by Embedded Processor

8.1.1 Processor Selection

Selection of an embedded processor is like selecting a microcontroller for a specific function. The main difference is that there are more options when it comes to embedded processors. These options allow a hardware designer to tailor his or her embedded processor and perform optimizations to best fit the hardware function being designed.

Generally optimizations available to a hardware designer include elimination of instructions that are not needed, use of just enough memory, memory mapping, data and address lengths, memory structure, use of proper cache size or elimination of it, and use or elimination of hardware processors like multipliers. Depending on a specific application, a hardware designer selects an embedded processor and tailors it to best satisfy his or her design constraints.

An embedded processor may come as a softcore, a hardcore, or you may design your own. A softcore may be available in pre-synthesis HDL description or as a post-synthesis description for a specific target. A hardcore is fixed in a chip or a layout and, like a microcontroller, offers very little, if any, customization or flexibility in the processor core hardware. An embedded processor can also be designed by a designer who is using it in a larger system. In this case, the designer uses VHDL or Verilog to design his or her processor, and has all the freedom to choose the functionality of the processor.

8.1.1.1 Softcores. Whether it comes as a pre-synthesis HDL description, or a post-synthesis target-specific description, a softcore offers certain customizations and options to choose from. These options make it a better fit for the specific application it is being designed for. A designer using an existing softcore is able to select instructions, hardware feature, or data and address size of a customizable softcore. This is usually done through software tools that the softcore vendor is providing. Tools are also available for helping the softcore user in the next two steps of an embedded system design, i.e., design of processor interfaces and its software.

For a softcore available in a pre-synthesis HDL, the designer is free to make any changes to the code that is necessary for speed and space optimizations. However, in this case, vendor provided software tools may no longer be useful in the next two steps of an embedded system design process.

8.1.1.2 Hardcores. A hardcore has a fixed architecture, a fixed instruction set, and usually a hardcore is designed for a specific application. For example, there as DSP hardcores for DSP applications or RISC hardcores for high-end data processing applications. A hardcore vendor provides all simulation and design tools that are needed for using the hardcore, configuring its interfaces, and developing its software.

8.1.1.3 HDL Processor Cores. Instead of using processor cores from core vendors, a hardware designer may choose to develop his or her own processor that can be used in many of the designs that the designer is involved in. In this case, the designer is able to modify the processor however it best implements the hardware function being designed.

With an HDL processor, there is no limit as to the target of the processor or implementation technology. Restructuring the processor for better power or space utilization is also possible when using an HDL processor core.

The biggest drawback in using your own HDL embedded core is lack of availability of a compiler and other software tools. Furthermore, for structuring the hardware of the processor to meet your requirements, you are on your own in developing and testing your VHDL or Verilog code of your processor. Because of this, use of homemade HDL processor cores is only recommended for simple and very specialized functions.

8.1.1.4 Embedded Processor Example. Altera Corporation offers the Nios II softcore processor for its solution to the embedded design methodology. Nios II is a softcore and is available for Altera FPGAs.

This processor comes in three flavors for small, medium and complex applications. In addition, each flavor of Nios II can be configured for instruction set, memory usage, and cache using Altera provided design tools. Altera's SOPC Builder Software is used for selecting and configuring Nios II processors.

8.1.2 Processor Interfacing

In an embedded design, the next step after selection of the processor is configuring its external bussing structure. This includes tasks such as design of memory and device selection logic, interrupt handling hardware, design of priority and bus arbitration, and other I/O and memory related hardware components. For large systems this part becomes so complex that shifts the focus of the hardware designer from implementing his or her hardware function of an embedded processor to designing the interface logic and external processor bus structure.

8.1.2.1 Simple Interfacing. Memory mapped I/O, use of fast single-cycle dedicated memories, and limiting a design to a single processor simplifies a design to the level that one can design an entire embedded system without needing complex hardware configuration tools.

Generous uses of interrupt, use of complex arbitration schemes, and using multiple bus masters, are factors that complicate design of an embedded system.

8.1.2.2 Embedded Processor Interface Example. A design Tool provided by Altera for design of embedded systems around their Nios II processor is their SOPC Builder Software. In addition to configuring the Nios II processor, this tool can be used for design of the interface of a Nios II based embedded system. The bus structure Altera uses for interfacing with Nios II is the Avalon switch fabric. The use of Avalon through SOPC Builder hides all handshakings, timings, arbitrations and other interfacing issues from the embedded system designer.

8.1.3 Developing Software

The last step in the embedded design of a hardware function is the development of the software to run on the embedded processor. This step involves writing the program, which is generally done in C or C++, and compiling it to the machine language of the newly configured embedded processor.

8.1.3.1 Basic Programming Task. If we were to use our own homemade processor for implementing a hardware function, our only choice of a language to program it would be the assembly language of our processor. Obviously for complex tasks, this is not an acceptable solution, and a high level language such as C or C++ should be used. Lack of other software utilities such as an instruction set simulator and a debugger for our homemade HDL embedded processor, limit utilization of such a processor.

8.1.3.2 Software Tools Example. Altera provides IDE integrated design environment for helping an embedded system designer develop software for Nios II. This environment has tools for C code development, a C compiler, and all the necessary debugging tools.

8.2 Filter Design

As an example for an embedded design we use an FIR filter. We will show how steps discussed in Section 8.1 and depicted in Figure 8.1 can be used to design a processor system that implements our specified digital filter.

In order to show the actual design steps and not overshadow these design steps with use of design tools and utilities, we will use our homemade processor of Chapter 4 (SAYEH) and perform all the design steps manually.

The SAYEH processor we are using for our embedded processor is available in behavioral pre-synthesis Verilog. We will use memory mapped I/O for its interfacing to keep its bussing structure simple. The filter program will be written in C, and we will show a hand translation of it into SAYEH Assembly Language.

8.2.1 Filter Concepts

This section gives a brief overview of filters. We will only discuss filter concepts to the point that we can discuss hardware implementation of digital filters.

8.2.1.1 Analog and Digital Filters.

A filter has the property of allowing some frequencies in its incoming waveform, and blocking others. An analog filter uses active or passive components for the job of filtering or eliminating certain frequencies from an incoming waveform. A simple passive analog filter is an RC circuit (Figure 8.2). This circuit is a low pass filter, meaning that it eliminates high frequency components of its input $f(t)$.

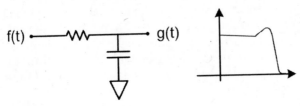

Figure 8.2 Passive Low Pass Filter

Because filters are mainly concerned with frequencies of an input signal, most filter discussions and representations are in the frequency domain. Filter representation in Figure 8.2 indicates that the filter allows lower frequencies to pass through, while higher frequencies are blocked.

An important application of filters is in audio and video processing. Since most such applications are done with digital computers, it is more appropriate to use digital techniques for filtering so that the job of filtering and other audio and video processing can all be done with the same integrated digital system.

8.2.1.2 Sampling.

An analog signal to be used with a digital filter must be turned into a series of digital data. This data is obtained by sampling the analog data and then digitizing the sample data as shown in Figure 8.3.

Figure 8.3 shows that an input analog signal is sampled with a sample-and-hold (S&H) circuit. This circuit outputs discrete data. The discrete analog data is then converted to digital data using an Analog-to-Digital converter (A/D). After filtering, and possibly other processings, the processed digital data is turned into analog (e.g., audio or video signals) by a Digital-to-Analog (D/A) converter.

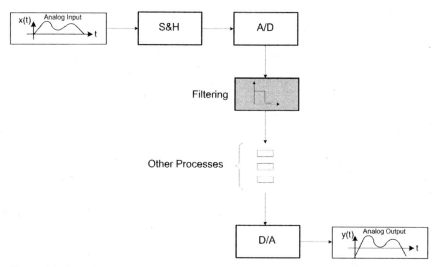

Figure 8.3 Sampling and Digitizing

According to Nyquist, in order to sample a signal without loss of information, the sampling frequency must be faster than twice the highest frequency of the incoming signal. This frequency is referred to as the Nyquist frequency. Signal sampling is done by a train of pulses at a frequency of at least the Nyquist frequency. Figure 8.4 shows the sampling process of an incoming $x(t)$ analog signal.

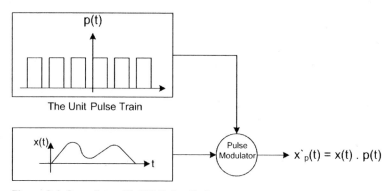

Figure 8.4 Sampling with P(t) Pulse Train

When modulated by a pulse modulator, each sample of an incoming signal becomes a pulse that is fed into the filter. The accumulation of the responses of a filter to the train of pulses that appear on its input becomes the output of the filter. The response of a filter to a unit pulse determines its functionality.

8.2.1.3 Impulse Response.

Digital filters are characterized by their response to an impulse. Figure 8.5 shows frequency response of low-pass, high-pass and band pass filters. Also shown here is the impulse response of a low-pass filter. The impulse response is described by a set of coefficients; the number of coefficients determines the order of the filter. The order of a filter determines how accurate filtering is done. This translates to a sharper frequency response for higher order filters.

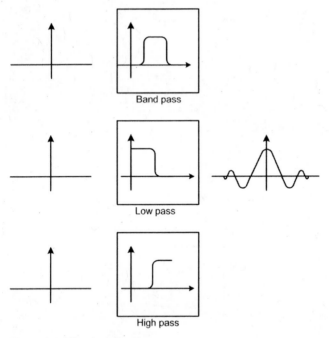

Figure 8.5 Filter Impulse Response

The response of a filter to a series of input samples can be considered as convolution of filter impulse responses to the individual samples. This convolution takes the present and past input samples and output responses into account.

8.2.1.4 Filter Equation.

What was described above will be put into an equation form from which the necessary hardware of a filter can be extracted. As discussed, as train of pulses from a sampled input source enter a digital filter, the response of the filter becomes the convolution of previous inputs and previous outputs. Equation below shows a filter output $y[n]$ in terms of its previous inputs, i.e., $x[n-k]$, and its previous outputs, $y[n-j]$. In this equation, input coefficients are indicated by $a[j]$ and output coefficients are $b[k]$.

$$y[n]= \sum a[j] * y[n\text{-}j] + \sum b[k] * x[n\text{-}k]$$

The block diagram for implementation of this equation is shown in Figure 8.6. This filter has a feedback that causes its impulse response to be infinite in duration. Such a filter is called an infinite impulse response (IIR) filter.

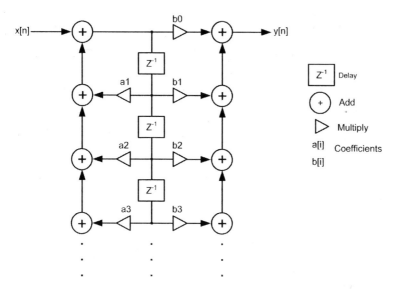

Figure 8.6 An IIR Filter

In Figure 8.6, the delay elements delay input values to be used in the next recursions. This filter is also called a recursive filter.

8.2.1.5 Finite Impulse Response (FIR) Filter. The IIR filter of Figure 8.6 can be simplified by eliminating its feedback. This eliminates output coefficients and simplifies calculation of remaining filter coefficients. Shown below is the resulting equation.

$$y[n]= \sum c[k] * x[n\text{-}k]$$

A filter made as such has finite impulse response and is referred to as an FIR or finite impulse response filter. In an FIR filter, filter coefficients can directly be extracted from its impulse response. An FIR filter block diagram is shown in Figure 8.7.

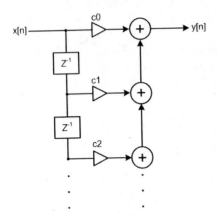

Figure 8.7 An FIR Filter

In the above sections we started with general filter concepts and ended with implementation of an FIR filter. The section that follows discusses alternatives for hardware implementation of this filter.

8.2.2 FIR Filter Hardware Implementation

The discussions of the previous section and filter block diagrams provide a good guideline for designing a digital system for a digital filter. This section shows two possibilities for implementing an FIR filter.

As discussed, filter coefficients are taken from the impulse response of a filter. These coefficients are the main factors in filter design. A generic hardware can be implemented using the structures that we discuss, and it can be configured for a specific filter by changing the coefficients.

The filter we are designing is a 4th order FIR filter with five coefficients that are shown in Figure 8.8. We will show an RTL and a CPU based hardware implementations.

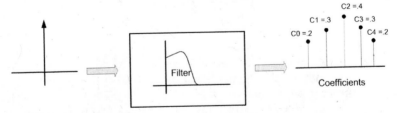

Figure 8.8 Filter to Design

8.2.2.1 FIR Filter RTL Design. Figure 8.9 shows the block diagram of the RT level hardware of a fourth order FIR filter. A series of regis-

ters provide delayed inputs that are multiplied by c_i coefficients and then added together to generate the output. The clock frequency for the registers must the same as the sampling frequency that is twice the largest frequency of the input signal.

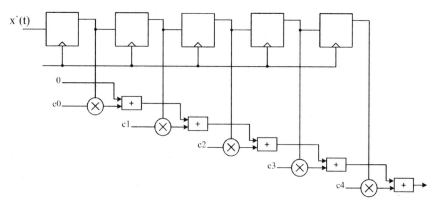

Figure 8.9 RTL FIR Filter

The hardware shown in Figure 8.9 is an iterative hardware and can be described using Verilog **generate** statements. This hardware can easily be implemented using registers, adders and multipliers. Of all these parts, the multipliers are the most complex and take more hardware than the other components.

8.2.2.2 Processor Core FIR Design. The algorithm presented in the block diagram of Figure 8.7 can be implemented with a processor running the iterative add and multiply procedure. A block diagram of this hardware implementation is shown in Figure 8.10.

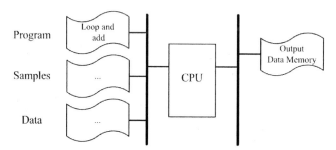

Figure 8.10 Processor Core for FIR Implementation

The data memory of the implementation of Figure 8.10 has a section for storing coefficients and another section for storing sampled data.

We assume that sampled data have been sampled and collected using a valid sampling rate for the input signal.

The program memory of the processor has a program that reads a new sample data performs shifting and five multiply and add operations, one for each coefficient of the filter. With each data read, the program outputs a new data for the circuit response.

8.2.3 FIR Embedded Implementation

The previous sections discussed operation and hardware of a digital filter. This section shows an embedded core implementation of a fifth order FIR filter. We will show the implementation of Figure 8.10 on an Altera FPGA.

Recall the steps we discussed in Section 8.1 for design of an embedded core. These steps are selection of the processor, design of the memory and I/O structure, and software development. The sections that follow will exercise these design steps.

Before an embedded design begins, the operation must be clearly defined. The operation of our filter is reading data, multiplying by all coefficients and outputting the result. The C code corresponding to this functionality is shown in Figure 8.11.

The program shown begins with header files and declarations. Following this part, it opens *input, parameter* and *output* files (lines 16 to 19). The *input* file is where sampled data are stored, and *output* is where result will be stored. The *parameter* file is where filter degree, and number of input samples are stored.

The loop that begins on line 26 and ends on line 47 (line numbers are shown in bold) reads data inputs calculates result and outputs data to the *output* file on line 46.

The loop that begins on line 37 and ends on line 43 performs multiplying data by filter coefficients and adding them as many times as there are coefficients. The result is collected in *temp*.

Note in this code that shifting of data that are multiplied (lines 32 and 40) results in using their eight most significant bits. This is done so that we will not require 16-bit multiplications in our implementation of this routine, which makes this C code conforms to the processor that we will be using for the implementation of this design. The processor we will use can only do 8-bit multiplications, which results in low accuracy of our filter design. This is a compromise we had to make to keep our design simple.

The program shown in Figure 8.11 works for any number of input data samples and FIR filters of any order.

```
01: #include <iostream.h>
02: #include <fstream.h>
03: #include <stdlib.h>
04: #include <stdio.h>
05: #include <string.h>
06:
07: int main ()
08: {
09: int history[16] = {0};
10: int i, j, n;
11: int temp;
12: int out[64] = {0};
13: int newInput, inputHigh, inputLow;
14: int newCoeff, coeffHigh, coeffLow;
15: int filterDegree, inputNo;
16: FILE * input;
17: FILE * coeff;
18: FILE * parameter;
19: FILE * output;
20: input = fopen("Input.bin","r");
21: parameter = fopen("Parameter.bin", "r");
22: output = fopen("Output.txt", "w");
23: filterDegree = fgetc(parameter);
24: inputNo = fgetc(parameter);
25: fclose(parameter);
26:     for (n = 0; n<inputNo; n++) {
27:         for (i = filterDegree-1; i>=0; --i) {
28:             history[i+1] = history[i];
29:         }
30:         inputHigh = fgetc(input);
31:         inputLow = fgetc(input);
32:         inputHigh = inputHigh << 8;
33:         newInput  = inputHigh | inputLow;
34:         history[0] = newInput;
35:         temp = 0;
36:         coeff = fopen("Coeff.bin","r");
37:         for (j = 0; j<filterDegree ; ++j) {
38:             coeffHigh = fgetc(coeff);
39:             coeffLow = fgetc(coeff);
40:             coeffHigh = coeffHigh << 8;
41:             newCoeff = coeffHigh | coeffLow;
42:             temp += history[j]*newCoeff;
43:         }
44:         fclose(coeff);
45:         out[n] = temp;
46:         fprintf(output,"%p\n", out[n]);
47:     }
48:     fclose(input);
49:     fclose(output);
50:     return 0;
51: }
```

Figure 8.11 FIR Filter C code

8.2.3.1 Processor Selection. We have many choices for the processor to perform the task of Figure 8.11. Because the purpose of this chapter is to present design steps and not design tools, we stay away from Nios II and other processors that require use of sophisticated SoC and embedded design tools. Our choice of the processor for this design is the SAYEH processor of Chapter 4.

SAYEH instruction set contains instructions that are not required for our simple task. On the other hand, SAYEH has an 8-bit multiplier that results in a 16-bit result. Although the use of this multiplication reduces the precision of our filter, we will not modify SAYEH or add a software multiplication routine for this introductory example. In other words, we will use SAYEH as is.

8.2.3.2 Memory and I/O Interfacing. As discussed in Section 8.1, the step after selection and/or configuration of the embedded processor is structuring memory and I/O of the processor and design of the CPU external busses. For a large system with many I/O devices and memory hierarchies this step involves design of address logic, I/O handshaking, arbitration, interrupt setting, priority encoding, etc. However, our system is much simpler than this.

Our embedded system needs a data memory for reading filter parameters, data input, filter coefficients, and writing filter outputs. In addition, the system needs an instruction memory for storing the filtering program to be read by the processor.

The bussing structure of our system only consists of processor data bus, address bus, and decoding logic for addressing these memory blocks. Our instruction memory begins at address 0000, and the data memory begins at 0100. We use FPGA on-chip clocked memory for the instruction memory, and fast signal cycle off-chip asynchronous RAM for the data memory. Figure 8.12 shows the bussing structure of our embedded system.

The decoder and read/write logic shown in Figure 8.12 has AND/OR logic for address decoding and issuing read/write signals. The select logic blocks in this figure are for connection of the bidirectional SAYEH data bus to the data busses of the data memory and instruction memory.

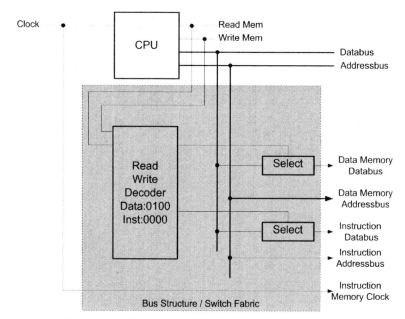

Figure 8.12 Embedded System Bus Structure

8.2.3.3 Filter Software.

The last step in the design of our FIR filter example is the development of its software. The algorithm for this software is that of Figure 8.11, the hardware structure that this software will be implemented in is shown in Figure 8.12, and the processor that the software runs on is SAYEH. In an automated environment, compiling the C program of Figure 8.11 with consideration of memory mappings, would be all that we needed to do for this step of the design. In our case, however, we have to develop our software in SAYEH assembly code.

Considering the memory structure of Figure 8.12 and requirements of our algorithm (Figure 8.11) as to filter parameters, data and coefficients, we decide on the memory mapping shown in Figure 8.13.

Filter program in SAYEH assembly is developed based on the memory map of Figure 8.13. We first read locations 0100 and 0101 for the degree of the filter and the number of input samples. Then each data sample that is read starting in 0140, is multiplied by its corresponding coefficient that being in 0130, stored in locations 0120 to 012f and added to previous data in these same locations. For each data read, an output is generated that is written starting in location 0180. Filter code in SAYEH assembly is shown in Figure 8.14.

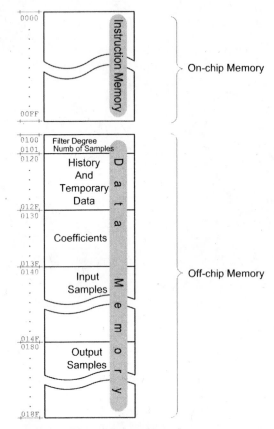

Figure 8.13 Filter Memory Map

The assembly code shown in this figure is translated to SAYEH machine language and becomes available for it to be loaded into the program memory (Figure 8.12). The task of assembly can be done manually, using the Verilog testbench *convert* of Chapter 4, or by writing an assembler for SAYEH.

```
0000 cwp
0001 mil r3 08
0002 mih r3 01 //0108: Out Loop Cntr
0003 mil r1 00
0004 mih r1 00
0005 sta r3 r1 // init Outer Loop
0006 mil r3 09
0007 mih r3 01 //0109: History Cntr
0008 sta r3 r1 // init History Cntr
0009 mil r3 0A
000A mih r3 01 //010A: Mult Loop Cntr
000B sta r3 r1 // init Mult Loop Cntr
000C awp 04
000D mil r3 00
000E mih r3 01
000F lda r2 r3 //0100:.Filter Degree
0010 mil r3 20
0011 mih r3 01 //0120: 1st Temp Data
0012 add r2 r3
0013 mil r3 01
0014 mih r3 00
0015 sub r2 r3
0016 lda r0 r2
0017 add r2 r3
0018 sta r2 r0 // shift history if n
0019 mil r2 09
001A mih r2 01
001B lda r0 r2
001C add r0 r3 // inct History cntr
001D sta r2 r0 // save hist cntr at 0109
001E mil r3 00
001F mih r3 01
0020 lda r2 r3
0021 sub r2 r0
0022 brz 07    // exit loop if hist=degree
0023 mil r3 20
0024 mih r3 01
0025 add r2 r3
0026 mil r3 00
0027 mih r3 00
0028 jpa r3 13
0029 mil r2 40
002A mih r2 01 //0104: 1st Input
002B mil r3 08
002C mih r3 01
002D lda r1 r3 // lda OuterLoopCntr
002E add r2 r1
002F lda r0 r2
0030 mil r2 20
0031 mih r2 01
0032 sta r2 r0 // new read data to 0120
0033 cwp
0034 awp 02
0035 mil r2 20
0036 mih r2 01
0037 mil r3 30
0038 mih r3 01 //0130:013F coefficients
0039 lda r0 r2
```

```
003A lda r1 r3
003B mul r0 r1
003C cwp
003D add r0 r2 //add MulOut to ACC
003E awp 02
003F mil r3 00
0040 mih r3 01
0041 lda r2 r3
0042 mil r3 0A
0043 mih r3 01
0044 lda r1 r3
0045 sub r2 r1
0046 brz 11    // quit MulLoop if degree
0047 mil r2 01
0048 mih r2 00
0049 add r1 r2 // inc MulLoop
004A sta r3 r1
004B awp 01
004C mil r1 20
004D mih r1 01
004E add r1 r0
004F mil r2 30
0050 mih r2 01
0051 add r2 r0
0052 cwp
0053 awp 02
0054 mil r1 00
0055 mih r1 00
0056 jpa r1 39
0057 cwp
0058 mil r1 80
0059 mih r1 01 // (0180:01BF) outputs
005A mil r2 08
005B mih r2 01
005C lda r3 r2
005D add r1 r3
005E sta r1 r0
005F mil r0 01
0060 mih r0 00
0061 add r0 r3 // inc outer Loop Counter
0062 sta r2 r0 // add Loop cuntr at 0108
0063 mil r3 08
0064 mih r3 01
0065 lda r2 r3
0066 sub r2 r0
0067 brz 0A    // quit main Loop if all
sample
0068 mil r0 00
0069 mih r0 00
006A mil r1 00
006B mih r1 00
006C mil r2 00
006D mih r2 00
006E mil r3 00
006F mih r3 00 // init registers
0070 jpa r3 06
0071 hlt
```

Figure 8.14 Filter Program Assembly Code

8.2.4 Building the FIR Filter

The previous section showed the complete design steps for an FIR filter. This section shows implementation of this design on an Altera FPGA. Figure 8.15 shows the block diagram of the filter in Quartus II. The processor code is available in Verilog. The *Decoder* block is built using basic gate primitives, and the rest of the bussing is just bus wires and tri-state gates.

The ROM shown in this figure is the program memory. The hex file that corresponds to the assembly code of Figure 8.14 goes into this memory at the initialization time. The *lpm_rom0* mega block is programmed to read *InstructionMem.mif* file upon start.

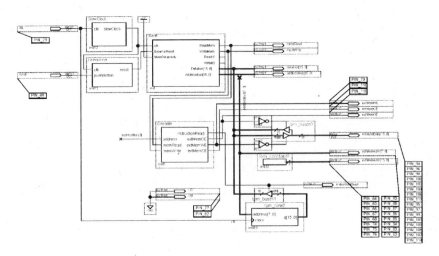

Figure 8.15 Quartus II Implementation of FIR

The external RAM that is used as data memory is not shown in the block diagram of Figure 8.15. Before the filter program starts, this RAM must be loaded with filter parameters, coefficients, and data. The DE2 development board has a control program and can be used to initialize its memories. Howver, for a board that does not have such a utility an FPGA program can handle the task of initializing board memories.

For example for the UP3 board, we have developed an FPGA hardware that reads data loaded into its internal memory and writes it into the selected external RAM. The FPGA internal memory is loaded with a *.mif* file. The FPGA hardware for performing this task is first programmed into the FPGA, and after it programs the external RAM, it is overwritten by the filter hardware of Figure 8.15. The RAM initializer Quartus II project is called *ROAM.bdf* and is available on the CD that accompanies this book.

The filter discussed here has been implemented on UP3 and DE2 Altera development boards. The filter design is generic and by changing its coefficient in the external data memory, it can be configured to implement any FIR filter for which an impulse response can be calculated. The filter can be modified to receive data via an external I/O port. Furthermore, controller hardware can be designed to load the external data memory from a PC connected to the development board that is being used for this filter.

8.3   Design of a Microcontroller

The previous section developed an application hardware on a development board using SAYEH. We selected the peripherals we needed and added memory and other necessary devices to our embedded processor system. This section tries to become more generic by designing a microcontroller system that can be programmed to perform various applications. This system has standard IO devices and memories connected to it.

Programming of this system is done from a PC running a C compiler. The PC is connected to the board through its serial or parallel printer port. An application program for our microcontroller is compiled into SAYEH assembly and after translation to SAYEH machine language it is programmed into the embedded SAYEH on the FPGA of our development board. This gives a high level programming interface for developing programs for our microcontroller. The overall structure of this system is shown in Figure 8.16.

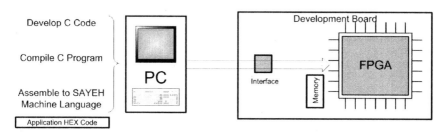

Figure 8.16 Programming Microcontroller

8.3.1   System Platform

The system being discussed here requires a C compiler and an assembler for our SAYEH processor. These programs run on a regular PC. The CD in the back of this book provides a preliminary version of these programs.

The microcontroller uses memory mapped IO devices and a memory for data and instructions. The user programming this system must be aware of memory and device locations. Programs developed on the PC (Figure 8.16) must consider these locations.

Interface logic for interfacing to the PC serial or parallel port is done on the FPGA of the development board using board connecters (serial for DE2 and parallel for UP3). Note that programming the FPGA to behave as a microcontroller is done through the regular programming pins (e.g., USB Blaster), but programming the memory of the microcontroller for a specific function of this microcontroller is done through our own designed interface logic on the FPGA.

8.3.2 Microcontroller Architecture

Figure 8.17 shows the architecture of our microcontroller. In addition to standard IO devices, this architecture has a dual port memory. While being programmed, the PC interface logic takes control of this memory and writes the application program into it. When this phase is complete, SAYEH will take over and uses this memory for its data and instructions.

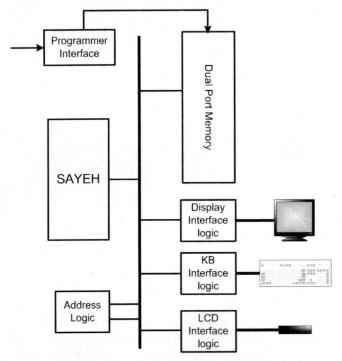

Figure 8.17 Microcontroller System Quartus II Implementation

The dual-port memory shown in Figure 8.17 is a 16-bit word 256K memory, and maps to locations 0000 to 00FF of SAYEH address space.

The keyboard shown here uses a buffer and it is mapped to locations 2000 and 2001. Reading from 2000 performs a read operation on the device FIFO, and reading from 2001 returns the status of FIFO.

An LCD driver is mapped to SAYEH at locations 3000 and 3001. Writing to 3000 and 3001 write command and data to the LCD, respectively. Reading these locations return the LCD's status word.

The VGA driver of our system is mapped to locations 1000 to 1FFF. Characters to be displayed can directly be written into these locations.

The complete implementation of this system in Quartus II is available on the CD on the back of this book. Figure 8.18. Although details of this diagram cannot be seen here, the relative positioning of its components is the same as that of Figure 8.18, which can be used to get more details about the wiring of the components of our microcontroller.

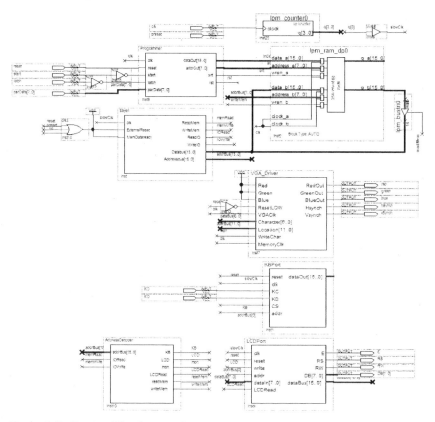

Figure 8.18 Quartus II Implementation

8.4 Summary

This chapter showed design of embedded systems. We discussed design steps that were needed for implementing a hardware function on an embedded processor. The key issue in this chapter was presentation of design steps without using SoC or SOPC design tools. The chapter focused on bare-bone hardware and software design.

The steps discussed were covered in a complete design; this design was an FIR filter that is a DSP application. The design was simple and used our homemade processor and its basic software utilities. Now that we are familiar with the actual steps of embedded system design, the next chapter shows a complete hardware/software environment for design of large scale applications using embedded systems.

9
Design of an Embedded System

The previous chapter showed how a software program and a processor that runs it could be used to implement a function that would otherwise require design of a special purpose hardware module. We also showed how such an implementation could be used along with other hardware functions for assembly and implementation and of a complete system that we refer to as an embedded system. The designs we used were generally small and therefore we did not require many sophisticated design tools or environments. Actually, the design of the structure of the hardware, and assembly of the program to run on our CPU were done manually.

An actual design of an embedded system cannot be done as easily as the designs of Chapter 8, and design tools and environments must come to the aid of the designer. This chapter shows tools and environments offered by Altera for design of embedded systems. Although the focus is on Altera's environments, but such tools are typical of most today's embedded system design environments. This chapter also shows an embedded system design example from specification to FPGA implementation.

Section 1 defines various pieces of an embedded system design environment. Discussion of the hardware elements of an embedded system begins with presenting an embedded processor in Section 2. This discussion continues with presenting the bus architecture of an embedded system in Section 3. The design tool for assembling the hardware parts of an embedded processor will be discussed in Section 4, while Section 5 is dedicated to the design tool for design of the software part of an embedded processor. Finally, Section 6 shows a complete design done with elements discussed in the preceding sections.

9.1 Designing an Embedded System

A typical embedded system consists of several processors connected to memories and other devices through a bussing structure. The first step in design of such systems is to decide what parts of a complete system are done in hardware using hardware blocks or Verilog code, and what parts are done by writing a program to run on a given processor. Once this decision is made, functions that are to be implemented in hardware will be designed using hardware design methods and synthesis tools, and functions implemented with a software program running on a processor will be designed using the C language and compilers and other software tools.

In addition to writing the C program, development of the software part of an embedded system requires general knowledge of the architecture of the processor the software runs on, and its bussing structure. Furthermore, considered as pieces of an embedded system design environment, an embedded system designer needs a tool for design and configuration of the hardware of the embedded processor and another for those of the software of the processor.

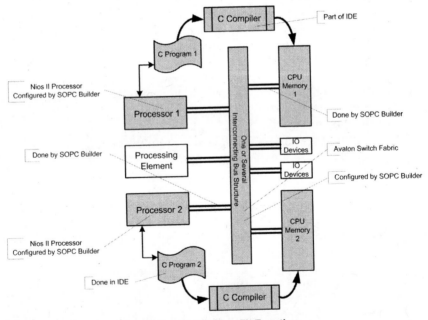

Figure 9.1 How pieces of an Embedded Design Fit Together

Altera's environment where decisions regarding hardware and software partitioning are done is the Quartus II software that we are already familiar with. This environment also provides tools and utilities for design and implementation of the hardware parts. Being the base of all our designs, we consider Quartus II as Altera's main platform of an embedded system design environment.

Taking off from Quartus II is the SOPC (System On Programmable Chip) Builder program that is used for putting together our embedded processors with their memories and IO devices through a given bus structure. Altera's processor for embedded designs is Nios II, and the bus structure that connects all system components is the Avalon switch fabric. Once the hardware part is in place, IDE (integrated Design Environment) takes off from SOPC Builder. IDE allows development of the C program that runs on the processor. It includes a C compiler, and necessary program entry and test utilities.

Embedded processors, interconnecting busses, hardware builders, and software tools are considers as pieces of an embedded system design environment. The relation between these facilities is shown in Figure 9.1. Altera's versions of these are the Nios II processor, Avalon bus, SOPC Builder, and IDE. Figure 9.1 is annotated with Altera's facilities for an embedded system design.

9.2 Nios II Processor

This section is an introduction to the Nios II embedded processor family. The Nios II processor is a general-purpose RISC processor core, providing:

- Full 32-bit instruction set, data path, and address space
- 32 general-purpose registers
- 32 external interrupt sources
- Single-instruction 32×32 multiply and divide producing a 32-bit result
- Dedicated instructions for computing 64-bit and 128-bit products of multiplication
- Floating-point instructions for single-precision floating-point operations
- Single-instruction barrel shifter
- Access to a variety of on-chip peripherals, and interfaces to off-chip memories and peripherals
- Hardware-assisted debug module enabling processor start, stop, step and trace under integrated development environment (IDE) control
- Software development environment based on the GNU C/C++ tool chain and Eclipse IDE
- Instruction set architecture (ISA) compatible across all Nios II processor systems

A Nios II processor system is equivalent to a microcontroller or "computer on a chip" that includes a CPU and a combination of peripherals and memory on a single chip. The term "Nios II processor system" refers to a Nios II processor core, a set of on-chip peripherals, on-chip memory, and interfaces to off-chip memory, all implemented on a single Altera chip. An example system including a processor, memories IO devices, and their interconnecting bus structure is shown in Figure 9.2. Like a microcontroller family, all Nios II processor systems use a consistent instruction set and programming model. The term "Nios II processor" or "Nios II CPU" refers to a Nios II processor core alone; we use these terms to discuss it from its architectural or programming point of view.

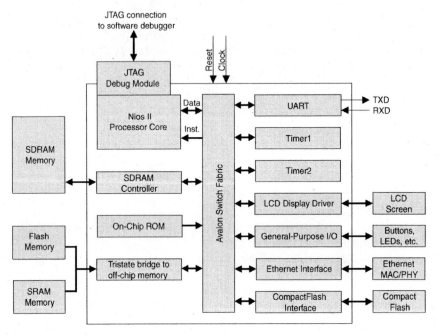

Figure 9.2 An Example Nios II System

9.2.1 Configurability Features of Nios II

This section introduces Nios II concepts that define a Nios II based system as configurable. These concepts are the major differences between an embedded system design using Nios II and discrete microcontrollers. For the most part, these concepts relate to the flexibility for hardware designers to fine-tune system implementation. Software programmers generally are not affected by the hardware implementation details, and can write programs without awareness of the configurable nature of the Nios II processor core.

9.2.1.1 Configurable Soft-Core Processor. The Nios II processor is a configurable soft-core processor, as opposed to a fixed, off-the-shelf microcontroller. In this context, "configurable" means that features can be added or removed on a system-by-system basis to meet performance or price goals. "Soft-core" means the CPU core comes in "soft" design form (i.e., not fixed in silicon), and can be targeted to any Altera FPGA family. Nios II comes in several versions as readymade processors. If these designs meet the system requirements, there is no need to configure the design further.

9.2.1.2 Flexible Peripheral Set & Address Map. A flexible peripheral set is one of the most notable differences between Nios II processor systems and fixed microcontrollers. Because of the soft-core nature of the Nios II processor, designers can easily build made-to-order Nios II processor systems with the exact peripheral set required for the target applications. A corollary of flexible peripherals is a flexible address map. Software constructs are provided to access memory and peripherals generically, independently of address location. Therefore, the flexible peripheral set and address map does not affect application developers.

Designers can choose from Altera standard peripherals, or design their own custom peripherals. Standard peripherals are those that are commonly used in microcontrollers, such as timers, serial communication interfaces, general-purpose I/O, SDRAM controllers, and other memory interfaces. Custom peripherals can also be deigned and integrated into Nios II processor systems.

For performance-critical systems that spend most CPU cycles executing a specific section of code, it is a common technique to create a custom peripheral that implements the same function in hardware. This approach offers performance benefit because the hardware implementation is faster than software. In addition, it frees the processor to perform other functions in parallel while the custom peripheral operates on data.

9.2.1.3 Custom Instructions. Like custom peripherals, custom instructions are a method to increase system performance by augmenting the processor with custom hardware. The soft-core nature of the Nios II processor enables designers to integrate custom logic into the arithmetic logic unit (ALU). Similar to native Nios II instructions, custom instruction logic can take values from up to two source registers and optionally write back a result to a destination register.

By using custom instructions, designers can fine tune the system hardware to meet performance goals. Because the processor is implemented on reprogrammable Altera FPGAs, a hardware/software co-design can consider tradeoffs between implementing various parts of a system in hardware or software.

From the software perspective, custom instructions appear as machine generated assembly macros or C functions, so programmers do not need to know assembly in order to use custom instructions.

9.2.1.4 Automated System Generation. Altera's SOPC Builder design tool fully automates the process of configuring processor features and generating a hardware design that can be programmed into an FPGA. The SOPC Builder graphical user interface (GUI) enables hardware designers to configure Nios II processor systems with any number of peripherals and memory interfaces. SOPC Builder can also import a designer's HDL design files, providing an easy mechanism to integrate custom logic into a Nios II processor system.

After system generation, the design can be programmed into a board, and software can be debugged executing on the board. Once the design is programmed into a board, the processor architecture is fixed. Software development proceeds in the same manner as for traditional, nonconfigurable processors.

9.2.2 Processor Architecture

This section describes the hardware structure of the Nios II processor, including a discussion of all the functional units of the Nios II architecture and the fundamentals of the Nios II processor hardware implementation.

The *Nios II architecture* describes an instruction set architecture (ISA). The ISA in turn necessitates a set of functional units that implement the instructions. A *Nios II processor core* is a hardware that implements the Nios II instruction set and supports the functional units described in this document. The processor core does not include peripherals or the connection logic to the outside world. It includes only the circuits required to implement the Nios II architecture. Figure 9.3 shows a block diagram of the Nios II processor core. User-visible functional units of the Nios II architecture are listed below.

- Register file
- Arithmetic logic unit and interface to custom instruction logic
- Resetting signals
- Exception and interrupt controller
- Instruction and data buses
- Instruction and data cache memories
- Tightly coupled memory interfaces for instructions and data
- JTAG debug module

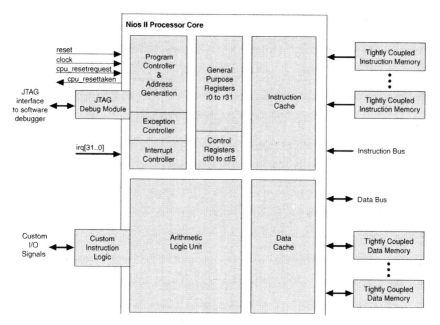

Figure 9.3 Nios II Processor Block Diagram

The functional units of the Nios II architecture form the foundation for the Nios II instruction set. However, this does not indicate that any unit is implemented in hardware. The Nios II architecture describes an instruction set, not a particular hardware implementation. A functional unit can be implemented in hardware, emulated in software, or omitted entirely.

A Nios II implementation is a set of design choices embodied by a particular Nios II processor core. All implementations support the instruction set defined in Altera's *Nios II Processor Reference Handbook*. Each implementation achieves specific objectives, such as smaller core size or higher performance. This allows the Nios II architecture to adapt to the needs of different target applications.

In the sub-sections that follow, we discuss hardware implementation details related to each functional unit.

9.2.2.1 Register File. The Nios II architecture supports a flat register file, consisting of thirty two 32-bit general-purpose integer registers, and six 32-bit control registers. The architecture supports supervisor and user modes that allow system code to protect the control registers from errant applications. The Nios II architecture allows for the future addition of floating point registers.

9.2.2.2 Arithmetic Logic Unit. The Nios II arithmetic logic unit (ALU) operates on data stored in general-purpose registers. ALU operations take one or two inputs from registers, and store a result back in a register. The ALU supports the data operations such as arithmetic, relational, logical, and shift and rotate operations. To implement any other operation, software computes the result by performing a combination of these fundamental operations.

Unimplemented Instructions. Some Nios II processor cores do not provide hardware to perform multiplication or division operations and are emulated in software. Instructions that the processor core may emulate in software: `mul`, `muli`, `mulxss`, `mulxsu`, `mulxuu`, `div`, `divu`. In such a core, these are known as unimplemented instructions. All other instructions are implemented in hardware.

The processor generates an exception whenever it issues an unimplemented instruction, and the exception handler calls a routine that emulates the operation in software. Therefore, unimplemented instructions do not affect the programmer's view of the processor.

Custom Instructions. The Nios II architecture supports user-defined custom instructions. The Nios II ALU connects directly to custom instruction logic, enabling designers to implement in hardware operations that are accessed and used exactly like native instructions.

Floating Point Instructions. The Nios II architecture supports single precision floating point instructions as specified by the IEEE Std 754-1985. These are implemented as custom instructions, and can be added to any Nios II processor core. The Nios II software development tools recognize C code that can take advantage of the floating point instructions when they are present in the processor core.

9.2.2.3 Reset Signals. The Nios II CPU core supports two reset signals. The global hardware reset signal (*reset*) forces the processor core to reset immediately. On the other hand, the *cpu_resetrequest* reset signal is a local reset signal that causes the CPU to reset without affecting other components in the Nios II system. With this reset, the processor finishes executing any instructions in the pipeline, and then enters the reset state and asserts the *cpu_resettaken* signal for one cycle. When in this state, the processor periodically checks if *cpu_resetrequest* remains asserted, and remains in reset for as long as *cpu_resetrequest* is asserted. The CPU does not respond to *cpu_resetrequest* when it is under the control of the JTAG debug module.

9.2.2.4 Exception & Interrupt Controller. The Nios II architecture provides a simple, non-vectored exception controller to handle all exception types. All exceptions, including hardware interrupts, cause the processor to transfer execution to a single *exception address*. The exception handler at this address determines the cause of the exception and dispatches an appropriate exception routine. The exception address is specified at system generation time.

The Nios II architecture supports thirty two external hardware interrupts. The processor core has 32 level-sensitive interrupt request (IRQ) inputs, *irq0* through *irq31*, providing a unique input for each interrupt source. IRQ priority is determined by software. The architecture supports nested interrupts. The software can enable and disable any interrupt source individually through the *ienable* control register, which contains an interrupt-enable bit for each of the IRQ inputs. Software can enable and disable interrupts globally using the PIE bit of the status control register.

A hardware interrupt is generated only if the PIE bit of the status register is **1**, an interrupt-request input, *irqn*, is asserted, and the corresponding bit *n* of the *ienable* register is **1**.

9.2.2.5 Memory & I/O Organization. The discussion of memory and I/O organization of this section covers both general concepts true of all Nios II processor systems, as well as features that may change from system to system.

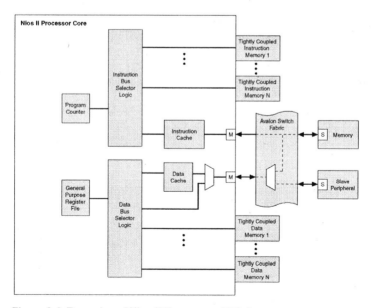

Figure 9.4 Examples of Nios II Memory and I/O Access

Figure 9.4 shows several ways a Nios II core can be connected to its memory and I/O devices. As shown, accessing memory and I/O devices can be done through instruction and data master ports of the Avalon switch fabric, tightly coupled instruction or data memory port, and fast cache memory internal to the Nios core. Connecting through instruction and data master ports implies use of Avalon master ports (shown by M in Figure 9.4) that connect to the corresponding memories. Such memories become slaves on the Avalon bus. The paragraphs that follow discuss these alternatives.

The Nios II architecture hides the hardware details from the programmer, so programmers can develop Nios II applications without awareness of the hardware implementation.

Instruction & Data Buses. The Nios II architecture supports separate instruction and data buses, classifying it as a Harvard architecture. Both the instruction and data buses are implemented as Avalon master ports that adhere to the Avalon interface specification. The data master port connects to both memory and peripheral components, while the instruction master port connects only to memory components.

The Nios II architecture provides memory-mapped I/O access. Both data memory and peripherals are mapped into the address space of the data master port. The Nios II architecture is little endian. Words and half words are stored in memory with the more-significant bytes at higher addresses.

The Nios II instruction bus is implemented as a 32-bit Avalon master port. The instruction master port performs a single function: Fetch instructions that will be executed by the processor. The instruction master port does not perform any write operations.

The instruction master port is a pipelined Avalon master port. Support for pipelined Avalon transfers minimizes the impact of synchronous memory with pipeline latency and increases the overall maximum frequency of the system. The instruction master port can issue successive read requests before data has returned from prior requests. The Nios II processor can pre-fetch sequential instructions and perform branch prediction to keep the instruction pipe as active as possible.

The instruction master port always retrieves 32 bits of data. The instruction master port relies on dynamic bus-sizing logic contained in the Avalon switch fabric. By virtue of dynamic bus sizing, every instruction fetch returns a full instruction word, regardless of the width of the target memory. Consequently, programs do not need to be aware of the widths of memory in the Nios II processor system.

The Nios II architecture supports on-chip cache memory for improving average instruction fetch performance when accessing slower memory. The Nios II architecture supports tightly coupled memory, which provides guaranteed low latency access to on-chip memory.

The Nios II data bus is implemented as a 32-bit Avalon master port. The two functions that the data master port performs are reading data from memory or a peripheral when the processor executes a load instruction, and writing data to memory or a peripheral when the processor executes a store instruction.

Byte-enable signals on the master port specify which of the four byte lane (s) to write during store operations. The data master port does not support pipelined Avalon transfers, because it is not meaningful to predict data addresses or to continue execution before data is retrieved. Consequently, any memory pipeline latency is perceived by the data master port as wait states. Load and store operations can complete in a single clock-cycle when the data master port is connected to zero wait-state memory.

The Nios II architecture supports on-chip cache memory for improving average data transfer performance when accessing slower memory. It also supports tightly coupled memory, which provides guaranteed low-latency access to on-chip memory.

Cache Memory. The Nios II architecture supports optional cache memories on both the instruction master port (instruction cache) and the data master port (data cache). Cache memory resides on-chip as an integral part of the Nios II processor core. The cache memories can improve the average memory access time for Nios II processor systems that use slow off-chip memory such as SDRAM for program and data storage.

A Nios II processor core may include one, both, or neither of the cache memories. Furthermore, for cores that provide data and/or instruction cache, the sizes of the caches are user-configurable.

Tightly Coupled Memory. Tightly coupled memory provides guaranteed low-latency memory access for performance-critical applications. Compared to cache memory, tightly coupled memory provides performance similar to cache memory, no cache overhead, and a guarantee that code or data is always available with the same latence.

Physically, a tightly coupled memory port is a separate master port on the Nios II processor core, similar to the instruction or data master port. A Nios II core can have zero, one, or multiple tightly coupled instruction and data memories. Each tightly coupled memory port connects directly to exactly one memory with guaranteed low, fixed latency. The memory is external to the Nios II core and is usually located on chip.

Tightly coupled memories occupy normal address space, the same as other memory devices connected via Avalon switch fabric.

The address ranges for tightly coupled memories (if any) are determined at system generation time. Software accesses tightly coupled memory using regular load and store instructions. From the software's perspective, there is no difference accessing tightly coupled memory compared to other memory.

9.2.2.6 Addressing Modes. The Nios II architecture supports register addressing, displacement addressing, immediate addressing, register indirect addressing, and absolute addressing.

In *register addressing*, all operands are registers, and results are stored back to a register. In *displacement addressing*, the address is calculated as the sum of a register and a signed, 16-bit immediate value. In *immediate addressing*, the operand is a constant within the instruction itself. *Register indirect addressing* uses displacement addressing, but the displacement is the constant 0. *Limited-range absolute addressing* is achieved by using displacement addressing with register *r0*, whose value is always 0x00.

9.2.2.7 JTAG Debug Module. The Nios II architecture supports a JTAG debug module that provides onchip emulation features to control the processor remotely from a host PC. PC-based software debugging tools communicate with the JTAG debug module and provide facilities, such as downloading programs to memory, starting and stopping execution, setting breakpoints and watchpoints, analyzing registers and memory, and collecting real-time execution trace data.

Soft-core processors such as the Nios II processor offer unique debug capabilities beyond the features of traditional, fixed processors. The softcore nature of the Nios II processor allows designers to debug a system in development using a full-featured debug core, and later remove the debug features to conserve logic resources. For the release version of a product, the JTAG debug module functionality can be reduced, or removed altogether.

9.2.3 Instruction Set

This section introduces the Nios II instructions categorized by type of operation performed. We will give a general overview of Nios II instructions. More details of the instructions of this machine can be found in Appendix A of this book and in the Nios II Processor Reference Handbook that is included in the CD in the back of this book. Instruction categories that we will discuss are: data transfer, arithmetic and logical, move, comparison, shift and rotate, program control, custom, and no-operation instructions.

9.2.3.1 Data Transfer Instructions. The Nios II architecture is a load-store architecture. Load and store instructions handle all data movement between registers, memory, and peripherals. Memories and peripherals share a common address space. Some Nios II processor cores use memory caching and/or write buffering to improve memory bandwidth. The architecture provides instructions for both cached and uncached accesses.

Data Transfer Instructions consists of Word Data Transfer Instructions (ldw, stw, ldwio & stwio), and Byte Data Transfer Instructions (ldb, ldbu, stb, ldh, ldhu, sth, ldbio, ldbuio, stbio, ldhio, ldhuio, sthio). Details of these instructions are discussed in Appendix A.

9.2.3.2 Arithmetic and Logical Instructions. Nios II logical instructions support AND, OR, XOR, and NOR operations. Arithmetic instructions support addition, subtraction, multiplication and division.

Arithmetic and Logical Instructions consist of *Standard Logical Instructions* (and, or, xor, nor), *Immediate Logical Instructions* (andi, ori, xori), *High Immediate Logical Instructions* (andhi, orhi, xorhi), *Standard Arithmetic Instructions* (add, sub, mul, div, divu), *Immediate Arithmetic Instructions* (addi, subi, muli), *Upper Multiplication Instructions* (mulxss, mulxuu), and *Long Multiplication Instruction* (mulxsu).

9.2.3.3 Move Instructions. Move instructions provide move operations to copy the value of a register or an immediate value to another register. This group of instructions consists of mov, movhi, movi, movui and, movia.

9.2.3.4 Comparison Instructions. The Nios II architecture supports a number of comparison instructions. All of these compare two registers or a register and an immediate value, and write either **1** (if true) or **0** to the result register. These instructions perform all the equality and relational operators similar to those of the C programming language.

Comparison Instructions consist of *Basic Comparison Instructions* (cmpeq, cmpne, cmpge, cmpgeu, cmpgt, cmpgtu, cmple, cmpleu, cmplt), and *Immediate Comparison Instructions* (cmpeqi, cmpnei, cmpgei, cmpgeui, cmpgti, cmpgtui, cmplei, cmpleui, cmplti).

9.2.3.5 Shift and Rotate Instructions. The Nios II architecture supports standard and immediate shift and rotate operations. Right and left versions of these instructions are provided. The number of bits to rotate or shift can be specified in a register or an immediate value.

Shift and Rotate Instructions consist of *Rotate Instructions* (rol, ror, roli), and *Shift Instructions* (rsll, slli, sra, srl, srai, srli).

9.2.3.6 Program Control Instructions.
The Nios II architecture supports the unconditional jump and call instructions. These instructions do not have delay slots.

Program Control Instructions subgroups are *Unconditional Jump and Call Instructions* (call, callr, ret, jmp, br) and *Conditional Branch Instructions* (bge, bgeu, bgt, bgtu, ble, bleu, blt, bltu, beq, bne).

9.2.3.7 Other Control Instructions.
In addition to the standard control instructions, Nios II supports instructions for debugging, status register manipulation, exception handling, and pipleline related instructions.

Exception Instructions (trap, eret), *Break Instructions* (break, bret), *Control Register Instructions* (rdctl, wrctl), *Cache Control Instructions* (flushd, flushi, initd, initi, flushp), and *Synchronization Instruction* (synch) are Other Control Instructions of Nios II.

9.2.3.8 Custom Instructions.
The custom instruction provides low-level access to custom instruction logic. The inclusion of custom instructions is specified at system generation time, and the function implemented by custom instruction logic is design dependent. Machine-generated C functions and assembly macros provide access to custom instructions, and hide implementation details from the user. Therefore, most software developers never use the custom assembly instruction directly.

9.2.3.9 No-Op Instruction.
The nop instruction is provided in the Nios II assembler, and is the no-operation instruction.

9.2.3.10 Potential Unimplemented Instructions.
Some Nios II processor cores do not support all instructions in hardware. In this case, the processor generates an exception after issuing an unimplemented instruction. The only instructions that may generate an unimplemented-instruction exception are: mul, muli, mulxss, mulxsu, mulxuu, div, divu. All other instructions are guaranteed not to generate an unimplemented instruction exception. An exception routine must exercise caution if it uses these instructions, because they could

generate another exception before the previous exception was properly handled.

9.2.4 Nios II Alternative Cores

The Nios II processor comes in three flavors that can be selected in the SOPC Builder configuration wizard. These alternative cores set a base for the exact processor upon which other configurations and customizations may be done. Currently, Altera offers three Nios II cores: *Nios II/f*, *Nios II/s*, and *Nios II/e*.

Nios II/f "fast" core is designed for fast performance. As a result, this core presents the most configuration options allowing you to fine-tune the processor for performance. Options available for this core include instruction and data cache, dynamic branch predication, hardware multiply and divide, and barrel shifter. This core uses between 1400 and 1600 FPGA logic elements. It also uses three 4K memory blocks plus whatever is needed for its caches.

Nios II/s—The Nios II/s "standard" core is designed for small size while maintaining performance. Options available for this core include instruction cache, branch predication, and hardware multiply and divide. This core uses between 1200 and 1400 FPGA logic elements. It also uses two 4K memory blocks plus whatever is needed for its caches.

Nios II/e—The Nios II/e "economy" core is designed to achieve the smallest possible core size. This core is a basic 32-bit RISC machine without any of the features mentioned for the other two versions of this processor. This core uses between 600 and 700 FPGA logic elements, and two 4k memory blocks.

9.3 Avalon Switch Fabric

Avalon switch fabric is a high-bandwidth interconnect structure that consumes minimal logic resources and provides greater flexibility than a typical shared system bus. This section describes the functions of Avalon switch fabric and the implementation of those functions.

9.3.1 Avalon Specification

Avalon switch fabric is the glue that binds together components in a system based on the Avalon interface. This switch fabric is the collection of interconnect and logic resources that connects Avalon master and slave ports on components in a system. Avalon switch fabric encapsulates the connection details of a system.

An example Avalon switch fabric is shown in Figure 9.5. This bus can be used for any number of master (M) and slave (S) components. The bus allows connection of any master to any slave that is on the bus. Various components on the bus can operate in different clock domains, have different data widths, and have big- or little-endian data orientation. Avalon facilitates a master writing (solid lines in Figure 9.5) and reading a slave port through the bus.

Some components in Figure 9.5 use multiple Avalon ports, i.e., processors and DMA. Because an Avalon component can have multiple Avalon ports, you can use Avalon switch fabric to create super interfaces that provide more functionality than a single Avalon port. For example, an Avalon slave port can have only one interrupt-request (IRQ) signal. However, by using three Avalon slave ports together, you can create a component that generates three separate IRQs. In this case, SOPC Builder generates the Avalon switch fabric to connect all ports.

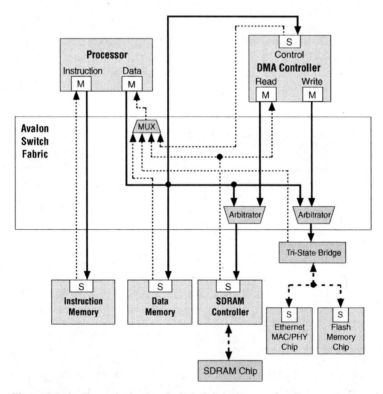

Figure 9.5 An Example Avalon Switch Fabric Connecting Components and Cores

Generating Avalon switch fabric is SOPC Builder's primary purpose. Because SOPC Builder generates Avalon switch fabric automatically, most users do not interact directly with it or the HDL that describes it. You do not need to know anything about the internal workings of Avalon switch fabric to take advantage of the services it provides. On the other hand, a basic understanding of how it works can help you optimize your components and systems. For example, knowledge of the arbitration mechanism can help designers of multi-master systems minimize the impact of arbitration on the system throughput.

9.3.1.1 Avalon Switch Fabric Implementation. Avalon switch fabric uses active logic to implement a switched interconnect structure that provides a dedicated path between master and slave ports. This bus consists of synchronous logic and routing resources inside an FPGA.

At each port interface, Avalon switch fabric manages Avalon transfers, responding to signals from the connected component. The signals that appear on the master port and corresponding slave port during a transfer can be very different, depending on how the Avalon switch fabric transports signals between the master-slave pair. In the path between master and slave ports, the Avalon switch fabric can introduce registers for timing synchronization, finite state machines for event sequencing, or nothing at all, depending on the services required by those ports.

9.3.1.2 Functions of Avalon Switch Fabric. Avalon switch fabric logic provides the following functions:

- Address Decoding
- Data-Path Multiplexing
- Wait-State Insertion
- Pipelining and Pipeline Management
- Endian Conversion
- Native Address Alignment & Dynamic Bus Sizing
- Arbitration for Multi-Master Systems
- Burst Management
- Clock Domain Crossing
- Interrupt Controller
- Reset Distribution

The behavior of these functions in a specific SOPC Builder system depends on the design of the components in the system and the settings made in the SOPC Builder GUI. The sections that follow describe how SOPC Builder implements each function. These sections can be skipped if a reader is not interested in the implementation details.

9.3.2 Address Decoding Logic

Address decoding logic in the Avalon switch fabric distributes an appropriate address and produces a chip-select signal for each slave port. Avalon selects a slave port whenever it is being addressed by a master. Slave components do not need to decode the address to determine when they are selected. Slave port addresses are always properly aligned for the data width of the slave port.

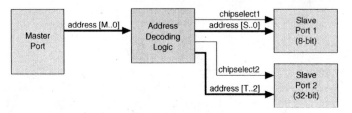

Figure 9.6 Address Decoding in Avalon

Figure 9.6 shows a block diagram of the address-decoding logic for one master and two slave ports. Separate address-decoding logic is generated for every master port. As shown, the address decoding logic handles the difference between the master address width (M) and slave address widths (S & T). It also maps only the necessary master address bits to access words in each slave port's address space.

9.3.3 Data-path Multiplexing

Data-path multiplexing logic in the Avalon switch fabric aggregates read-data signals from multiple slave ports during a read transfer, and presents the signals from only the selected slave back to the master port. Figure 9.7 shows a block diagram of the data-path multiplexing logic for one master and two slave ports.

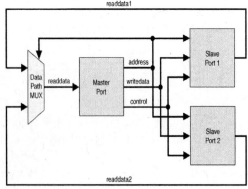

Figure 9.7 Data Path Multiplexing

Data-path multiplexing is not necessary in the write-data direction for write transfers. The write-data signals are distributed equally to all slave ports, and each slave port ignores write-data except for when the address-decoding logic selects that port.

9.3.4 Wait-state Insertion

Wait states extend the duration of a transfer by one or more cycles for the benefit of components with special synchronization needs.

Wait-state insertion logic accommodates the timing needs of each slave port, and coordinates the master port to wait until the slave can proceed. Avalon switch fabric inserts wait states into a transfer when the target slave port cannot respond in a single clock cycle. Avalon switch fabric also inserts wait states in cases when slave read-enable and write-enable signals have setup or hold time requirements.

Wait-state insertion logic is a small finite-state machine that translates control signal sequencing between the slave side and the master side. Figure 9.8 shows a block diagram of the wait-state insertion logic between one master and one slave.

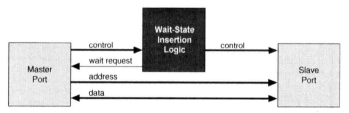

Figure 9.8 Wait State Block Diagram

9.3.5 Pipelining

SOPC Builder can pipeline the Avalon switch fabric by inserting stages of registers between master-slave pairs, and appropriate pipeline management logic for taking advantage these registers. Adding pipeline registers can increase the performance of the system and ensure that the critical timing path does not occur inside the Avalon switch fabric.

The pipeline registers introduce one or more clock cycles of latency between master-slave pairs, which creates a trade-off between transfer latency and maximum frequency of operation. The pipeline registers can also increase logic utilization considerably, depending on the complexity of the system. Components that support pipelined Avalon transfers minimize the effects of the pipeline latency.

9.3.6 Endian Conversion

In general, an Avalon-based system can contain both big and little endian components. The endianness of an Avalon port depends on the component design. Endianness affects the order a master port expects individual bytes to be arranged within a larger word. If all master ports in the system use the same endianness, then all master ports' perception of byte addresses is consistent within the system. In this case there is no further endian-related design consideration required.

Avalon switch fabric provides endian-conversion functionality to allow master ports of differing endianness to share memory. When several master ports of equal data widths access a common memory, the Avalon endian-conversion logic hides the endian difference of master ports for as long as the master ports read and write the memory using only native width units (e.g., always 32-bit) of data.

9.3.7 Native Address Alignment and Dynamic Bus Sizing

SOPC Builder generates Avalon switch fabric to accommodate master and slave ports with unmatched data widths. Address alignment affects how slave data is aligned in a master port's address space, in the case that the master and slave data widths are different. Address alignment is a property of each slave port, and it may be different for each slave port in a system. A slave port can declare itself to use either native address alignment, or dynamic bus sizing.

Slave ports that access address-mapped registers inside the component generally use native address alignment. Native address alignment is when each slave offset (i.e., word) is smaller than and maps to exactly one master word, regardless of the data width of the ports, and one transfer on the master port generates exactly one transfer on the slave port.

Slave ports that access memory devices generally use dynamic bus sizing. Dynamic bus sizing hides the details of interfacing a narrow memory device to a wider master port, and vice versa. When an N-bit master port accesses a slave port with dynamic bus sizing, the master port operates exclusively on full N-bit words of data, without awareness of the slave data width.

9.3.8 Arbitration for Multi-Master Systems

Avalon switch fabric supports systems with multiple master components. In a system with multiple master ports, such as the system pictured in Figure 9.5, the Avalon switch fabric provides shared access to slave ports using a technique called slave-side arbitration. Slave-side arbitration determines which master port gains access to a specific slave port in the event that multiple master ports attempt to access the same slave port at the same time.

9.3.9.1 Slave-Side Arbitration. The multi-master architecture used by Avalon switch fabric does not have shared bus lines; instead Avalon master-slave pairs are connected by dedicated paths. A master port never waits to access a slave port, unless a different master port attempts to access the same slave port at the same time. As a result, multiple master ports can simultaneously transfer data with independent slave ports.

A multi-master Avalon system requires arbitration, but only when two masters contend for the same slave port. This arbitration is called slave-side arbitration, because it is implemented at the point where two (or more) master ports connect to a single slave. Master ports contend for individual slave ports, not for a shared bus resource. For example, the example of Figure 9.5 shows a system with two master ports (CPU and DMA), sharing a slave port (SDRAM controller). Arbitration is performed on the SDRAM slave port; the arbitrator dictates which master port gains access to the slave port if both master ports initiate a transfer with the slave at the same time.

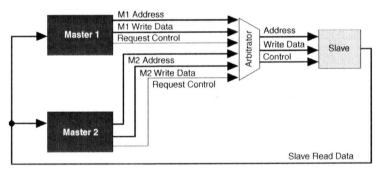

Figure 9.9 Multi Master Connections

Figure 9.9 focuses on the two master ports and the shared slave port, and shows additional detail of the data, address, and control paths. The arbitrator logic multiplexes all address, data, and control signals from a master port to a shared slave port.

9.3.9.2 Arbitrator Details. SOPC Builder generates an arbitrator for every slave port connected to multiple master ports, based on arbitration parameters specified in the SOPC Builder GUI.

The arbitrator logic evaluates the address and control signals from each master port at every clock cycle when a new transfer can begin, and determines which master port, if any, is requesting access to the slave. It then selects the mater port that is to gain access to the slave next. Other master ports, not granted access, are forced to wait. The arbitrator logic uses multiplexers to connect address, control, and data paths between the multiple master ports and the slave port. An example arbitration logic in a system with two master ports, each connected to two slave ports, is shown in Figure 9.10.

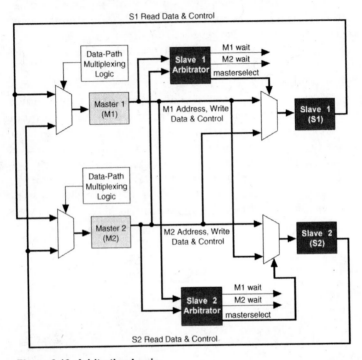

Figure 9.10 Arbitration Logic

Arbitration rules based on which the arbitrator grants access to master ports when they contend are *Fairness-Based Shares, Round-Robin Scheduling, Burst Transfers*, and *Minimum Share Value*. In a fairness-based arbitration scheme, each master-slave port pair has an integer value of transfer shares. In case of content, the arbitrator grants access to the master with smaller share. In a Minimum Share Value scheme, a component can declare the minimum number of transfers it needs when given access in a Round-Robin cycle.

9.3.9 Burst Management

Avalon switch fabric provides burst management logic to accommodate the burst capabilities of each port in the system, including ports that do not support burst transfers. Burst management logic is a finite state machine that translates the sequencing of address and control signals between the slave side and the master side.

The maximum burst length for each port is determined by the component design and is independent of other ports in the system. Therefore, a particular master port might be capable of initiating a burst longer than a slave port's maximum supported burst length. In this case, the burst management logic translates the master burst into smaller slave bursts, or into individual slave transfers if the slave port does not support bursts. Until the master port completes the burst, the Avalon arbitrator logic prevents other master ports from accessing the target slave port.

For example, if a master port initiates a burst of 16 transfers to a slave port with maximum burst length of 8, the burst management logic initiates two bursts of length 8 to the slave port. If a master port initiates a burst of 16 transfers to a slave port that does not support bursts, the burst management logic initiates 16 separate transfers to the slave port.

9.3.10 Clock Domain Crossing

SOPC Builder generates clock-domain crossing (CDC) logic that hides the details of interfacing components operating in asynchronous clock domains. The Avalon switch fabric upholds the Avalon protocol with each port independently, and therefore each Avalon port need only be aware of its own clock domain. Avalon switch fabric logic propagates transfers across clock domain boundaries transparently to the user.

The CDC logic in Avalon switch fabric allows component interfaces to operate at a different clock frequency than system logic. It eliminates the need to design CDC hardware manually. It allows each Avalon port to operate in only one clock domain, which reduces design complexity of components. With CDC, master ports can access any slave port without awareness of the slave clock domain.

9.3.10.1 Clock Domain-Crossing Logic. The CDC logic consists of two finite state machines (FSM), one in each clock domain, which use a simple hand-shaking protocol to propagate transfer control signals (read request, write request, and the master wait-request signals) across the clock boundary.

9.3.10.2 Master-Slave Communication. With CDC, transfers proceed as normal on the slave and the master side, without a special protocol to handle crossing clock domains. From the perspective of a slave port, there is nothing different about a transfer initiated by a master port in a different clock domain. From the perspective of a master port, a transfer across clock domains simply takes extra clock cycles. Similar to other transfer delay cases (for example, arbitration delay and/or wait states on the slave side), the Avalon switch fabric simply forces the master port to wait until the transfer terminates. As a result, latency-aware master ports do not benefit from pipelining when performing transfers to a different clock domain.

9.3.11 Interrupt Controller

In systems with one or more slave ports that generate IRQs, the Avalon switch fabric includes interrupt controller logic. A separate interrupt controller is generated for each master port that accepts interrupts. The interrupt controller aggregates IRQ signals from all slave ports, and maps slave IRQ outputs to user-specified values on the master IRQ inputs.

Each slave port optionally produces an IRQ output signal. There are two master signals related to interrupts: irq and irqnumber. SOPC Builder generates the interrupt controller in one of two configurations, software priority or hardware priority, depending on the interrupt signals present on the master port.

9.3.11.1 Software Priority. In the software priority configuration, the Avalon switch fabric passes IRQs directly from slave to master port, without making any assumptions about IRQ priority. In the event that multiple slave ports assert their IRQs simultaneously, the master logic (presumably under software control) determines which IRQ has highest priority, then responds appropriately.

Using software priority, the interrupt controller can handle up to 32 slave IRQ inputs. The interrupt controller generates a 32-bit signal $irq[31..0]$ to the master port, and maps slave IRQ signals to bits of $irq[31..0]$. Any unassigned bits of $irq[31..0]$ are permanently disabled.

9.3.11.2 Hardware Priority. In the hardware priority configuration, in the event that multiple slaves assert their IRQs simultaneously, the Avalon switch fabric passes the IRQ of highest priority to the master port. An IRQ of lesser priority is undetectable until a master port clears all IRQs of higher priority.

Using hardware priority, the interrupt controller uses a priority encoder and can handle up to 64 slave IRQ signals. The interrupt controller generates a 1-bit irq signal to the master port, signifying that one or more slave ports have generated an IRQ. The controller also generates a 6-bit $irqnumber$ signal, which outputs the encoded value of the highest pending IRQ.

9.3.12 Reset Distribution

The Avalon switch fabric generates and distributes a system-wide reset pulse to all logic in the system module. The switch fabric distributes the reset signal conditioned for each clock domain. The duration of the reset signal is at least one clock period.

9.4 SOPC Builder Overview

SOPC Builder is a system development tool for creating systems based on processors, peripherals, and memories. This software is included in Altera's Quartus II and is for defining and generating a complete system-on-a-programmable-chip (SOPC).

SOPC Builder is a general-purpose tool for creating arbitrary SOPC designs that may or may not contain a processor. This tool automates the task of integrating hardware components into a larger system. Using SOPC Builder, you specify the system components in a graphical user interface (GUI), and SOPC Builder generates the interconnect logic automatically. SOPC Builder outputs HDL files that define all components of the system, and a top-level HDL design file that connects all the components together. SOPC Builder can generate Verilog HDL or VHDL.

In addition to its role as a hardware generation tool, SOPC Builder also serves as the starting point for system simulation and embedded software creation. SOPC Builder provides features to ease writing software and to accelerate system simulation.

9.4.1 Architecture of SOPC Builder Systems

This section introduces the architectural structure of systems built with SOPC Builder, and describes its primary functions. This program is used for building a system of processors and interfaces to be programmed into an Altera FPGA. Figure 9.2 is an example of a system that has been put together with SOPC Builder.

An SOPC Builder *component* is a design module that SOPC Builder recognizes and can automatically integrate into a system. SOPC Builder connects multiple components together to create a top-level HDL file called the *system module*. SOPC Builder generates *Avalon switch fabric* that contains logic to manage the connectivity of all components in the system.

9.4.1.1 SOPC Builder Components.
SOPC Builder components are the building blocks of the system module. The components use the Avalon interface for the physical connection of components. This tool can be used to connect any logical device (either on-chip or off-chip)

that has an Avalon interface. The Avalon interface uses an address-mapped read/write protocol that allows master components to read and/or write any slave component.

A component can be a logical device that is entirely contained within the system module, such as a processor component. Alternately, a component can act as an interface to an off-chip device, such as an SRAM interface component. In addition to the Avalon interface, a component can have other signals that connect to logic outside the system module. Non-Avalon signals can provide a special-purpose interface to the system module.

Altera and third-party developers provide ready-to-use SOPC Builder components, such as:

- Microprocessors, such as the Nios II processor
- Microcontroller peripherals
- Timers
- Serial communication interfaces, such as a UART and a serial peripheral interface (SPI)
- General purpose I/O
- Digital signal processing (DSP) functions
- Communications peripherals
- Interfaces to off-chip devices
    o Memory controllers
    o Buses and bridges
    o Application-specific standard products (ASSP)
    o Application-specific integrated circuits (ASIC)
    o Processors

SOPC Builder Ready components are those intellectual property (IP) designs that have plug–and–play integration with SOPC Builder. These functions may be accompanied by software drivers, low level routines, or other software design files.

Users can also define their own components to be used in an SOPC design. SOPC Builder provides an easy method to develop and connect new custom components. With the Avalon interface, user-defined logic need only adhere to a simple interface based on address, data, read enable, and write enable signals. The following summarizes steps needed for integrating custom logic into an SOPC Builder system:

1. Define the interface to the user-defined component.
2. If the component logic resides on-chip, write HDL files describing the component in either Verilog HDL or VHDL.
3. Use the SOPC Builder component editor wizard to specify

the interface and optionally package your HDL files into an SOPC Builder component.

4. Instantiate your component in the same manner as other SOPC Builder Ready components.

Once an SOPC Builder component is created, it can be reused in other SOPC Builder systems, and shared with other design teams.

9.4.1.2 Avalon Switch Fabric. The Avalon switch fabric is the glue that binds SOPC Builder-generated systems together. As discussed in the previous section, Avalon is the collection of signals and logic that connects master and slave components, including address decoding, data-path multiplexing, wait-state generation, arbitration, interrupt controller, and data-width matching. SOPC Builder generates this switch fabric automatically, to free designers from manually performing the tedious, error-prone task of connecting hardware modules.

SOPC Builder abstracts the complexity of interconnect logic, allowing designers to focus on the details of their custom components and the high-level system architecture. Automatically generating the Avalon switch fabric is the keystone to achieving this purpose.

9.4.2 Functions of SOPC Builder

The purpose of the SOPC Builder GUI is to allow designers to define the structure of a hardware system, and then generate the system. This involves generating the hardware, creating a memory map for software development, and creating a simulation model.

9.4.2.1 Hardware Generation. The GUI of SOPC Builder is designed for the tasks of adding components to a system, configuring the components, and specifying how they connect together.

After all components are added and all necessary system parameters are specified, SOPC Builder is ready to generate the Avalon switch fabric and output the HDL files that describe the system. During system generation, SOPC Builder outputs an HDL file for the top-level system module and for each component in the system. In addition, it generates a Block Symbol File (**.bsf**) representation of the top-level system module for use in Quartus II Block Diagram Files (**.bdf**). Optionally, SOPC Builder generates software files for embedded software development, such as a memory-map header file and component drivers, as well as a testbench for the system module and ModelSim simulation project files.

After a system module is generated, it can be compiled directly by the Quartus II software, or instantiated in a larger FPGA design.

9.4.2.2 Memory Map for Software Development.

For each microprocessor in the system, SOPC Builder optionally generates a header file that defines the address of each slave component. In addition, each slave component can provide software drivers and other software functions and libraries for the processor.

The process for writing software for the system depends heavily on the nature of the processor in the system. For example, Nios II processor systems use Nios II processor-specific software development tools. These tools are separate from SOPC Builder, but they do use the output of SOPC Builder as the foundation for software development.

9.4.2.3 Creating a Simulation Model.

During system generation, SOPC Builder optionally outputs a push-button simulation environment for system simulation. SOPC Builder generates both a simulation model and a testbench for the entire system. The testbench instantiates the system module, assigns values and drives all clocks and resets appropriately, and optionally instantiates simulation models for off-chip devices.

9.5 IDE Integrated Development Environment

As discussed in Section 9.1, another piece necessary for putting an embedded system together is an environment for development and testing of the software programs that run on the processors of the embedded system. Altera uses IDE for this purpose. This section discusses the main features of this development environment. The main features of this design tool are its project definition utility, editor and compiler, debugger, and flash programmer. These tools will be discussed here.

9.5.1 IDE Project Manager

The Nios II IDE provides several project management tasks that speed up the development of embedded applications. A project wizard is used to automate the set-up of the C/C++ application project and system library projects.

Additionally, the Nios II IDE provides software code examples that are available in the form of project templates. These templates help software engineers use tested templates instead of starting from scratch. Each template is a collection of software files and project settings. Designers can add their own source code to the project by placing the code in the project directory or importing the files into the project.

Another form of ready-made software modules provided in IDE is its software components, also known as "system software". System software components provide designers with an easy way to configure their system for their specific target hardware. Components included are Nios II run-time library (also known as the hardware abstraction layer (HAL)), Lightweight IP TCP/IP stack, MicroC/OS-II real-time operating system (RTOS), and Altera Zip file system.

9.5.2 Source Code Editor

Altera's Nios II IDE provides a full-featured C/C++ source editor. The editor has the syntax highlighting feature, and is linked to the C/C++ compiler of IDE for debugging error correcting. Standard features of C/C++ source editors are available in IDE's editor.

9.5.3 C/C++ Compiler

IDE includes a C/C++ compiler for compiling programs that are to run on the Nios II core of an embedded system. This compiler is based on the industry-standard GNU tool chain.

The Nios II IDE provides a graphical user interface to the GCC compiler. The Nios II IDE build environment is designed to facilitate software development for Altera's Nios II processors, providing an easy-to-use push-button flow, while also allowing designers to manipulate advanced build settings.

The Nios II IDE build environment automatically produces a makefile based on the user's specific system configuration (the SOPC Builder-generated PTF file). Changes made in the Nios II IDE compiler/linker settings are automatically reflected in this auto-generated makefile. These settings can include options for the generation of memory initialization files (MIF), flash content, simulator initialization files (DAT/HEX), and profile summary files.

9.5.4 Debugger

The Nios II IDE contains a software debugger based on the GNU debugger, GDB. The debugger provides basic debug features, as well as several advanced features. The basic features include the following:

- Run control
- Call stack view
- Software breakpoints
- Disassembly code view
- Debug information view
- Instruction set simulator (ISS) target

In addition to the basic debug features, the Nios II IDE debugger also has several advanced debugging capabilities, such as Hardware breakpoints for debugging code in ROM or flash, Data triggers, and Instruction trace.

The Nios II IDE debugger connects to the target hardware using a JTAG debug module. Altera UP3 and DE series development boards are acceptable debugger targets. The debug information view provides the users with access to local variables, registers, memory, breakpoints, and expression evaluation functions.

9.5.5 Flash Programmer

Many designs that utilize Nios II processors also incorporate flash memory on the board as a means to store an FPGA configuration and/or Nios II program data. The Nios II IDE includes a convenient method of programming this flash. Any common flash interface (CFI) compliant flash device connected to the FPGA can be programmed using the Nios II IDE flash programmer. In addition to CFI flash, the Nios II IDE flash programmer can program any Altera serial configuration device connected to the FPGA.

The Nios II IDE flash programmer is pre-configured to work with all of the boards available with the Nios II development kits, but can be easily ported to any custom hardware.

9.6 An Embedded System Design: Calculator

Material presented in the previous sections showed all the pieces that are needed for putting an embedded system together. This section uses what has been presented to design an embedded system using the Nios II processor and its related tools.

The example embedded system we design here is a calculator that we will implement it an Altera FPGA. The calculator design involves a processor, its interfaces, a software program and its IO devices. This example shows the complete flow of a design, as well as the methodology and design tools.

In addition to being a comprehensive example, this design also utilizes some of the hardware cores discussed in the previous chapter. The methodology presented here complements the material of Chapter 6 in presenting design of a complete hardware/software system.

9.6.1 System Specification

The system we are designing has a keyboard a processor and an LCD, as shown in Figure 9.11. Data and operations to be performed are entered on the keyboard. The keyboard driver prepares parallel data for the processor to read. The processor reads this information through its interface, performs the specified operations, and puts it on its LCD interface for display. When alerted, the LCD driver takes data from its parallel port and displays it on its LCD.

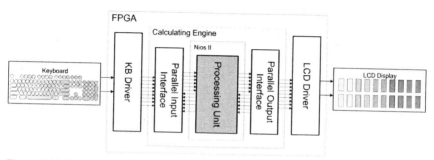

Figure 9.11 Calculator Components and Interfaces

The keyboard and LCD of Figure 9.11 are the physical devices that are attached to our FPGA pins. Keyboard and LCD drivers will be implemented on the FPGA and they have already been discussed in Chapter 7. The calculating engine is a Nios II based system that includes a Nios II processor, necessary memories, and IO devices. The design of this part is discussed next.

9.6.2 Calculating Engine

A general schematic diagram of the calculating engine part of our calculator is shown in Figure 9.12. The central part of this engine is a Nios II processor with a program that handles communication with the keyboard driver, collection of data, performing calculation tasks, and displaying results on the LCD through its driver.

The program memory of this system will be programmed with the compiled C/C++ program that we will develop for performing IO operations and calculation tasks. This program is the software part of our embedded processor system. The data memory is used for storing intermediate data used and needed by the program in the instruction memory. All systems shown here connect to the Nios II processor through the Avalon bus system.

The keyboard and the LCD are connected to the processor through PIO interfaces of the interconnecting bus of this system.

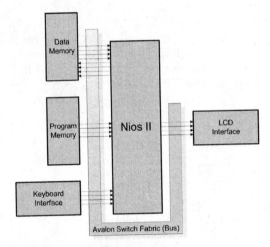

Figure 9.12 Calculating Engine

9.6.3 Calculator IO Interface

The keyboard driver of Chapter 7 reads keyboard codes from the keyboard and generates ASCII codes. When a key is released it puts the ASCII code of the character that corresponds to the uppercase character of the key pressed on its output and issues the *KeyReleased* signal for one clock duration.

The Nios II processor of Figure 9.12 has two PIO (Parallel IO) interfaces that handle the keyboard. Through a 1-bit PIO, Nios II continuously monitors *KeyReleased*. When detected, through another input PIO (an 8-bit one), it reads the parallel output of the keyboard driver. When this is read, the Nios II program decides on its next step depending on whether byte read is data or operation.

The LCD display of Chapter 7 has an 8-bit input, a write enable (*E*) and a reset input (*RS*). Through the LCD PIO interface, the Nios II processor puts the 8-bit ASCII code of a character to be displayed on its output port, and issues write enable to the LCD driver through another PIO. When resetting needs to be done, Nios II issues *reset* to LCD through another of its PIO interfaces.

Summarizing the above, through its Avalon bus, the Nios II processor in Figure 9.12 has two input PIOs for the keyboard. One is 1-bit and the other is an 8-bit PIO. This processor has three output PIOs on its LCD side. One is an 8-bit one, and the other two are 1-bit PIOs.

9.6.4 Design of Calculating Engine

We use the SOPC Builder discussed in Section 9.4 for the design of our Calculating Engine. The complete design including keyboard and

LCD drivers is defined as a Quartus II project, and the calculating engine becomes an SOPC Builder design file. Using screen shots, this section shows all necessary steps for definition of this design file.

9.6.4.1 Calculator Project. We define a *calculator* project in Quartus II. The central part of this is the calculating engine which we define as an SOPC file named: *NIOSII_CPU*.

A new SOPC file can be created from the New item of Quartus II File menu just like defining a new schematic or a Verilog file. Figure 9.13 shows the New menu in which SOPC Builder System is selected.

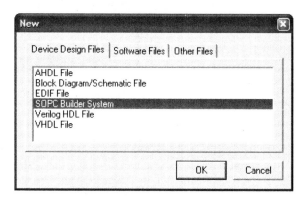

Figure 9.13 A New SOPC System Design File

The next window in creating a new SOPC system asks whether we want Verilog or VHDL for our target HDL. In this window (Figure 9.14), we select Verilog.

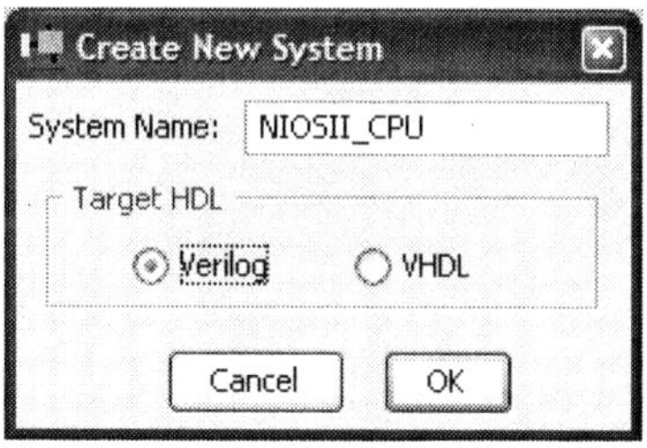

Figure 9.14 Target Language Selection

After completion of this task, a new blank SOPC Builder window (shown in Figure 9.15) opens. In this window, for the target board, we specify Nios Evaluation Board, Cyclone (EP1C12), or any other development board we are using or developing this example.

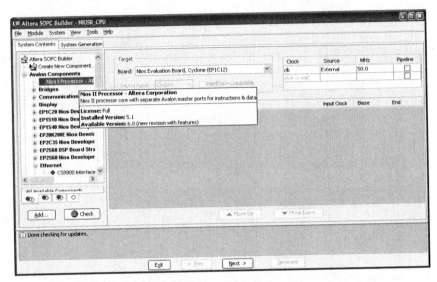

Figure 9.15 New Blank SOPC System

9.6.4.2 Processor Specification.
The first component to place in our new *NIOSII_CPU* SOPC system is a Nios II processor. This is done by double-clicking "NIOS II Processor" in the window of Figure 9.15.

When this is done, in a series of windows, the SOPC Builder helps us decide on the options we want for our new processor. Figure 9.16 shows the first such window.

Nios II Core	Caches & Tightly Coupled Memories	JTAG Debug Module	Custom Instructions

Select a Nios II core:

	⊙ Nios II/e	○ Nios II/s	○ Nios II/f
Nios II Selector Guide Family: Cyclone f_{system}: 50 MHz	RISC 32-bit	RISC 32-bit **Instruction Cache** **Branch Prediction** **Hardware Multiply** **Hardware Divide**	RISC 32-bit Instruction Cache Branch Prediction Hardware Multiply Hardware Divide **Barrel Shifter** **Data Cache** **Dynamic Branch Prediction**
Performance at 50 MHz	Up to 7 DMIPS	Up to 29 DMIPS	Up to 49 DMIPS
Logic Usage	600-700 LEs	1200-1400 LEs	1400-1800 LEs
Memory Usage	Two M4Ks	Two M4Ks + cache	Three M4Ks + cache

Figure 9.16 Processor Type Selection

Any of the three Nios II processor types, /e, /s, or /f can be selected. Nios II/e is the basic processor. Nios II/s is more advanced than /e, and has hardware divide and multiply plus an instruction cache. Nios II/f is the most advanced Nios II processor type. Options of these three processor types are shown in Figure 9.16. We choose Nios II/e for the processor of our calculator design.

In the next three windows after that of Figure 9.16, processor cache, JTAG Debug, and custom instructions can be specified. Because our processor does not have a cache, nothing can be selected under the cache tab. To keep our design simple, we choose "No Debugger" in the window that appears under the JTAG tab. The last tab of Figure 9.16 brings up Custom Instructions. For our choice of processor we have a very limited set of custom instructions, from which we select none.

After completion of this customization, the Nios II processor of our choice will be added to our *NIOSII_CPU* SOPC Builder page. Part of this page is shown in Figure 9.17. Note here that, placement of the processor automatically causes placement of an Avalon bus master for the instruction memory and one for the data memory. Connections to the memory will be shown when we place our system's memories.

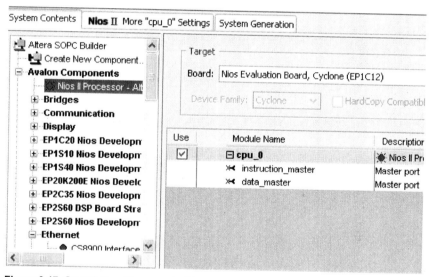

Figure 9.17 Completing Placement of CPU

9.6.4.3 Memory Specification.

For selection and addition of memory to our system, scroll down in the system contents part of the main SOPC Builder window until memory choices appear. As shown in Figure 9.18 we can select various types of SDRAM, SRAM, flash, or on-chip memories. Double-clicking on the memory type selects it, and as indicated in Figure 9.18 we have selected on-chip memory for this processor's memory. This means that the memory the processor uses will be that of the FPGA.

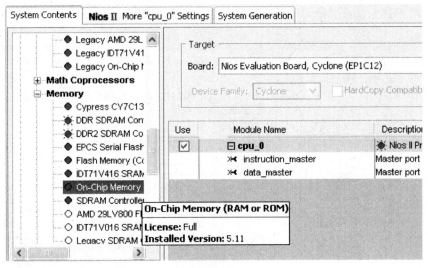

Figure 9.18 Memory Selection

When this memory is selected, a memory specification window appears, in which memory details can be specified. We are using this first memory for our program memory. As shown in Figure 9.19, we use a memory width of 32 bits and total size of 8 K bytes.

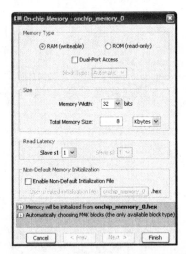

Figure 9.19 Memory Specification Window

When we click on Finish in Figure 9.19, the specified memory will be added to our *NIOSII_CPU* SOPC system. We will use this memory for the program memory. We follow the same exact procedure to add a 32-bit 4 K byte memory for the data memory. Both these memories will now appear on our SOPC system page, as shown in Figure 9.20.

When both program and data memories are placed, several changes must be made to their default configurations. The first thing is to use names that indicate which memory is used for data and which one is for program. This can be done by right-clicking on the memory entry and selecting the rename option.

We change the names as shown in Figure 9.20. The first memory (the 8 K memory) is used for data and it is named *onchip_prog_ram*, and the second memory is *onchip_data_ram*.

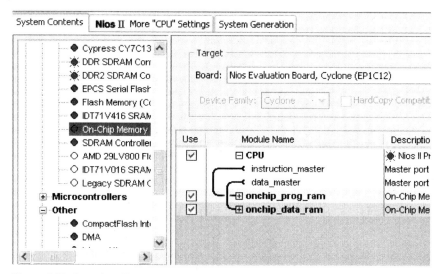

Figure 9.20 Complete Memory Specification

Next in the design process is specification of the connection of Avalon *instruction_master* and *data_master* to the new memories. Since the SOPC Builder does not know which memory is used for data or program, it connects *instruction_master* and *data_master* to all memories that are placed. This may cause an extra overhead that we will never utilize. To avoid this extra hardware we can remove the bus connections that we are not using. To do this we can move the screen curser to the connection that is to be removed and click it off. See in Figure 9.20 that *instruction_master* of Avalon bus only connects to the memory we are using for program, and the *data_master* only connects to the data RAM.

Another item we may need to change in the so-far configuration of our system is the Base address of our memories. The SOPC Builder uses some default base addresses starting from 0x00000000. This address serves fine for starting the program RAM. However, since we are yet unsure of the size of our program, we may need to leave some room for increasing the size of the program memory. This means that the next base address default used for the data memory may be too close to our program memory. Therefore we have changed the starting address of data RAM to 0x00003000. Since Figure 9.20 only shows a portion of the complete window, this change is not shown. This new address will show in another screen-shot later in this section. With this change, if at a later time, after all memories and interfaces have been placed, we decide that we need more program memory, we will not be forced to change base addresses of all our Avalon switch fabric connections.

9.6.4.4 Parallel IO Devices. The next step in this design is selection and placement of IO interfaces. For this purpose we can choose from one of the existing interfaces and configure it, or develop our own custom interface using the **Create New Component** option of **System Contents** tab of the main SOPC window.

For this design, we use Parallel I/O (PIO) for all one-bit, 8-bit, input and output ports of our system. PIO is a simple IO interface that can be configured for size, input or output. This interface has status and data registers, as well as address decoders and multiplexers. The hardware of this interface is already part of the Avalon switch fabric.

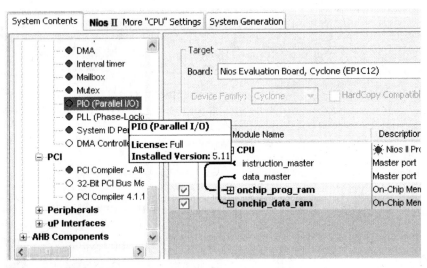

Figure 9.21 PIO Definition

Our design needs the following PIOs:

- A 1-bit input PIO for keyboard KeyReleased output
- A 8-bit input PIO for keyboard data output
- A 1-bit output PIO for LCD reset input
- A 1-bit output PIO for LCD write enable input
- An 8-bit output PIO for LCD data input

For specifying a PIO (one of the above), in the **System Contents** tab of the main SOPC Builder window, select PIO and double-click it (see Figure 9.21).

In the next series of windows that appear, details of the PIO that is being configured are specified. The first such window is shown in Figure 9.22. In this window size and IO direction of the port are specified. Depending on the choices here, other configuration windows may appear.

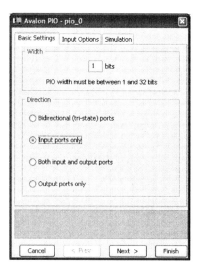

Figure 9.22 PIO Specification

We are using the PIO shown in Figure 9.22 for the *KeyReleased* input from the keyboard. This is a 1-bit input port. In our case, this input is being polled by the CPU. If needed, in one of the configuration windows, we could specify this as an interrupt input port.

9.6.4.5 Completing the Design. Figure 9.23 shows our *NIOSII_CPU* that is our calculating engine of our *calculator* project after placement of all necessary PIOs, and completing our system contents. As shown, the base location of various components are automatically set according to their required registers and address decoding.

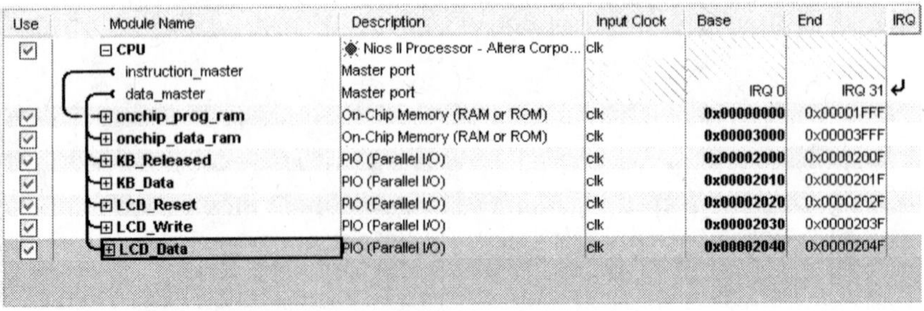

Use	Module Name	Description	Input Clock	Base	End	IRQ
☑	⊟ CPU	☀ Nios II Processor - Altera Corpo...	clk			
	◄ instruction_master	Master port				
	◄ data_master	Master port			IRQ 0	IRQ 31 ↵
☑	⊞ onchip_prog_ram	On-Chip Memory (RAM or ROM)	clk	0x00000000	0x00001FFF	
☑	⊞ onchip_data_ram	On-Chip Memory (RAM or ROM)	clk	0x00003000	0x00003FFF	
☑	⊞ KB_Released	PIO (Parallel I/O)	clk	0x00002000	0x0000200F	
☑	⊞ KB_Data	PIO (Parallel I/O)	clk	0x00002010	0x0000201F	
☑	⊞ LCD_Reset	PIO (Parallel I/O)	clk	0x00002020	0x0000202F	
☑	⊞ LCD_Write	PIO (Parallel I/O)	clk	0x00002030	0x0000203F	
☑	⊞ LCD_Data	PIO (Parallel I/O)	clk	0x00002040	0x0000204F	

Figure 9.23 System Contents, Complete

Now that specification of our system contents is complete, we go
on to the next step by clicking Next in the main SOPC Builder window (see bottom of Figure 9.15). This next step is for the CPU settings. As shown in Figure 9.24, this window allows us to change reset
address, exception address, and break location. For larger Nios II options (/s or /f) more choices would be available in this window. We
keep these values as they are and will not modify them for this design.

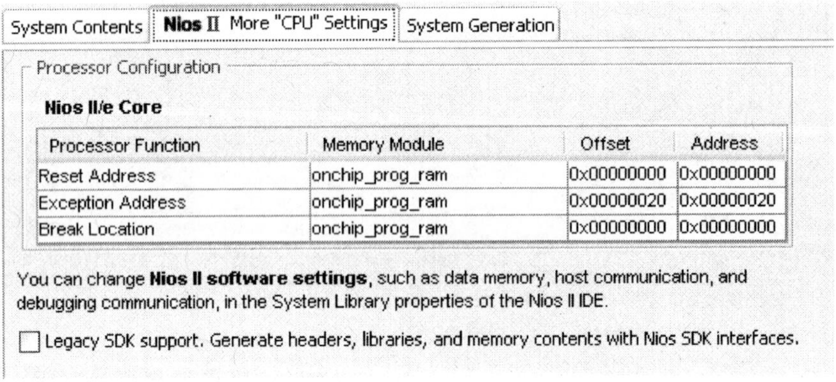

System Contents **Nios** II More "CPU" Settings System Generation

Processor Configuration

Nios II/e Core

Processor Function	Memory Module	Offset	Address
Reset Address	onchip_prog_ram	0x00000000	0x00000000
Exception Address	onchip_prog_ram	0x00000020	0x00000020
Break Location	onchip_prog_ram	0x00000000	0x00000000

You can change **Nios II software settings,** such as data memory, host communication, and
debugging communication, in the System Library properties of the Nios II IDE.

☐ Legacy SDK support. Generate headers, libraries, and memory contents with Nios SDK interfaces.

Figure 9.24 CPU Settings Window

Clicking Next in the SOPC Builder brings up the System Generation window from which selecting the Generate button begins the generation of the system for which contents and options have been set.
Figure 9.25 shows a portion of the SOPC Builder window after the
generation of the system. Note in this window directories of files that
are generated for simulation and testing of our design. With a successful completion of the system, a symbol will be generated for our
SOPC system.

Figure 9.26 shows the *NIOSII_CPU* symbol after some editing. Like any library component, this component can be placed in a Quartus II block diagram file and used in an upper level design.

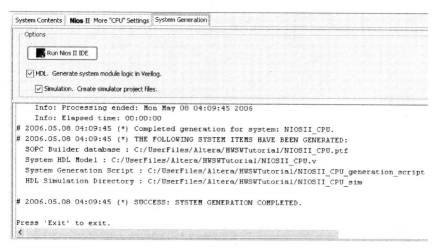

Figure 9.25 Completion of System Generation

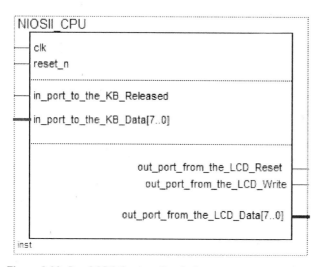

Figure 9.26 Our SOPC System Symbol

Going back to Figure 9.25, see the button for running the Nios II IDE on the upper left side of this window. Clicking on this starts the Altera Nios II IDE that we use for entering the C/C++ code that runs in our *NIOSII_CPU* system.

9.6.5 Building Calculator Software

The environment for editing and running C/C++ code in our system is the Nios II IDE. This can run from the SOPC Builder as shown in Figure 9.25, or independently. This section shows how a software project is started, how it is linked with the hardware that we have designed and how a complete hardware/software project is built.

9.6.5.1 Software Project Definition.   Figure 9.27 shows the menu part of the first window that opens after the welcome window of the Nios II IDE. From the File menu of this window select New and in the menu that opens select C/C++ Application.

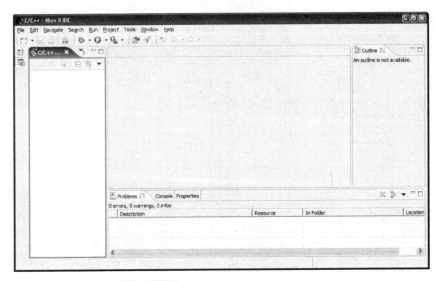

Figure 9.27 Nios II IDE First Window

This will allow you to start a new C/C++ application project in a selected directory. The window that opens as a result of this selection is shown in Figure 9.28.

In Figure 9.28 we use *CalculatorSoftware* for the name of the software part of our calculator. To link this software with the *NIOSII_CPU* SOPC system that we have built, under Select Target Hardware, select this SOPC system as the target hardware. Note in this figure that our SOPC system has *.ptf* file extension. As shown here, *CPU* is the name we used for the processor part of *NIOSII_CPU* SOPC system, which is consistent with the name shown in Figure 9.23.

We have now defined the CPU and the CPU system of our software project. The program for this CPU system will be defined next.

Before entering the code, we have to define our program and data memories. For this purpose, as shown in Figure 9.29, right click on *CalculatorSoftware_syslib* to bring up the properties window.

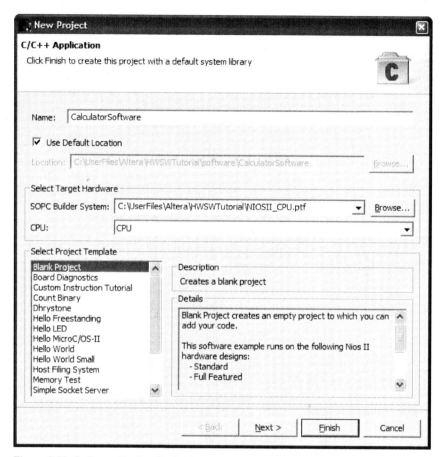

Figure 9.28 Software Project Definition

In the properties window that opens after Properties is selected you can specify memories for data and program. Referring back to Figure 9.23 note that our *instruction_master* is only connected to *on-chip_prog_ram* and *data_master* of the Avalon switch fabric is only connected to *onchip_data_ram*. This means that we only have one choice for our data memory and only one for the program memory. Had we had multiple data and instruction master connections, we had to decide on the specific memories for our data and program. Figure 9.30 shows the part of the Properties window that allows selection of memories for data and program.

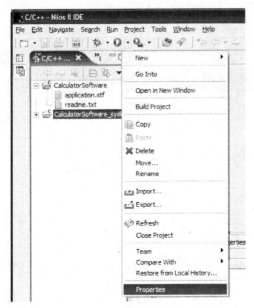

Figure 9.29 Properties of Software Project

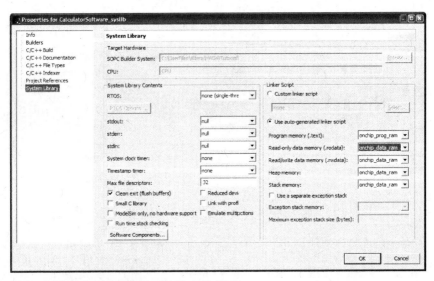

Figure 9.30 Data and Program Memory Selection

9.6.5.2 C program Hardware Interaction. The software program we are developing reads keyboard status and data, processes data, and writes the results to the LCD output. There are two parts to the C program that we are developing. One is interaction with hardware, i.e., keyboard and LCD; and the other is performing proper operations on data that is read. The latter part is a C program that will be discussed in the next section. The former, however, must consider parameters of the SOPC system that the software is going into.

Note in Figure 9.23 that we have defined PIO ports for our keyboard and LCD. Our software program must do all readings and writings to these ports. For this purpose we define C pointers for reading and writing these locations. Definitions shown in Figure 9.31 define pointers to keyboard and LCD ports.

```
#define KB-RELEASED    ((int *) 0x2000)
#define KB-DATA        ((int *) 0x2010)
#define LCD-RESET      ((int *) 0x2020)
#define LCD-WRITE      ((int *) 0x2030)
#define LCD-DATA       ((int *) 0x2040)
```

Figure 9.31 Hardware Interface Definitions

With the definitions shown, reading data from the keyboard and writing data to the LCD display can be done by the C statements shown below:

New_data_taken_from_keyboard = * KB-DATA;

*LCD-WRITE = New_data_to_display_on_LCD;

Definitions shown in Figure 9.31 can go in a header file, or at the beginning of the C program.

When the SOPC builder builds our system, it creates a file named *system.h* in its directory that contains the above definitions and other system related parameters. Instead of defining our own parameters, we could use the *system.h* header file. However, for a small example like ours, using our own definitions is not a complex task. If DMAs were used, more complex parameters would have been defined, which would necessitate the use of the *system.h* header file.

The C/C++ Nios II IDE (Figure 9.27) allows us to define our C/C++ files using its syntax-highlighted editor. For defining our new calculator C file, go to the **File** menu of the IDE window and select a new file as shown in Figure 9.32.

Once this is done, a new file, which we name *program.c* will open in our *CalculatorSoftware* directory. Figure 9.33 shows the header part of our C program for the calculator.

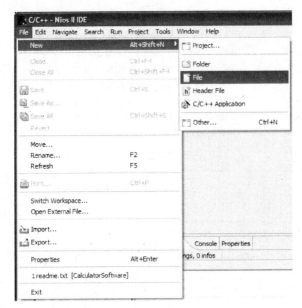

Figure 9.32 Defining a New C File

```
       {k = k + 2;}}


void  WriteNumOnLCD(int number);

int main()
{
   unsigned int key_released;
   unsigned int key;
   int k,i;

   int operand1;
   int operand2;
   int result;
   int opcode;
```

Figure 9.33 Calculator C Program

9.6.5.3 Project Built. Building the software part is the last part of building our calculator. This phase compiles our C program and generates memory contents for the program memory of *NIOSII_CPU* SOPC project.

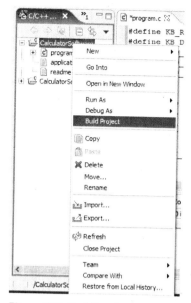

Figure 9.34 Building Software Part

As shown in Figure 9.34, the *CalculatorSoftware* project is built by right-clicking it in the Nios II IDE and selecting **Build Project**. After completion of project build, a report of the software is created in the environment's console. This report indicates the memory usage and if our allocated memory was sufficient for our software program. Had it not been, an error would be issued alarming us of this. Figure 9.35 shows the build report in the **Console** window.

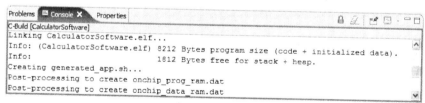

Figure 9.35 Build Report

9.6.6 Calculator Program

The program that is loaded in our Nios II processor memory reads keyboard data through its keyboard PIO, performs calculations and displays the result on the LCD through its LCD PIO. This section discusses the details of this program, which is shown in Figure 9.36.

```
001: #define KB_RELEASED ((int *)0x2000)
002: #define KB_DATA ((int *)0x2010)
003: #define LCD_RESET ((int *)0x2020)
004: #define LCD_WRITE ((int *)0x2030)
005: #define LCD_DATA  ((int *) 0x2040)
006:
007: #define ADD (65)
008: #define SUB (83)
009: #define MUL (77)
010: #define DIV (68)
011: #define EQ  (61)
012: #define NEW_OPERATION  (67)
013:
014: #define READ_KEY(key_value,key_ready,i,k) { \
015:     key_ready = *KB_RELEASED; \
016:     while(key_ready == 1) { \
017:     key_ready = *KB_RELEASED;} \
018:     while(key_ready == 0) { \
019:     key_ready = *KB_RELEASED;} \
020:     for(i=0;i<100000;i++) {k=k+2;} \
021:     (key_value) = *KB_DATA; }
022: #define WRITE_LCD(data){\
023:         *LCD_DATA =  data;\
024:         *LCD_WRITE = 1;\
025:         *LCD_WRITE = 0;}
026: #define DELAY(i,k) {\
027:         for(i=0;i<10000;i++)\
028:         {k = k + 2;}}
029: #define RESET_LCD(i,k)  {*LCD_RESET = 0; \
030:         *LCD_RESET = 1; DELAY(i,k);  \
031:         *LCD_RESET = 0;   }
032:
033: void  WriteNumOnLCD(int number);
034:
035: int main()
036: {  unsigned int key_released;
037:    unsigned int key;
038:    int k,i;
039:
040:    int operand1;
041:    int operand2;
042:    int result;
043:    int opcode;
044:    int read_operand_flag;
045:
046:    //...................//
047:
048:    operand1 = 0;
049:    operand2 = 0;
050:    result = 0;
051:    opcode = ADD;
052:    key=48;
053:    key_released = 0;
054:    read_operand_flag = 0;
055:
056:    while(1){
057:      READ_KEY(key,key_released,i,k);
058:      if(read_operand_flag == 0)
059:      { if((key< 58) && (key>47))
060:        {  operand1 = (operand1*10)+(key-48);
061:           WRITE_LCD(key); DELAY(i,k);
062:        }else if(key==ADD||key==SUB||key==MUL||key==DIV)
063:        { read_operand_flag = 1;
064:           opcode = key;
065:           WRITE_LCD(key); DELAY(i,k);
066:        }
067:      }
```

```
068:        else if(read_operand_flag == 1)
069:        {   if((key< 58) && (key>47))
070:            {   operand2 = (operand2*10)+(key-48);
071:                WRITE_LCD(key); DELAY(i,k);
072:            }else if (key == EQ)
073:            {   RESET_LCD; DELAY(i,k);
074:                if((opcode == DIV) && (operand2 == 0))
075:                {   WRITE_LCD(69); DELAY(i,k);
076:                }
077:                else
078:                { if(opcode == ADD)
079:                      result = operand1 + operand2;
080:                  else if (opcode == SUB)
081:                      result = operand1 - operand2;
082:                  else if (opcode == MUL)
083:                      result = operand1 * operand2;
084:                  else
085:                      result = operand1 / operand2;
086:                  operand1 = 0;
087:                  operand2 = 0;
088:                  WriteNumOnLCD(result);
089:                  result = 0;
090:                }
091:                read_operand_flag = 2;
092:            }
093:        }
094:        else
095:        { if((key == NEW_OPERATION))
095:          { RESET_LCD; DELAY(i,k);
096:            read_operand_flag = 0;
097:          }
099:        }
100:    }
101:    return 0;
102: }
103: void  WriteNumOnLCD(int number)
104: { int digit;
105:   int dig_count;
106:   int num;
107:   int base;
108:   int i,k;
109:
110:   digit = 0;
111:   dig_count = 1;
112:   if(number < 0)
113:   {   WRITE_LCD((45));
114:       DELAY(i,k);
115:       num = 0 - (number);
116:   }
117:   else num = number;
118:   base = 10;
119:   while(num > base) {base *= 10; dig_count++;}
120:   base = base / 10;
121:   while(dig_count > 0)
122:   { dig_count--;
123:     digit = num / base;
124:     WRITE_LCD((digit+48)); DELAY(i,k);
125:     num = num % base;
126:     base = base / 10;
127:   }
```

Figure 9.36 Calculation C Program

Lines in the listing of Figure 9.36 are numbered and we will refer to them accordingly. Lines 001 to 005 define pointers for keyboard and LCD PIO ports. As mentioned in the preceding sections, these addresses are defined by our hardware setup by SOPC Builder.

Lines 007 to 012 define ASCII characters for operations of the calculator. We are using A, S, M, D and = for add, subtract, multiply, divide and equal. Since our keyboard interface only reads keys without the shift-key held, we are using A, S, M, and D for +, -, * and /, respectively.

The *READ_KEY* definition begins on line 014. This parameterized definition continues reading *KB_RELEASED* until a positive pulse is detected. When this happens, *KB_DATA* PIO is read into the *key_value* parameter.

Definition beginning on line 022 writes the *data* parameter into LCD. The DELAY definition on line 026 is for generating real time delays. This definition is used for the LCD reset- and write-operations. The delay value is generated by k=k+2 operation that takes approximately 10 clock cycles. This repeats for 10,000 times giving a total delay of approximately 100,000 clocks. The definition on line 029 resets the LCD by putting a 1 into its reset PIO and holding it for a given delay value.

The main code of our calculation program begins on line 035. The first part of this code declares the necessary variables, which is then followed by initial values. The *read_operand_flag* is 0 when the program is reading the first operand of an operation, and it is 1 when reading the second operand.

The main calculation loop begins on line 056. Lines 058 to 067 read the first operand and the operator code. For the operand, 0 to 9 ASCII characters are read and *operand1* is calculated. As characters are read, they are written into the LCD. As shown on line 061 writing into the LCD must be followed by a delay period that is achieved by DELAY. Completion of the first operand in detected on line 062 when a key for one of the operators is pressed. The operator code is captured in *opcode* on line 064. Note the delay after writing operator code to the LCD.

Lines 068 to 072 read the second operand and calculate *operand2*. Completion of the second operand in detected on line 072 when the = sign code is detected, in which case the LCD is reset, and made ready for display of the results.

Lines 074 to 076 display an Error if the operation is divide and *operand2* is 0.

Lines 078 to 085 calculate the result based on the operation being performed. When done, function *WriteNumOnLCD* on line 088 writes the multi-character *result* to the LCD port.

Following this, on lines 095 to 099, if character C is pressed, a new operation begins and the LCD is reset.

The remaining part of code of Figure 9.36 is the *WriteNumOnLCD* function. This function counts the number of digits of its *number* argument and displays them one by one. If the result is negative, lines 112 to 116 write a minus sign to the LCD before writings of the result digits begin. Lines 121 to 127 calculate result digits and display them on the LCD. Note that with each LCD write there is a wait delay defined by DELAY.

9.6.7 Completing the Calculator System

The sections above described how the processing unit of our calculator example is formed. We showed its hardware part and its software part and tools and utilities for formation of the complete machine. As discussed above, and as shown in Figure 9.26, a Quartus II symbol is generated for the calculating machine.

To complete this system, we use the calculator symbol in Quartus II to connect it to the keyboard and the LCD drivers of Chapter 7. This configuration is done according to the diagram of Figure 9.11. The complete Quartus II schematic diagram is shown in Figure 9.37.

The complete system shown in this diagram, has serial keyboard inputs and generates parallel data and the corresponding handshaking signals for writing to the LCD. As an alternative, the output could be displayed on a VGA display monitor using the VGA driver that we developed in Chapter 7.

9.7 Summary

This chapter showed the complete top-down design of an embedded system with hardware and software components. We showed how a design that started from a schematic capture environment such as Quartus II could be used to define, not only hardware components, but also components that were implemented as a software program running on a processor. We presented SOPC Builder and IDE as tools that branch out from Quartus II for aiding the design of an embedded processor. The first part of this chapter discussed Altera tools for putting an embedded system together. We then used an example to exercise the methodology of creating such a system. Although Altera environment was discussed, an embedded system design environment generally includes tools similar to those discussed in this chapter.

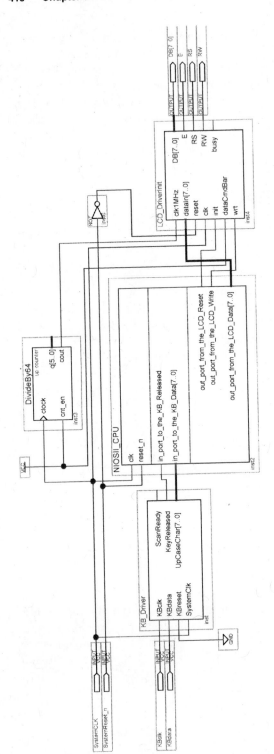

Figure 9.37 Calculator System Schematic

A ▲ Nios II Instruction Set

This appendix introduces the Nios II instructions categorized by type of operation performed. More details of the instructions of this machine can be found in the Nios II Processor Reference Handbook that is included in the CD in the back of this book. Instruction categories that we will discuss are: data transfer, arithmetic and logical, move, comparison, shift and rotate, program control, custom, and no-operation instructions.

A.1 Data Transfer Instructions

The Nios II architecture is a load-store architecture. Load and store instructions handle all data movement between registers, memory, and peripherals. Memories and peripherals share a common address space. Some Nios II processor cores use memory caching and/or write buffering to improve memory bandwidth. The architecture provides instructions for both cached and uncached accesses.

Word Data Transfer Instructions (ldw, stw, ldwio & stwio).
The ldw and stw instructions load and store 32-bit data words from/to memory. The effective address is the sum of a register's contents and a signed immediate value contained in the instruction. Memory transfers may be cached or buffered to improve program performance. This caching and buffering may cause memory cycles to occur out of order, and caching may suppress some cycles entirely.

Data transfers for I/O peripherals should use ldwio and stwio. ldwio and stwio instructions load and store 32-bit data words from/to peripherals without caching and buffering. Access cycles for ldwio and stwio instructions are guaranteed to occur in instruction order and never will be suppressed.

Byte Data Transfer Instructions (ldb, ldbu, stb, ldh, ldhu, sth, ldbio, ldbuio, stbio, ldhio, ldhuio, sthio). Load instructions ldb, ldbu, ldh and ldhu load a byte or half-word from memory to a register. ldb and ldh sign-extend the value to 32 bits, and ldbu and ldhu zero-extend the value to 32 bits. stb and sth store byte and half-word values, respectively. Memory accesses may be cached or buffered to improve performance. To transfer data to I/O peripherals, use the "io" versions of the instructions, i.e., ldbio, ldbuio, stbio, ldhio, ldhuio, and sthio.

A.2 Arithmetic and Logical Instructions

Nios II logical instructions support AND, OR, XOR, and NOR operations. Arithmetic instructions support addition, subtraction, multiplication and division.

Standard Logical Instructions (and, or, xor, nor). Instructions, and, or, xor, and nor are the standard 32-bit logical operations. These operations take two register values and combine them bit-wise to form a result for a third register.

Immediate Logical Instructions (andi, ori, xori). Instructions, andi, ori, and xori are the immediate versions of and, or, and xor instructions.

High Immediate Logical Instructions (andhi, orhi, xorhi). Instructions, andhi, orhi, and xorhi take 16-bit data, right jusatify them to a 32-bit word and use them as immediate data.

Standard Arithmetic Instructions (add, sub, mul, div, divu). Instructions, add, sub, mul, div, and divu are the standard 32-bit arithmetic operations. These operations take two register values as input and store the result in the third register. The divu instruction performs the divide unsigned operation.

Immediate Arithmetic Instructions (addi, subi, muli). Instructions, addi, subi, and muli are the immediate versions of add, sub, and mul. The instruction word includes a 16-bit signed value.

Upper Multiplication Instructions (`mulxss`, `mulxuu`). Instructions, `mulxss`, and `mulxuu`, and `muli` provide access to the upper 32 bits of a 32×32 multiplication operation. Choose the appropriate instruction depending on whether the operands should be treated as signed or unsigned values. It is not necessary to precede these instructions with a `mul`.

Long Multiplication Instructions (`mulxsu`). The `mulxsu` instruction used in computing a 128-bit result of a 64×64 signed multiplication.

A.3 Move Instructions

Move instructions provide move operations to copy the value of a register or an immediate value to another register.

Move Instructions (`mov`, `movhi`, `movi`, `movui`, `movia`). The `mov` instruction copies the value of one register to another register. `movi` moves a 16-bit signed immediate value to a register, and sign-extends the value to 32 bits. `movui` and `movhi` move an immediate 16-bit value into the lower or upper 16-bits of a register, inserting zeros in the remaining bit positions. Use `movia` to load a register with an address.

A.4 Comparison Instructions

The Nios II architecture supports a number of comparison instructions. All of these compare two registers or a register and an immediate value, and write either 1 (if true) or 0 to the result register. These instructions perform all the equality and relational operators of the C programming language.

Basic Comparison Instructions (`cmpeq`, `cmpne`, `cmpge`, `cmpgeu`, `cmpgt`, `cmpgtu`, `cmple`, `cmpleu`, `cmplt`). All compare instructions begin with "cmp". Following this, "eq", "ne", "ge", "gt", "le", and "lt" are for equal, not-equal, greater-than-or-equal, greater-than, less-than-or-equal, and less-than, respectively. The use of "u" at the end of the instruction is for unsigned comparison, versus signes 2's complement. In this case, the arguments are treated as 32-bit unsigned numbers.

Immediate Comparison Instructions (`cmpeqi`, `cmpnei`, `cmpgei`, `cmpgeui`, `cmpgti`, `cmpgtui`, `cmplei`, `cmpleui`, `cmplti`). Compare instructions ending with "i" are immediate versions of the comparison operations. They compare the value of a register and a 16-bit immediate value. Signed operations sign-extend the immediate value to 32-bits. Unsigned operations fill the upper bits with zero.

A.5 Shift and Rotate Instructions

The Nios II architecture supports standard and immediate shift and rotate operations. Right and left versions of these instructions are provided. The number of bits to rotate or shift can be specified in a register or an immediate value.

Rotate Instructions (`rol`, `ror`, `roli`). The `rol` and `roli` instructions provide left bit-rotation. `roli` uses an immediate value to specify the number of bits to rotate. The `ror` instruction provides right bit-rotation. There is no immediate version of `ror`, because `roli` can be used to implement the equivalent operation.

Shift Instructions (`rsll`, `slli`, `sra`, `srl`, `srai`, `srli`). The shift instructions implement the << and >> operators of the C programming language. The `sll`, `slli`, `srl`, `srli` instructions provide left and right logical bit-shifting operations, inserting zeros. The `sra` and `srai` instructions provide arithmetic right bit-shifting, duplicating the sign bit in the most significant bit. `slli`, `srli` and `srai` use an immediate value to specify the number of bits to shift.

A.6 Program Control Instructions

The Nios II architecture supports the unconditional jump and call instructions. These instructions do not have delay slots.

Unconditional Jump and Call Instructions (`call`, `callr`, `ret`, `jmp`, `br`). The `call` instruction calls a subroutine using an immediate value as the subroutine's absolute address, and stores the return address in register *ra*. `callr` instruction calls a subroutine at the absolute address contained in a register, and stores the return address in register `ra`. This instruction serves the roll of dereferencing a C function pointer. The `ret` instruction is used to return from subroutines called by `call` or `callr`. This instruction loads and executes the instruction specified by the address in register `ra`. The `jmp` instruction jumps to an absolute address contained in a register. This instruction is used to implement switch statements of the C programming language. The `br` instruction is for branch relative to the current instruction. With this instruction, a signed immediate value gives the offset of the next instruction to execute.

Conditional Branch Instructions (bge, bgeu, bgt, bgtu, ble, bleu, blt, bltu, beq, bne). The conditional-branch instructions compare register values directly, and branch if the expression is true. The conditional branches support the equality and relational comparisons of the C programming language. Notations: "eq", "ne", "ge", "gt", "le", "lt", and "u" used with these instructions have the same meanings as those of compare instructions. These instructions provide relative branches that compare two register values and branch if the expression is true.

A.7 Other Control Instructions

In addition to the standard control instructions, Nios II supports instructions for debugging, status register manipulation, exception handling, and pipleline related instructions. These instructions are discussed below.

Exception Instructions (trap, eret). The trap and eret instructions generate and return from exceptions. These instructions are similar to the call/ret pair, but are used for exceptions. trap saves the status register in the *estatus* register, saves the return address in the *ea* register, and then transfers execution to the exception handler. eret returns from exception processing by restoring status from *estatus*, and executing the instruction specified by the address in *ea*.

Break Instructions (break, bret). The break and bret instructions generate and return from breaks. break and bret are used exclusively by software debugging tools. Programmers never use these instructions in application code.

Control Register Instructions (rdctl, wrctl). The rdctl and wrctl instructions read and write control registers, such as the status register. The value is read from or stored to a general-purpose register.

Cache Control Instructions (flushd, flushi, initd, initi, flushp). The instructions flushd, flushi, initd, initi are used to manage the data and instruction cache memories. The flushp instruction flushes all pre-fetched instructions from the pipeline. This is necessary before jumping to recently-modified instruction memory.

Synchronization Instructions (synch). The `sync` instruction ensures that all previously-issued operations have completed before allowing execution of subsequent load and store operations.

A.8 Custom Instructions

The `custom` instruction provides low-level access to custom instruction logic. The inclusion of custom instructions is specified at system generation time, and the function implemented by custom instruction logic is design dependent. Machine-generated C functions and assembly macros provide access to custom instructions, and hide implementation details from the user. Therefore, most software developers never use the custom assembly instruction directly.

A.9 No-Op Instruction

The `nop` instruction is provided in the Nios II assembler, and is the no-operation instruction.

A.10   Potential Unimplemented Instructions

Some Nios II processor cores do not support all instructions in hardware. In this case, the processor generates an exception after issuing an unimplemented instruction. The only instructions that may generate an unimplemented-instruction exception are: `mul`, `muli`, `mulxss`, `mulxsu`, `mulxuu`, `div`, `divu`. All other instructions are guaranteed not to generate an unimplemented instruction exception. An exception routine must exercise caution if it uses these instructions, because they could generate another exception before the previous exception was properly handled.

B Additional Resources

The early chapters of this book provided the necessary background material for understanding design of embedded systems. For the readers requiring more in-depth reading material, or alternative ways of presentations, this appendix provides a list of references and books sorted by their subject areas.

Listed below are two comprehensive books on logic design. The web site shown provides simple review of many electrical and computer engineering topics, including logic design.

- Brown, S., and Z. Vranesic, *Fundamentals of Digital Logic with Verilog Design*, McGraw-Hill, New York, 2002, ISBN: 0-07-283878-7.
- Nelson, V. P., H. T. Nagle, B. D. Carroll, et al., *Digital Logic Circuit Analysis & Design*, Prentice-Hall, Inc., New Jersey, 1996, ISBN: 0134638948..
- www.play-hookey.com

Many books and references are available on Verilog and RTL design with hardware description languages. Listed below are several books and IEEE references.

- Bening, L., and H. D. Foster, *Principles of Verifiable RTL Design Second Edition–A Functional Coding Style Supporting Verification Processes in Verilog*, Springer, Boston, 2001, ISBN: 0792373685.

- IEEE Std 1364-2001, *IEEE Standard Verilog Language Reference Manual*, SH94921-TBR (print) SS94921-TBR (electronic), ISBN 0-7381-2827-9 (print and electronic), 2001.
- IEEE Std 1076-2002, *IEEE Standard VHDL Language Reference Manual*, SH94983-TBR (print) SS94983-TBR (electronic), ISBN 0-7381-3247-0 (print) 0-7381-3248-9 (electronic), 2002.
- Navabi, Z., *Digital Design and Implementation with Field Programmable Devices*, Kluwer Academic Publishers, Boston, 2005, ISBN: 1-4020-8011-5.
- Navabi, Z., "*Verilog Computer-Based Training Course*"; CBT CD with hardcopy user's manual, McGraw-Hill, New York, 2002, ISBN 0-07-137473-6.
- Navabi, Z., *VHDL: Analysis and Modeling of Digital Systems (Series in Electrical and Computer Engineering)*, McGraw-Hill College Division, New York, 1992, ISBN: 0070464723.
- Navabi, Z., *Verilog Digital System Design: Register Transfer Level Synthesis, Testbench, and Verification (Series in Electrical and Computer Engineering)*, McGraw-Hill College Division, New York, 2006, ISBN: 007144564-1.
- Palnitkar, S., *Verilog HDL*, 2d ed, Prentice Hall PTR, New Jersey, 2003, ISBN: 0130449113.
- Thomas, P. R., and P. Moorby, *The Verilog Hardware Description Language*, Springer, Boston, 2002, ISBN: 1402070896.

The following textbook is a comprehensive book on computer architectures.

- Patterson, D.A., J.L. Hennessy, P.J. Ashenden, et al., *Computer Organization and Design: The Hardware/Software Interface, Third Edition*, Morgan Kaufmann, San Francisco, 2004, ISBN: 1558606041.

Index

#define, 150
#endif, 150
#if, 150
#include, 150
#undef, 150
$fflush, 185, 187
$finish, 127
$fopen, 184, 187
$fscanf, 185, 186, 187
$fseek, 185, 186, 187
$random, 127
$stop, 127, 128, 138, 184
.mif file, 362
?:, 88
2's complement, 17, 18, 45, 91, 422
4-value logic, 78

A/D, 350
abstraction level, 2, 3
AC, 171, 177, 182
accumulator, 163
activity levels, 43
add-and-shift, 128
address alignment, 386
addressing mode, 161
ALU, 165, 171, 177, 182, 190, 374
always block, 94, 98, 102

analog filter, 350
and, 80
AND-plane, 198, 199, 200, 201, 202, 203, 206, 207, 208, 209, 210, 211, 214
arbitration, 387
Arithmetic instructions, 379, 420
arrays, 158
artistic abilities, 272
ASIC, 7, 8, 392
assembler, 11, 147, 161, 183, 194
assembly code, 359
assembly language, 147, 349
assign statement, 77, 87, 88
asynchronous RAM, 236, 358
asynchronous set, 58, 105
Avalon, 369, 382, 404
Avalon interface, 392
Avalon switch fabric, 382, 383, 393

BCD. See Binary Coded Decimal
begin, 94
bidirectional, 64, 128
binary addition, 16
Binary Coded Decimal, 35, 42